Communications
in Computer and Information Science 2483

Series Editors

Gang Li, *School of Information Technology, Deakin University, Burwood, VIC, Australia*

Joaquim Filipe, *Polytechnic Institute of Setúbal, Setúbal, Portugal*

Zhiwei Xu, *Chinese Academy of Sciences, Beijing, China*

Rationale

The CCIS series is devoted to the publication of proceedings of computer science conferences. Its aim is to efficiently disseminate original research results in informatics in printed and electronic form. While the focus is on publication of peer-reviewed full papers presenting mature work, inclusion of reviewed short papers reporting on work in progress is welcome, too. Besides globally relevant meetings with internationally representative program committees guaranteeing a strict peer-reviewing and paper selection process, conferences run by societies or of high regional or national relevance are also considered for publication.

Topics

The topical scope of CCIS spans the entire spectrum of informatics ranging from foundational topics in the theory of computing to information and communications science and technology and a broad variety of interdisciplinary application fields.

Information for Volume Editors and Authors

Publication in CCIS is free of charge. No royalties are paid, however, we offer registered conference participants temporary free access to the online version of the conference proceedings on SpringerLink (http://link.springer.com) by means of an http referrer from the conference website and/or a number of complimentary printed copies, as specified in the official acceptance email of the event.

CCIS proceedings can be published in time for distribution at conferences or as post-proceedings, and delivered in the form of printed books and/or electronically as USBs and/or e-content licenses for accessing proceedings at SpringerLink. Furthermore, CCIS proceedings are included in the CCIS electronic book series hosted in the SpringerLink digital library at http://link.springer.com/bookseries/7899. Conferences publishing in CCIS are allowed to use Online Conference Service (OCS) for managing the whole proceedings lifecycle (from submission and reviewing to preparing for publication) free of charge.

Publication process

The language of publication is exclusively English. Authors publishing in CCIS have to sign the Springer CCIS copyright transfer form, however, they are free to use their material published in CCIS for substantially changed, more elaborate subsequent publications elsewhere. For the preparation of the camera-ready papers/files, authors have to strictly adhere to the Springer CCIS Authors' Instructions and are strongly encouraged to use the CCIS LaTeX style files or templates.

Abstracting/Indexing

CCIS is abstracted/indexed in DBLP, Google Scholar, EI-Compendex, Mathematical Reviews, SCImago, Scopus. CCIS volumes are also submitted for the inclusion in ISI Proceedings.

How to start

To start the evaluation of your proposal for inclusion in the CCIS series, please send an e-mail to ccis@springer.com

Zaharuddin Mohamed · Fazilah Hassan ·
Gary Tan · Anita Ahmad · Leow Pei Ling ·
Salinda Buyamin
Editors

Systems Modelling and Simulation

First International Symposium, SMS 2024
Johor Bahru, Malaysia, December 16–17, 2024
Proceedings

Editors
Zaharuddin Mohamed
Universiti Teknologi Malaysia
Johor Bahru, Johor, Malaysia

Gary Tan
National University of Singapore
Singapore, Singapore

Leow Pei Ling
Universiti Teknologi Malaysia
Johor Bahru, Malaysia

Fazilah Hassan
Universiti Teknologi Malaysia
Johor Bahru, Malaysia

Anita Ahmad
Universiti Teknologi Malaysia
Johor Bahru, Malaysia

Salinda Buyamin
Universiti Teknologi Malaysia
Johor Bahru, Malaysia

ISSN 1865-0929   ISSN 1865-0937 (electronic)
Communications in Computer and Information Science
ISBN 978-981-96-4612-8   ISBN 978-981-96-4613-5 (eBook)
https://doi.org/10.1007/978-981-96-4613-5

© The Editor(s) (if applicable) and The Author(s), under exclusive license
to Springer Nature Singapore Pte Ltd. 2025

This work is subject to copyright. All rights are solely and exclusively licensed by the Publisher, whether the whole or part of the material is concerned, specifically the rights of translation, reprinting, reuse of illustrations, recitation, broadcasting, reproduction on microfilms or in any other physical way, and transmission or information storage and retrieval, electronic adaptation, computer software, or by similar or dissimilar methodology now known or hereafter developed.
The use of general descriptive names, registered names, trademarks, service marks, etc. in this publication does not imply, even in the absence of a specific statement, that such names are exempt from the relevant protective laws and regulations and therefore free for general use.
The publisher, the authors and the editors are safe to assume that the advice and information in this book are believed to be true and accurate at the date of publication. Neither the publisher nor the authors or the editors give a warranty, expressed or implied, with respect to the material contained herein or for any errors or omissions that may have been made. The publisher remains neutral with regard to jurisdictional claims in published maps and institutional affiliations.

This Springer imprint is published by the registered company Springer Nature Singapore Pte Ltd.
The registered company address is: 152 Beach Road, #21-01/04 Gateway East, Singapore 189721, Singapore

If disposing of this product, please recycle the paper.

Preface

The International Symposium on Systems Modelling and Simulation (SMS 2024) was held in Johor Bahru, Johor, Malaysia on 16–17th December 2024. This symposium was organized by the Malaysian Simulation Society and the technical co-sponsor the Society of Simulation and Gaming Singapore.

This symposium provided a platform for scientists, academicians, and professionals from around the world to present and discuss their work and new emerging simulation technologies. SMS 2024 focused on important topics in Modelling and Simulation that are related to current and emerging technologies. In particular, applications of modelling and simulation to advanced systems such as robotics, smart manufacturing, intelligent systems, and machine learning were of interest. Moreover, SMS 2024 encouraged articles that apply theoretical work in practice.

The papers presented at SMS 2024 have been compiled in this volume of Communications in Computer and Information Science (CCIS). A total of 65 papers were submitted to the symposium's three main tracks, namely: Emerging Trends in Modelling and Simulation, Artificial Intelligence and Computing, and Interdisciplinary Applications of Systems Modelling. Following a single-blind review process, 27 papers were accepted for oral presentation.

As editorial members of the SMS 2024 symposium, we would like to express our gratitude to all the authors who chose this symposium as a venue for their publications. We are also very grateful for the support and tremendous effort of the Program and Organizing Committee members for soliciting and selecting research papers with a balance of high quality and new ideas and applications. Thank you from us.

February 2025

Zaharuddin Mohamed
Fazilah Hassan
Gary Tan
Anita Ahmad
Leow Pei Ling
Salinda Buyamin

Organization

Organizing Committee

Chair

Anita Ahmad — Universiti Teknologi Malaysia, Malaysia

Deputy Chair

Herman Wahid — Universiti Teknologi Malaysia, Malaysia

Secretary

Nurul Adilla Mohd Subha — Universiti Teknologi Malaysia, Malaysia

Treasurer

Nurulaqilla Khamis — Universiti Teknologi Malaysia, Malaysia

Technical

Gary Tan	National University of Singapore, Singapore
Liang Li	Ritsumeikan University, Japan
Mohd Fua'ad Rahmat	Universiti Teknologi Malaysia, Malaysia
Norhaliza Abd Wahab	Universiti Teknologi Malaysia, Malaysia
Xiao Song	Beihang University, China
Zaharuddin Mohamed	Universiti Teknologi Malaysia, Malaysia

Publication

Fazilah Hassan	Universiti Teknologi Malaysia, Malaysia
Leow Pei Ling	Universiti Teknologi Malaysia, Malaysia

| Mohd Shahrizal Rusli | Universiti Teknologi Malaysia, Malaysia |
| Yasmin Abdul Wahab | Universiti Malaysia Pahang Al-Sultan Abdullah, Malaysia |

Programme

Hassrizal Hassan Basri	Universiti Malaysia Perlis, Malaysia
Mohd Ariffanan Mohd Basri	Universiti Teknologi Malaysia, Malaysia
Salinda Buyamin	Universiti Teknologi Malaysia, Malaysia

Local Arrangements

Mohd Saiful Azimi Mahmud	Universiti Teknologi Malaysia, Malaysia
Muhamad Fadli Ghani	Universiti Teknologi Malaysia, Malaysia
Noor Hanis Izuddin Mat Lazim	Universiti Sains Islam Malaysia, Malaysia

Website and Publicity

Hazriq Izzuan Jaafar	Universiti Teknikal Malaysia Melaka, Malaysia
Herman Wahid	Universiti Teknologi Malaysia, Malaysia
Noorhazirah Sunar	Universiti Teknologi Malaysia, Malaysia

International Program Committee

Abul K. M. Azad	Northern Illinois University, USA
Aduwati Sali	Universiti Putra Malaysia, Malaysia
Andi Andriansyah	Universitas Mercu Buana, Indonesia
Asmarashid Ponniran	Universiti Tun Hussien Onn Malaysia, Malaysia
Awanis Romli	Universiti Malaysia Pahang Al-Sultan Abdullah, Malaysia
Bidyadhar Subudhi	National Institute of Technology Rourkela, India
Byeong-Yun Chang	Ajou University, South Korea
Firdaus	Universitas Islam Indonesia, Indonesia
Gary Tan	National University of Singapore, Singapore
He Chen	Hebei University, China
Jie Huang	Beijing Institute of Technology, China
Le Anh Tuan	Vietnam Maritime University, Vietnam
Liang Li	Ritsumeikan University, Japan

Lin Zhang	Beihang University, China
Maurizio Palesi	University of Catania, Italy
Mehmet Önder Efe	Hacettepe University, Turkey
Mohammad Hasan Shaheed	Queen Mary University of London, UK
Mohammad Osman Tokhi	London South Bank University, UK
Muhammad Abid	COMSATS University Islamabad, Pakistan
Muhammad Nasiruddin Mahyuddin	Universiti Sains Malaysia, Malaysia
Mustapha Muhammad	Bayero University, Nigeria
Naomie Salim	Universiti Teknologi Malaysia, Malaysia
Ning Sun	Nankai University, China
Pawel Kwiaton	Częstochowa University of Technology, Poland
Sarvat M. Ahmad	King Fahd University of Petroleum and Minerals, Saudi Arabia
Satoshi Tanaka	Ritsumeikan University, Japan
Shaharin Anwar bin Sulaiman	Universiti Teknologi Petronas, Malaysia
Siti Fauziah Toha	Universiti Islam Antarabangsa Malaysia, Malaysia
Subhas Mukhopadhya	Macquarie University, Australia
Sumeet S. Aphale	University of Aberdeen, UK
Ramon Vilanova Arbos	University of Barcelona, Spain
Tan Ai Hui	Multimedia Universiti, Malaysia

Contents

A Refined First-Order Sparse TGV Model with L1 Norm Data Fidelity
for Enhanced Image Denoising 1
   Cheng Zhang and Kin Sam Yen

Application of Underwater Flexible Manipulator in Teaching Nonlinear
Systems and Control Courses 14
   Zixing Zhao and Jie Huang

A Spring-Pendulum Educational Device for Teaching Dynamics
and Control Courses 27
   Chenglei Yang and Jie Huang

3D Scene Reconstruction Using Lidar Point Clouds and Images 41
   *Sulaiman Sabikan, Kenneth Edmond Ntende,
   Sophan Wahyudi Nawawi, and Shahrudin Zakaria*

Dissolved Oxygen Concentration Modelling Based on Nanobubble
Technology in Recirculating Aquaculture System 57
   *Indra Sakti, Nadiatulhuda Zulkifli, Sofia, Hilman Syaeful Alam,
   Mohd Fua'ad Rahmat, Sevia Mahdaliza Idrus Nameh,
   Hanif Fakhrurroja, Anto Tri Sugiarto, and Ahmad Aminudin*

Analysis of Ultrasonic Signal Arrival Time in Concrete with Multiple
Defects 72
   *Farah Aina Jamal Mohamad, Anita Ahmad, Ruzairi Abdul Rahim,
   Juliza Jamaludin, and Syarfa Najihah Raisin*

Predictive Maintenance in Aerospace: Leveraging Machine Learning
for Optimized Filter Lifespan 87
   *Syaimak Abdul Shukor, Jia Hui Tang, Iqraq Kamal,
   Mohd Faizzuan Mohammad, Zalinda Othman, Shahnorbanun Sahran,
   Nor Samsiah Sani, Mohd Ridzwan Yaakub, and Azizi Abdullah*

Safety Control with Conflict Resolution for Multi Robots in Flexible
Manufacturing System 102
   *Jakaisa Riskhalifah Bhuwana, Muhammad Zakiyullah Romdlony,
   Angga Rusdinar, Norhaliza Abdul Wahab, and Muhammad Azhar Ismail*

Investigating Interdisciplinary Research Impact: A Framework
for Integrating Linguistic and Citation Information 118
   Masanao Ochi, Masanori Shiro, Jun'ichiro Mori, and Ichiro Sakata

Identification and Control of Industrial Hydraulic Actuators
with a Hammerstein-Wiener Model Approach 131
   *Nur Husnina Mohamad Ali, Hazriq Izzuan Jaafar, Rozaimi Ghazali,
Muhamad Fadli Ghani, Chong Chee Soon, and Zulfatman Has*

Machine Learning and NLP-Based Approach for Constraint Acquisition
Problems 146
   *Abdessalam Bahafid, Zakarya Erraji, Naira Abdou Mohamed,
Anass Allak, Kamel Gaanoun, and Imade Benelallam*

Improving Twitter Sentiment Analysis with a Hybrid BERT and Graph
Neural Network Ensemble 159
   Amar Taggu and Nabam Teyi

Autonomous Quadcopter UAV Waypoint Navigation: Simulation
in Mission Planner 171
   Mohd Yusuf Amran and Mohd Ariffanan Mohd Basri

Emotions Divergence in Online Social Network During Pandemic:
A Sentiment Analysis 182
   *Mohd Sharul Nizam Mohd Danuri, Nur Hidayatullah Rashidia,
and Rohizah Abd Rahman*

Modeling of Low Salinity Waterflooding Process in Laboratory
Experiments 195
   *Nguyen Le-Khoi, Lac Tran-Hoang-Gia, Viet Pham-Tuan,
and Lan Mai-Cao*

Multimodal Biometric Recognition Using Fuzzified BTC with a Novel
Hybrid JSO-CSO Algorithm 210
   *Indu Singh, Siddartha Aggarwal, Shrey Gupta, Vansh Khandelwal,
and Abhishek Verma*

Delayed Transitions in GPenSIM 224
   Yuming Feng, Slawomir Samolej, and Reggie Davidrajuh

Application Interface for GPenSIM 237
   Yuming Feng, Slawomir Samolej, and Reggie Davidrajuh

Analyzing Production Line Schedules with Tecnomatix Plant Simulation 251
   *Jocelyn Yeoh, Li-Pei Wong, Mohd. Zulkifli Che Wanik,
and Siang Kok Chia*

Modelling of Electrical Capacitance Tomography Sensor Array
for Air-Wax Detection 268
   *Shahrulnizahani Mohammad Din, Pei Ling Leow,
Jaysuman Pusppanathan, Ruzairi Abdul Rahim, Xian Feng Hor,
Wen Pin Gooi, Yasmin Abdul Wahab, and Suzanna Ridzuan Aw*

Modeling and Simulation of an ECT Sensor for Non-invasive Agarwood
Inspection 281
   *Muhammad Aiqil Sarudin, Yasmin Abdul Wahab,
Nurhafizah Abu Talip Yusof, Mohd Mawardi Saari,
Suzanna Ridzuan Aw, Mohd Shafie Bakar, Nurul Wahidah Arshad,
and Sia Yee Yu*

Hardware Accelerator Simulation for Colour Correction Algorithm 295
   *Shu Ting Loh, Mohd Shahrizal Rusli, Muhammad Nadzir Marsono,
Ab Al-Hadi Ab Rahman, Norlina Paraman, Shahidatul Sadiah,
Michael Loong Peng Tan, Izam Kamisian, Jaysuman Pusppanathan,
and Nur Diyana Kamarudin*

Analysis of the Impact of Watermarking Technique in Neural Network
Models to Predict Lung Diseases 308
   Tuan Nguyen-Thanh and Kiet Vo-Tuan

Adaptive Genetic Algorithm Based LQR for Optimal Control of Nonlinear
Double Pendulum Gantry Crane 322
   *Mohamed O. Elhabib, Herman Wahid, Zaharuddin Mohamed,
and Hussein Shutari*

Performance Evaluation Customer Access Network Using Difference
Modulation Technique 338
   *Sivaguru Mugunthan, Juwairiyyah Abdul Rahman,
Hasliza Abu Hassan, Norliana Muslim,
and Mohammad Syuhaimi Ab Rahman*

Extracting Narrative Events in Andersen's Fairy Tales Using a Hybrid
BERT-LSTM Model 355
   Erna Daniati, Aji Prasetya Wibawa, and Wahyu Sakti Gunawan Irianto

Galvanic Corrosion Progression Analysis at Aluminium-Carbon Fiber
Interfaces in ACCC/TW Conductors 373
  *Mohamad Izzat Nawawi, Muhammad Iqbal Shafei,
  Shahnurriman Abdul Rahman, and Konstantinos Kopsidas*

Author Index 387

A Refined First-Order Sparse TGV Model with L1 Norm Data Fidelity for Enhanced Image Denoising

Cheng Zhang and Kin Sam Yen(✉)

School of Mechanical Engineering, Universiti Sains Malaysia, Engineering Campus, Nibong Tebal, Penang, Malaysia
meyks@usm.my

Abstract. The total generalized variation (TGV) method effectively reduces noise and mitigates the staircase effect commonly observed in total variation (TV) approaches. However, its capacity to preserve structural features and fine details within images remains limited, making it less effective in certain applications. To address these shortcomings, the paper introduce overlapping group sparsity (OGS) regularization into the first-order gradient of the TGV model. This integration significantly enhances denoising performance while further minimizing staircase artifacts. By leveraging the sparse structural information embedded within images, our method achieves superior noise suppression compared to conventional TGV models. Furthermore, we replace the commonly used L2 norm in the data fidelity term with the L1 norm, which better preserves undistorted image regions, particularly in cases of complex noise. To efficiently solve the resulting optimization problem, we employ a split Bregman method, ensuring computational efficiency. Experimental evaluations highlight the superiority of our approach, showing that it consistently outperforms state-of-the-art TV- and TGV-based methods in both noise reduction and detail preservation, delivering significant advancements in image denoising techniques.

Keywords: Total Generalized Variation · Staircase Effect · Overlapping Group Sparsity

1 Introduction

Image denoising is a vital process in image analysis, aimed at removing noise while preserving key attributes such as contrast, clarity, and texture. Maintaining these attributes is essential for subsequent tasks like image segmentation, feature extraction, and target recognition. Therefore, the effectiveness of denoising significantly impacts the success of image processing workflows.

The dual objectives of noise reduction and detail preservation are paramount in image restoration. Since both detail features and noise components reside in the high-frequency spectrum, conventional denoising methods often misinterpret essential details as noise, leading to detail loss and blurring [1]. Variational methods, however, have shown effectiveness in balancing noise removal and detail preservation [2]. These methods involve minimizing an energy function to achieve a smoother image state, functioning as a regularization technique. The well-known total variation (TV) approach [3] exemplifies this strategy, using the L2 norm for the data fidelity term.

The TV model, which assumes images are piecewise smooth, excels in edge preservation but suffers from the staircase effect. To address this, Bredies et al. proposed the total generalized variation (TGV) model [4]. This model integrates first- and higher-order regularizations to enhance smoothness and gradient consistency while reducing staircase artifacts. Nevertheless, TGV models lack explicit consideration of structural similarity in images.

Further advancements have incorporated overlapping group sparsity (OGS) into TV regularization (TV-OGS), as demonstrated by Liu et al. [5]. OGS introduces neighborhood gradient constraints and sparsity in the image difference domain, enhancing the differentiation between smooth and boundary regions while reducing staircase artifacts. However, sparse gradients within overlapping groups have limited effectiveness in fully mitigating the staircase effect. Zhang et al. [6] achieved promising results by applying OGS to simultaneously constrain the first- and second-order components of TGV, but this approach led to a reduction in computational efficiency.

In image restoration, the choice of data fidelity term has been relatively limited, resulting in insufficient research diversity. The traditional TV model, using the L2 norm, is well-suited for images corrupted by Gaussian noise. Chan et al. [7] introduced the TV-L1 model, which integrates an L1 norm-based data fidelity term. This model enhances signal sparsity and better preserves image contrast after restoration, distinguishing it from the traditional TV model. However, it exacerbates the staircase effect compared to traditional TV methods.

This paper introduces an enhanced image denoising technique: the TGV regularization term with a first-order OGS convergence operator based on the L1-norm data fidelity term (TGV-F1OGS-L1). The primary goal is to reduce noise while effectively preserving edge structures. Adjusting the data fidelity or regularization terms in variational models often involves trade-offs, making it challenging to balance performance. By leveraging the L1 norm's ability to preserve geometric texture, we modify the data fidelity term in the proposed model from the traditional L2 norm to the L1 norm. Although this modification exacerbates the staircase effect, we counteract it by combining the TGV and OGS regularization terms, leveraging their ability to mitigate staircase artifacts.

High-order gradient constraints are crucial for image reconstruction, as they control the continuity of low-order gradients [8]. To streamline the proposed model's complexity, we apply the OGS convergence technique exclusively to the first-order gradient within the regularization term. This approach explores image edge continuity and pixel-level neighborhood information, achieving the overarching goal of eliminating noise and staircase effects while safeguarding edges.

To improve efficiency, we utilize the split Bregman method, recognized for its fast convergence in handling L1 norm-based partial differential equations [9], to solve the proposed energy model. This technique divides the problem into sub-problems, solved using the fast Fourier transform (FFT) and the majorization-minimization (MM) method [10] integrated with the OGS convergence technique.

The proposed TGV-F1OGS-L1 model stands out by using the L1-norm data term and the OGS regularization term. This novel synergy addresses the shortcomings of traditional TGV and TV-OGS models by preserving finer edge details while reducing the staircase effect and optimizing computational efficiency via the split Bregman method.

2 Related Work

2.1 The Fundamental of Total Generalized Variation

In contrast to traditional TV regularization, TGV adapts the gradient order to specific information and local edge features across different image regions. This adaptability allows TGV to perform well in areas with relatively rapid changes within smooth regions. However, a notable drawback is its tendency to induce edge blurring, which can lead to the loss of essential features in the processed image. Efforts have been made to propose enhancements based on the TGV model [11].

The traditional TGV_λ^2 regularization term is defined as follows:

$$\text{TGV}_\lambda^2(f) = \min_{\omega \in BD(\Omega)} \alpha_0 \int_\Omega |\nabla f - v| dx + \alpha_1 \int_\Omega |\varepsilon(v)| dx. \tag{1}$$

where $\varepsilon(v)$ constrains the second-order gradient within the processed image and is a symmetric gradient derivative.

$$\varepsilon(v) = \frac{\nabla v + \nabla v^T}{2} = \begin{bmatrix} \nabla_x v_x & \frac{1}{2}(\nabla_x v_y + \nabla_y v_x) \\ \frac{1}{2}(\nabla_x v_y + \nabla_y v_x) & \nabla_y v_y \end{bmatrix}. \tag{2}$$

The weakly symmetric derivative $\varepsilon(v)$ functions as a matrix-valued Radon metric with a regularity weaker than that of $\nabla^2 f$. In the equation, α_0 and α_1 are non-negative parameters that serve as gradient weights, balancing the influence of the different orders-gradients. The first term represents the total first-order gradient amplitude, and the second term corresponds to the second-order gradient amplitude.

Unlike TV approach, the TGV method introduces the auxiliary parameter v, which can adaptively adjust its value based on the local details and edges of the image. The gradient operator $\nabla = \begin{bmatrix} \nabla_x \\ \nabla_y \end{bmatrix}$ comprises horizontal (∇_x) and and vertical (∇_y) differential operator. The vector $v = \begin{pmatrix} v_x \\ v_y \end{pmatrix}$ limits the sparsity of the first-order gradient, with v_x and v_y corresponding to the components of v.

2.2 Overlapping Group Sparsity Regularization Term

In the context of a two-dimensional image f, the overlapping group matrix $\widetilde{f}_{i,j,k,k}$ is defined as follows:

$$\tilde{f}_{i,j,k,k} = \begin{pmatrix} f_{i-K_l,j-K_l} & f_{i-K_l,j-K_l+1} & \cdots & f_{i-K_l,j+K_r} \\ f_{i-K_l+1,j-K_l} & f_{i-K_l+1,j-K_l+1} & \cdots & f_{i-K_l+1,j+K_r} \\ \vdots & \vdots & \ddots & \vdots \\ f_{i+K_r,j-K_l} & f_{i+K_r,j-K_l+1} & \cdots & f_{i+K_r,j+K_r} \end{pmatrix}. \quad (3)$$

where $K_l = \lfloor \frac{K-1}{2} \rfloor$, $K_r = \lfloor \frac{K}{2} \rfloor$, and $\lfloor x \rfloor$ denotes the rounding operator. As seen from Eq. (3), the OGS regularization term utilizes $K \times K$ pixel neighborhood gradients within the image, where K denotes the group size. The overlapping group gradient for each pixel is defined as:

$$\varphi(f) = \sum_{i=1}^{K} \sum_{j=1}^{K} \|\tilde{f}_{i,j,k,k}\|_2 \quad (4)$$

3 The Proposed Method

3.1 TGV Regularization Term with First-Order OGS Based on L1 Norm Data Fidelity Term

For simplicity, the image is treated as a square matrix, and the second-order TGV denoising model is outlined in Eq. (5).

$$\begin{aligned} f = & \arg\min_{f} \{ \tfrac{1}{2}\|f-g\|_2^2 + \mu \mathrm{TGV}_\alpha^2(f) \} = \\ & \arg\min_{f,v} \{ \tfrac{1}{2}\|f-g\|_2^2 + \mu\alpha_0 \|\nabla f - v\|_1 + \mu\alpha_1 \|\varepsilon(v)\|_1 \} \end{aligned} \quad (5)$$

where $\tfrac{1}{2}\|f-g\|_2^2$ refers to the data fidelity term, while $\mathrm{TGV}_\alpha^2(f)$ indicates the second-order TGV. The Lagrange multiplier μ effectively balances the regularization and data fidelity terms. The parameters α_0 and α_1 are non-negative gradient weights that balance first- and second-order gradients. Here, f represents the denoised image, and f is the discrete vectorized form of the image. The noisy observation image is denoted as g. The gradient operator $\nabla f = [\nabla_x f; \nabla_y f]$ comprises horizontal ($\nabla_x f$) and and vertical ($\nabla_y f$) differential components. The vector $v = [v_x; v_y]$ enforces sparsity in the first-order gradient, where v_x and v_y denote its horizontal and vertical components. From this, we can derive the following equation:

$$\begin{aligned} (f,v) = \arg\min_{f,v} & \{ \tfrac{1}{2}\|f-g\|_2^2 + \mu\alpha_0(\|\nabla_x f - v_x\|_1 + \|\nabla_y f - v_y\|_1) + \\ & \tau\alpha_1(\|\nabla_x v_x\|_1 + \|\nabla_x v_y + \nabla_x v_y\|_1 + \|\nabla_y v_y\|_1) \} \end{aligned} \quad (6)$$

Although TGV has shown improvements [4], it has limitations in fully utilizing neighborhood information for each pixel, often overlooking valuable structural details and inadequately protecting weak edges and fine details [6]. In contrast, OGS regularization enhances the distinction between regions with smooth and edge gradients. In the

TGV model, the L2 norm effectively removes noise in smooth regions but frequently causes edge blurring [12]. Conversely, while the L1 norm-based regularization term preserves edge features, the L2 norm is highly sensitive to outliers, often resulting in excessive edge smoothing. Additionally, when dealing with non-Gaussian noise, the L2 norm is less effective at suppressing noise-induced outliers, limiting its denoising capability. On the other hand, the L1 norm provides sparser weights, benefiting sparse data and improving robustness against noise outliers in inverse problems [13].

To overcome these limitations, we have developed an extended TGV model, termed the TGV-F1-OGS-L1 model. This approach incorporates OGS within the first-order gradient component of the TGV regularization term and employs an L1 norm-based data fidelity term. By including both high-order gradients and first-order OGS, our method effectively eliminates staircase effects, compensating for the influence of the L1 norm within the data term. Furthermore, this approach retains the L1 norm's ability to protect edges, thereby fully achieving the goal of removing staircase artifacts while preserving edge details. The proposed model is defined in Eq. (7):

$$f = arg\min_{f} \{\frac{1}{2}\|f - g\|_1 + \mu TGV^2_{\alpha-F1-OGS}(f)\}$$

$$(f, v) = arg\min_{u,v} \{\frac{1}{2}\|f - g\|_1 + \mu\alpha_0[\varphi(\nabla_x f - v_x) + \varphi(\nabla_y f - v_y)] + \tau\alpha_1[(\nabla_x v_x) + (\nabla_x v_y + \nabla_x v_y) + (\nabla_y v_y)]\} \quad (7)$$

The proposed model in vector form potentially requires extensive matrix computations. To reduce computational complexity, the proposed model is reformulated in matrix form as:

$$(F, V) = \min_{F,V} \frac{1}{2}\|F - G\|_1 + \mu TGV^2_{\alpha-F1OGS}(F)$$
$$= \frac{1}{2}\|F - G\|_1 + \mu\{\alpha_0[\varphi(K_x * F - V_x) + \varphi(K_y * F - V_y)] \cdot \quad (8)$$
$$+ \alpha_1(K_x * V_x + K_y * V_y + K_y * V_x + K_x * V_y)\}.$$

where $TGV^2_{\alpha-F1OGS}(f)$ denotes the second-order TGV regularization combined with first-order OGS. The matrices F, G, V_x, V_y are defined, with $K_x = [1, -1]$ and $K_y = [\begin{smallmatrix}1\\-1\end{smallmatrix}]$. The symbol $*$ denotes the 2D convolution operation. The differential matrix satisfies the relation $\nabla_x f = vec(K_x * F)$ and $\nabla_y f = vec(K_y * F)$, where vec is the vectorization operator.

3.2 Numerical Algorithm

Solving Eq. (8) entails optimization challenges due to multiple L1 norms. The L1 norm's non-smooth nature poses significant challenges. To tackle these issues, a fast split Bregman framework is employed for solving Eq. (8).

First, we introduce auxiliary variables $S_x, S_y, L_{xx}, L_{xy}, L_{yy}$, and rewrite Eq. (8) as follows:

$$(F, V, S, L) = \min_{F.V.S.L} \tfrac{1}{2}\|F - G\|_1 + \mu TGV^2_{\alpha-F1OGS}(F)$$
$$= \min_{F,V,S,L} \tfrac{1}{2}\|F - G\|_1 + \mu\{\alpha_0[\varphi(S_x) + \varphi(S_y)] + \alpha_1(L_{xx} + L_{yy} + L_{xy})\}$$
$$\text{s.t.} \ S_x = K_x * F - V_x, \ S_y = K_y * F - V_y \quad (9)$$
$$L_{xx} = K_x * V_x, \ L_{yy} = K_y * V_y$$
$$L_{xy} = K_y * V_x + K_x * V_y$$

In the reformulated equations, we define the vectors $S = [S_x; S_y]$, $L = [l_{xx}; l_{xy}; l_{yy}]$, and $d = [d_x, d_y, d_{xx}; d_{xy}; d_{yy}]$. The split Bregman framework [9] is used to simplify the proposed model by dividing it into subproblems. Intermediate variables are introduced to decouple these subproblems.

$$L(F, S_x, S_y, L_{xx}, L_{yy}, L_{xy}, d_x, d_y, d_{xx}, d_{yy}, d_{xy}) =$$
$$\tfrac{1}{2}\|F - G\|_1 + \mu a_0[\varphi(S_x) + \varphi(S_y)] +$$
$$\mu a_1(L_{xx} + L_{yy} + L_{xy}) + \tfrac{\lambda_1}{2}\|S_x - (K_x * F - V_x) - d_x\|_2^2 +$$
$$\tfrac{\lambda_1}{2}\|S_y - (K_y * F - V_y) - d_y^k\|_2^2 + \tfrac{\lambda_2}{2}\|L_{xx} - K_x * V_x - d_{xx}^k\|_2^2 +$$
$$\tfrac{\lambda_2}{2}\|L_{yy} - (K_y * V_y) - d_{yy}^k\|_2^2 + \tfrac{\lambda_2}{2}\|L_{xy} - (K_y * V_x + K_x * V_y) - d_{xy}^k\|_2^2\} \quad (10)$$

Non-negative penalty parameters λ_1 and λ_2 are established following the principles outlined in [9], ensuring that as the auxiliary variable d approaches zero, the optimization problem in Eq. (10) converges to the solution of Eq. (8).

To effectively solving the optimization problem in Eq. (10), the split Bregman iteration begins by alternately updating each variable F, V, S, L, and d via minimization. This iterative process breaks down into several optimization subproblems. Specifically, for the F subproblem, the objective function defined as:

$$F^{k+1} = \tfrac{1}{2}\|F - G\|_1 + \tfrac{\lambda_1}{2}(\|S_x^{(k)} - (K_x * F - V_x) - d_x^k\|_2^2$$
$$+ \|S_y^{(k)} - (K_y * F - V_y) - d_y^k\|_2^2) \quad (11)$$

The objective function of V_x sub-problem is:

$$V_x^{k+1} = \tfrac{\lambda_1}{2}(\|V_x + S_x^{(k)} - K_x * F - d_x^k\|_2^2 +$$
$$\|L_{xx}^{(k)} - K_x * V_x - d_{xx}^k\|_2^2) +$$
$$\tfrac{\lambda_2}{2}\|L_{xy}^{(k)} - (K_y * V_x + K_x * V_y) - d_{xy}^k\|_2^2. \quad (12)$$

The objective function of V_y sub-problem is:

$$V_y^{k+1} = \tfrac{\lambda_1}{2}(\|V_y + S_y^{(k)} - K_y * F - d_y^k\|_2^2) +$$
$$\tfrac{\lambda_2}{2}[\|L_{yy}^{(k)} - K_y * V_y - d_{yy}^k\|_2^2 + \|L_{xy}^{(k)} - (K_y * V_x + K_x * V_y) - d_{xy}^k\|_2^2] \quad (13)$$

The objective function of L sub-problem can be solved using a nonlinear contraction operator, denoted as $\text{shrink}_u()$.

$$\begin{cases} L_{xx}^{(k+1)} = \text{shrink}_{\frac{\mu\alpha_1}{\lambda_2}}\left(K_x * V_x^{(k+1)} + d_{xx}^k\right) \\ L_{yy}^{(k+1)} = \text{shrink}_{\frac{\mu\alpha_1}{\lambda_2}}\left(K_y * V_y^{(k+1)} + d_{yy}^k\right) \\ L_{xy}^{(k+1)} = \text{shrink}_{\frac{\mu\alpha_1}{\lambda_2}}\left(K_y * V_x^{(k+1)} + K_x * V_y^{(k+1)} + d_{xy}^k\right) \end{cases} \quad (14)$$

Auxiliary variable update:

$$\begin{aligned} d_x^{k+1} &= d_x^k + \left(K_x * F^{k+1} - V_x^{k+1} S_x^{(k)} - S_x^{k+1}\right) \\ d_y^{k+1} &= d_y^k + \left(K_y * F^{k+1} - V_y^{k+1} - S_y^{k+1}\right) \\ d_{xx}^{k+1} &= d_{xx}^k + K_x * V_x^{k+1} - L_{xx}^{k+1} \\ d_{xy}^{k+1} &= d_{xy}^k + \left(K_x * V_y^{k+1} + K_y * V_x^{k+1}\right) - L_{xy}^{k+1} \\ d_{yy}^{k+1} &= d_{yy}^k + K_y * V_y^{k+1} - L_{yy}^{k+1} \end{aligned} \quad (15)$$

3.3 Experimental Setup

Two experiments were conducted to assess the proposed model's performance under Gaussian white noise. Three 512×512-resolution grayscale images, illustrated in Fig. 1, were used in the experiments. The red rectangles in these images highlight the regions of interest for observing the results after denoising. The images tested the model's ability to remove Gaussian white noise and retain structural details.

Before denoising, the group size K in Eq. (3) needed careful selection to balance between avoiding excessive size that introduces unstructured data and ensuring the inclusion of relevant neighborhood structural information. To investigate the effect of K on the model, all other parameters were fixed, and the model was used to denoise the "Barbara" image contaminated with zero-mean Gaussian noise ($\sigma = 30$). The resulting denoised images were analyzed to determine the optimal selection of K. The 30% zero-mean Gaussian white noise level was chosen as it represents a challenging scenario for image denoising, allowing a rigorous evaluation of the proposed method's ability to suppress noise while preserving fine image details. This noise level is also commonly used in the literature, ensuring a fair comparison with existing methods and providing a consistent baseline for performance evaluation.

The proposed model's denoising and staircase elimination capabilities were compared with state-of-the-art models, including TV [2], TV-OGS [5], and TGV [4]. All models were implemented in MATLAB 2022b using a 4.6 GHz CPU and 16 GB RAM laptop.

To evaluate denoising quality quantitatively, three metrics were used: Peak signal-to-noise ratio (PSNR), structural similarity index (SSIM), and processing time [14]. PSNR quantifies the ratio of peak signal energy to average noise energy, where higher values reflect reduced noise or distortion in the denoised image.

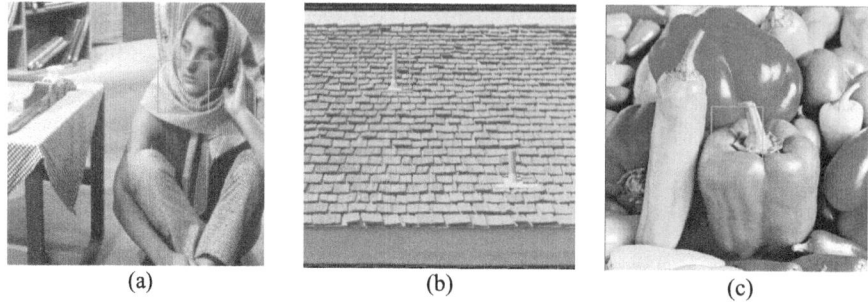

Fig. 1. (a) "Barbara" (b) "Roof" (c) "Peppers" images.

4 Results and Discussion

The first experiment emphasizes the significance of choosing an optimal group size K to enhance the proposed model's denoising performance. As shown in Fig. 2, the highest PSNR and SSIM values are achieved when K is set to 3, establishing $K = 3$ as the optimal group size for our model.

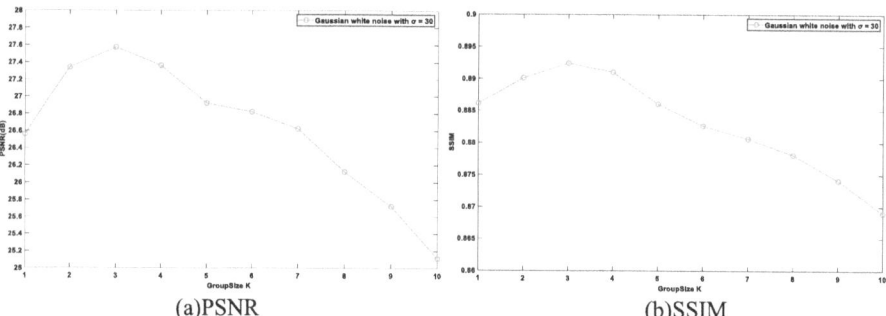

(a)PSNR   (b)SSIM

Fig. 2. PSNR and SSIM values after the "Barbara" image denoising using various group sizes K.

To illustrate the processing effects, we applied 30% Gaussian white noise and presented the denoising outcomes for the "Barbara" image at various values of K in Fig. 5. The figure demonstrates that reducing K results in substantial residual noise, while increasing K leads to excessive smoothing of texture details. Optimal texture preservation, with no apparent residual noise, is achieved at $K = 3$. The selection of parameters, such as regularization weights and group size K, should be adapted to the characteristics of the image, including noise level and texture complexity. For high noise levels, larger regularization weights can enhance noise suppression, while for images with fine textures, smaller weights help preserve details. The group size K should balance noise reduction and detail preservation, with $K = 3$ found to be optimal in this study for typical textured images.

Gaussian white noise was applied at noise levels 10, 20 and 30. We compared denoising methods including TV, TV-OGS, TGV, and our proposed method. The results for the

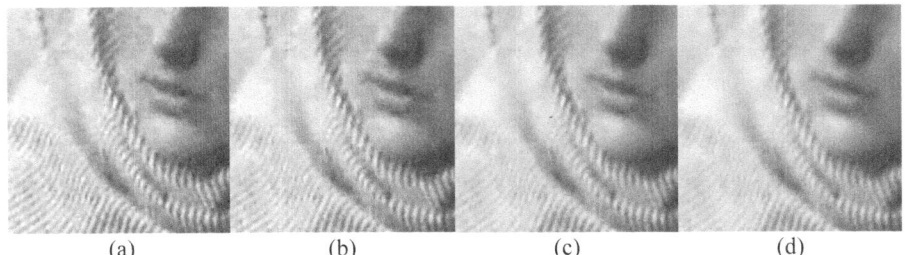

Fig. 3. The denoising performance of the proposed model on the "Barbara" image with 30% Gaussian noise at different K values: (a) $K = 2$, (b) $K = 3$, (c) $K = 5$, and (d) $K = 7$.

three images are presented in Table 1, where bold font highlights optimal performance indicators.

Table 1. Denoising performance of different methods.

Noise	Method	"Barbara" image			"Roof" image			"Peppers" image		
		PSNR	SSIM	Time	PSNR	SSIM	Time	PSNR	SSIM	Time
10%	TV	34.159	0.968	**0.58**	28.434	0.941	**0.63**	31.271	0.937	**0.51**
	TV-OGS	34.911	0.977	0.73	29.361	0.963	0.91	32.151	0.945	0.89
	TGV	35.075	0.973	1.27	29.365	0.960	1.49	32.239	0.949	1.38
	Our	**36.153**	**0.987**	1.82	**30.982**	**0.984**	2.21	**33.387**	**0.961**	2.18
20%	TV	28.337	0.925	**0.49**	26.219	0.935	**0.55**	28.561	0.908	**0.67**
	TV-OGS	39.867	0.942	0.97	27.035	0.651	0.85	29.356	0.915	1.01
	TGV	39.694	0.939	1.43	27.191	0.949	1.34	29.234	0.917	1.56
	Our	**31.615**	**0.957**	2.07	**28.464**	**0.969**	1.98	**31.005**	**0.939**	2.47
30%	TV	26.141	0.870	**0.58**	24.716	0.919	**0.56**	28.556	0.870	**0.47**
	TV-OGS	27.003	0.881	1.11	25.572	0.927	0.91	29.704	0.885	0.94
	TGV	26.816	0.884	1.35	25.566	0.929	1.11	29.243	0.886	1.50
	Our	**27.769**	**0.892**	2.19	**26.234**	**0.943**	2.04	**30.051**	**0.898**	2.28

The table shows that our proposed model achieves higher PSNR and SSIM values than other methods, demonstrating its superior denoising performance under Gaussian white noise. Notably, TGV emerges as the second-best performer, reflecting the benefits of incorporating the second-order gradient. However, TGV requires more computational time than TV-OGS. Due to the complexity of our proposed model, it has the slowest computational speed among the compared algorithms.

To provide a clearer understanding of the denoising efficacy, we conducted a comparative analysis of the three images using various models. Figures 4, 5 and 6 present the performance outcomes after processing with both the proposed model and other methods. These images were subjected to 30% Gaussian white noise, and denoising results were obtained using TV, TV-OGS, TGV, and our proposed method, enabling a comprehensive comparison.

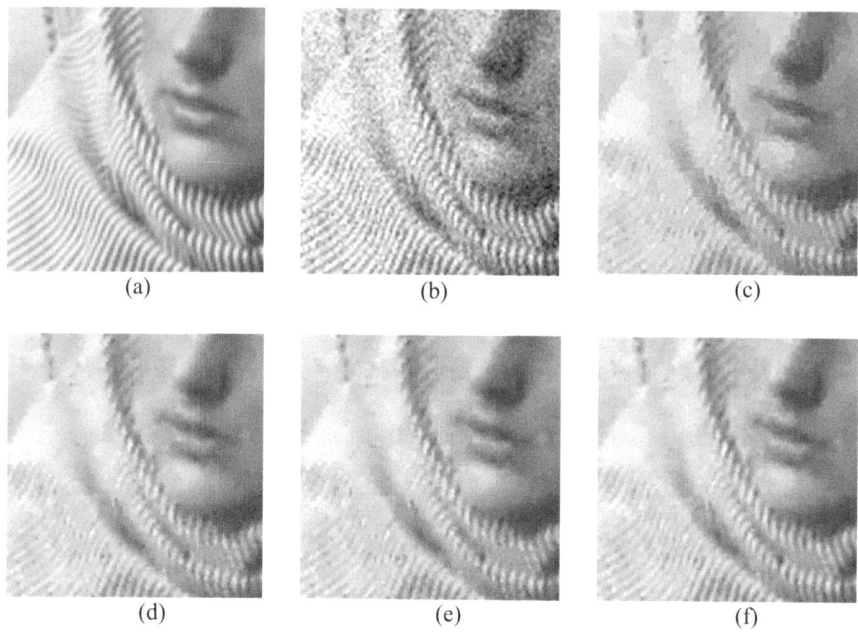

Fig. 4. (a) Enlarged patch of the "Barbara" image, (b) with noise, (c) denoised using TV, (d) denoised using TV-OGS, (e) denoised using TGV, (f) denoised using the proposed model.

Figure 4 presents enlarged patches of the "Barbara" image before and after denoising. In Fig. 4(c), the TV model introduces a noticeable staircase effect, particularly on the face, where block-like structures appear that were not present in the original image. Figures 4(d) and 4(e) show significant improvement over 4(c), with a noticeable reduction in staircase effects. However, minor residual staircase effects are still visible in Figs. 4(d) and 4(e), especially around the mouth and nose. A closer inspection of the cheek texture reveals that Figs. 4(d) and 4(e) exhibit some blurring compared to Fig. 4(f).

Figure 5 shows enlarged patches of the "Roof" image before and after denoising. Noticeable staircase effects are apparent in Fig. 5(c), where the denoised images have undergone significant changes compared to the original image due to the staircase effect. Additionally, several white spots, representing residual noise, are observable in Figs. 5(c) through 5(e). These staircase effects and residual noise are significantly reduced in Fig. 5(g). A close examination of the tile edges in Figs. 5(c) through 5(f) reveals that only the edges in Fig. 5(f) exhibit the sharpest features, highlighting the superior edge protection capability of the proposed model.

Figure 6 presents enlarged patches of the "Peppers" image before and after denoising. A detailed examination of the pepper stems reveals a distinct staircase effect in Figs. 6(c), accompanied by slight edge blurring. While no obvious staircase effect is present in

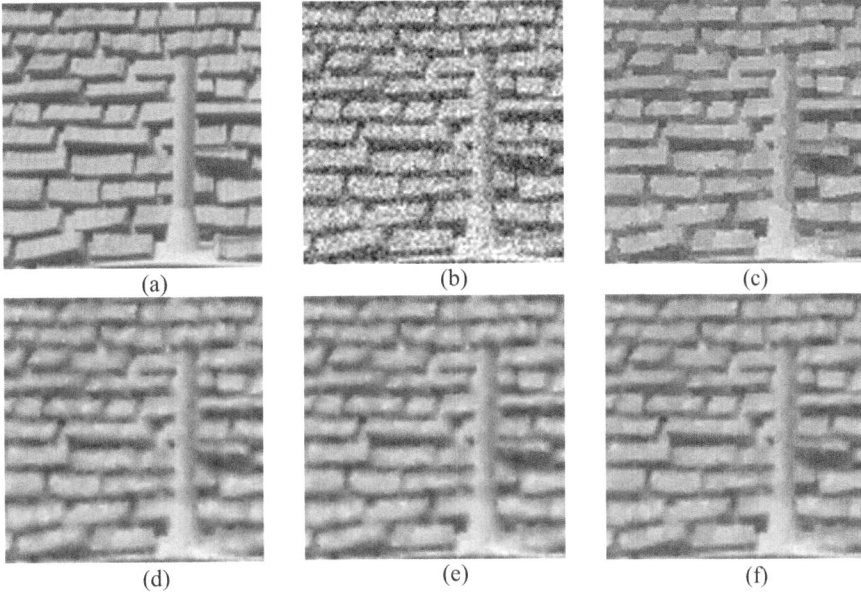

Fig. 5. (a) Enlarged patch of the "Roof" image, (b) with noise, (c) denoised using TV, (d) denoised using TV-OGS, (e) denoised using TGV, (f) denoised using the proposed model.

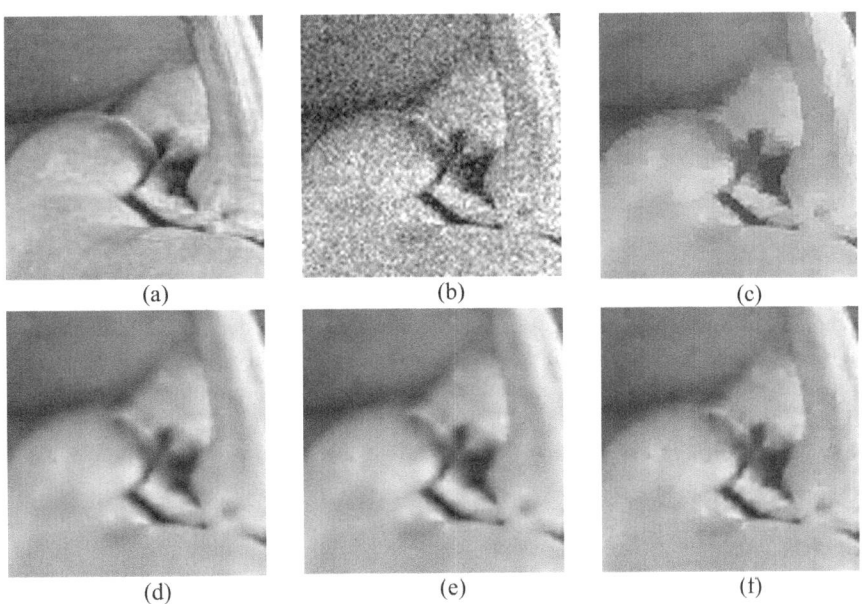

Fig. 6. (a) Enlarged patch of the "Peppers" image, (b) with noise, (c) denoised using TV, (d) denoised using TV-OGS, (e) denoised using TGV, (f) denoised using the proposed model.

Figs. 6(d) and 6(e), the edges of the pepper stalks show some blurring. Notably, only the relevant portion of Fig. 6(f) displays the sharpest edge and the most effective edge protection.

In terms of visual quality, the TV method exhibited residual noise, and some edges showed noticeable blurring during the denoising process. In contrast, the images reconstructed by TV-OGS and TGV demonstrated superior denoising performance compared to TV. However, traces of noise remained in the reconstructed images produced by TV-OGS and TGV, along with some loss of edge detail, indicating limitations in noise removal and detail preservation. In contrast, the proposed model effectively eliminated residual noise while maintaining the integrity of the image's texture and edges.

The proposed method's computational efficiency is enhanced by employing the Split Bregman method for L1-based problems. Our previous work [6] systematically compared the Split Bregman method and ADMM in models with L1 data fidelity terms. Results showed that the Split Bregman method converges faster with similar or higher accuracy by simplifying the optimization problem into subproblems.

This efficiency is particularly significant in handling sparse structures and regularization terms, as the Split Bregman method leverages fast updates and Fourier-based operations, reducing overall computational time. While ADMM is more versatile and can handle a wider range of constraints, its slower convergence in scenarios dominated by L1-based fidelity terms makes it less suitable for the proposed model, which heavily relies on L1 regularization.

By adopting the Split Bregman framework, the proposed method achieves efficient computation while maintaining high-quality denoising performance. The proposed model demonstrates excellent denoising performance but has some limitations. Its computational cost is higher compared to simpler models like TV or TV-OGS, primarily due to the inclusion of both TGV and OGS terms. Additionally, the model may exhibit sensitivity to non-Gaussian noise types, requiring careful parameter tuning to maintain optimal performance.

5 Conclusion

This paper introduces a denoising model integrating second-order TGV, first-order OGS regularization, and an L1 norm data fidelity term. To efficiently address L1-norm regularization in the energy equation, a fast split Bregman method is employed. Comparative analyses with state-of-the-art models show that the model removes Gaussian noise effectively while excelling in preserving image details. However, it should be noted that the model requires manual adjustment of the regularization parameter to suit the specific characteristics of the input images. Future research may focus on developing adaptive parameter adjustments based on a comprehensive analysis of different noise profiles across various images.

References

1. Sagheer, S.V.M., George, S.N.: A review on medical image denoising algorithms. Biomed. Signal Process. Control **61**, 102036 (2020)

2. Zhou, Y., Guo, Z., Li, Y., Yao, W., Wu, B.: A variational model to remove multiplicative noise based on SAR image feature preservation. Inverse Probl. Imaging **19**(2), 253–281 (2024)
3. Rudin, L.I., Osher, S., Fatemi, E.: Nonlinear total variation based noise removal algorithms. Physica D **60**(1–4), 259–268 (1992)
4. Bredies, K., Kunisch, K., Pock, T.: Total generalized variation. SIAM J. Imag. Sci. **3**(3), 492–526 (2010)
5. Liu, J., Huang, T.Z., Selesnick, I.W., Lv, X.G., Chen, P.Y.: Image restoration using total variation with overlapping group sparsity. Inf. Sci. **295**, 232–246 (2015)
6. Zhang, C., Yen, K.S.: Denoising on textured image using total generalized variation with overlapping group sparsity based on fast split bregman method. IEEE Access **12**, 19145–19157 (2024)
7. Chan, T.F., Esedoglu, S.: Aspects of total variation regularized L 1 function approximation. SIAM J. Appl. Math. **65**(5), 1817–1837 (2005)
8. Ningshan, X., Chen, W., Guoqiang, R., Yongmei, H.: Blind image restoration method regularized by hybrid gradient sparse prior. Opto-Electron. Eng. **48**(6), 210040–1 (2021)
9. Goldstein, T., Osher, S.: The split Bregman method for L1-regularized problems. SIAM J. Imag. Sci. **2**(2), 323–343 (2009)
10. Sun, Y., Babu, P., Palomar, D.P.: Majorization-minimization algorithms in signal processing, communications, and machine learning. IEEE Trans. Signal Process. **65**(3), 794–816 (2016)
11. Shu, Q.: Spatially adaptive TGV-regularized variational model for single image dehazing. J. Electron. Imaging **30**(6), 063018 (2021)
12. Liu, X., Sun, T.: Mixed Gaussian-impulse noise removal using non-convex high-order TV penalty. Appl. Numer. Math. **201**, 72–84 (2024)
13. Calvetti, D., Pragliola, M., Somersalo, E.: Sparsity promoting hybrid solvers for hierarchical Bayesian inverse problems. SIAM J. Sci. Comput. **42**(6), A3761–A3784 (2020)
14. Wang, Z., Bovik, A.C., Sheikh, H.R., Simoncelli, E.P.: Image quality assessment: from error visibility to structural similarity. IEEE Trans. Image Process.ans. Image Process. **13**(4), 600–612 (2004)

Application of Underwater Flexible Manipulator in Teaching Nonlinear Systems and Control Courses

Zixing Zhao and Jie Huang(✉)

School of Mechanical Engineering, Beijing Institute of Technology, Beijing 100081, China
bit_huangjie@bit.edu.cn

Abstract. This paper introduces an innovative teaching approach for the nonlinear systems and intelligent control course, leveraging underwater flexible manipulator to enhance both theoretical comprehension and practical application. The proposed approach integrates fundamental knowledge of mechanical systems with simulation experiments, providing master's candidates with a comprehensive grasp of modeling, analysis, identification, nonlinear dynamics, and control. By utilizing an underwater flexible manipulator model, students gain a deeper insight into dynamic characteristics and vibration control, particularly regarding the interaction between fluid and the flexible structure, which produce unique dynamic phenomena. This novel teaching method, adaptable to both on-campus and online formats, incorporates modeling of fluid and manipulator dynamics, frequency and modal shape analyses, as well as time-domain and parameter variation simulations. Furthermore, this approach extends to physical experimental platforms such as PLCs and cranes. The paper details course assignments, tests, and an advanced study segment, offering a robust framework for engaging students in nonlinear systems analysis and control.

Keywords: Education · Curriculum · Customized Teaching Approach · Underwater Flexible Manipulator · Simulation

1 Introduction

The nonlinear systems and intelligent control course includes theoretical lectures to teach fundamental knowledge and simulation experiments to verify dynamic characteristics and achieve vibration control [1, 2]. Key topics cover modeling, analysis, identification, nonlinear dynamics, optimization, and intelligent control of mechanical systems [3, 4]. Targeted at master's candidates, the course builds upon foundational concepts from prerequisite courses such as fluid mechanics and mechanical vibrations. Simulation experiments using an underwater flexible manipulator model allow students to develop an intuitive and profound understanding of the learned concepts.

This paper presents a new teaching method based on theoretical lectures and simulation experiments using the underwater flexible manipulator model. The vibration of the flexible manipulator, characterized by infinite modes, has gathered significant attention from researchers studying its dynamic properties and control [5, 6]. In underwater environments, flexible manipulators interact with the fluid, exhibiting unique dynamic phenomena [7, 8]. The proposed teaching method is also applicable to physical experimental platforms, such as PLCs and cranes [9, 10].

The main contribution of this paper lies in proposing an adaptive, customized and spatially unrestricted teaching method that can be conducted both on-campus and online. Section 2 introduces the course homework, which includes fluid and manipulator dynamics, frequency analysis, and modal shape analysis. Section 3 details the course tests, covering time-domain simulations, parameter variation simulations, drive distance variation simulations, and controlled system simulations. Section 4 provides an advanced study segment for students to explore the course content further.

2 Homework

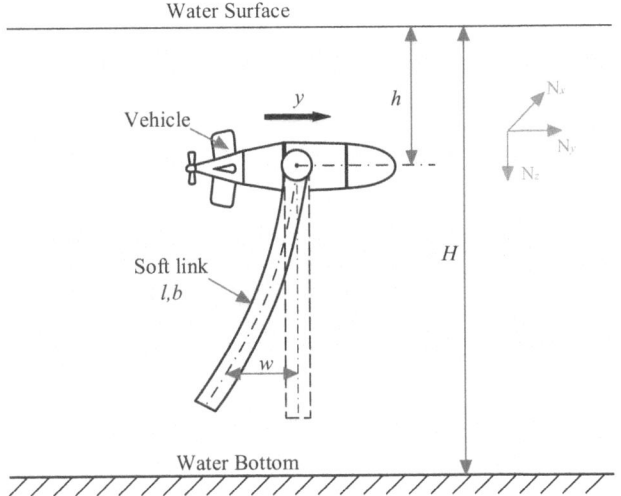

Fig. 1. Model of underwater flexible manipulator.

A series of assignments can progressively enhance students' understanding of theoretical courses and reinforce acquired knowledge. Students are required to complete assignments related to fluid dynamics modeling, manipulator dynamics modeling, and frequency analysis as part of this study. Through these tasks, students consolidate their grasp of the course material.

The first assignment requires students to extract a physical model diagram from real-world models. Specifically, students are to draw the model of an underwater flexible manipulator as depicted in Fig. 1. Based on the model diagram, the inertial coordinate system is defined as $N_x N_y N_z$. Students can observe that the vehicle moves under the water. The system's input is the acceleration of the vehicle, denoted as d^2y/dt^2, while the output consists of the deflection of the flexible manipulator, denoted as w, and the velocity potential of the fluid, denoted as ϕ. The cylindrical coordinate system, represented as $R_r R_\theta R_z$, is established with the center of contact between the vehicle and the flexible manipulator's surface as the origin, rigidly attached to the vehicle. The angle between the R_r-axis of the cylindrical coordinate system and the N_x direction is θ. The radial distance from the origin is r. The height in the direction perpendicular to the r-θ plane is z. The liquid depth is denoted as H, the depth of the vehicle from the water surface as h, and the length and radius of the flexible manipulator as l and b, respectively.

2.1 Modeling

Fluid Dynamics. Employing the **mode-superposition method**, the fluid flow's velocity potential can be represented as:

$$\phi(r, \theta, z, t) = \sum_{s=1}^{+\infty} [R_s(r, \theta, z) T_s(t)] \quad (1)$$

where $R_s(r,\theta,z)$ is the mode shape for the sth order mode of velocity potential, and $T_s(t)$ is the corresponding time function.

To simplify the complex physical model, reasonable assumptions are introduced: The fluid in the flow field is an ideal, incompressible, inviscid fluid. The fluid motion is irrotational. The bottom of the flow field is rigid without elastic deformation. There are no fluctuations on the surface of the flow field. The mass of the vehicle is sufficiently large that the motion of the manipulator and the fluid does not affect it. Additionally, the vehicle does not cause any disturbance to the flow field. Under these assumptions, the expression for the mode shape function of the velocity potential can be derived as:

$$R_s(r, \theta, z) = E_s K_1(mr) \sin\theta \sin(mz + mh) \quad (2)$$

where K_1 is the second kind of first-order Bessel function. Based on the aforementioned assumptions, at the interface between the flow field and the manipulator, normal velocity continuity conditions are applied to determine the coefficient E_s. The expression of the coefficient, m, is:

$$m = \frac{(2s-1)\pi}{2H} \quad (3)$$

Disregarding the coupling between different modes of velocity potential, students can yield the velocity potential function:

$$\phi(r, \theta, z, t) = \sum_{s=1}^{+\infty} \left(K_1(mr) \sin\theta \sin(mz + mh) \left[D_1 \int_{z=0}^{l} \frac{\partial w}{\partial t} \sin(mz + mh) dz + D_2 \frac{dy}{dt} \right] \right) \quad (4)$$

where w represents the deflection of the flexible manipulator, and the expressions for coefficients D_1 and D_2 are:

$$D_1 = \frac{1}{m \frac{\partial K_1(mb)}{\partial r} \int_{z=0}^{l} \sin^2(mz + mh) dz} \quad (5)$$

$$D_2 = \frac{\int_{z=0}^{l} \sin(mz + mh)dz}{m\frac{\partial K_1(mb)}{\partial r} \int_{z=0}^{l} \sin^2(mz + mh)dz} \tag{6}$$

Based on Euler's fluid dynamics equation and neglecting higher-order terms, students can obtain:

$$\frac{p}{\rho_w} + \frac{\partial \phi}{\partial t} + gz - \mathbf{V}_e \cdot \nabla \phi = 0 \tag{7}$$

where p is the water pressure, ρ_w represents the liquid density, indicates the dot product operator, ∇ represents the gradient operator, and $\mathbf{V}_e$ denotes the vibrational velocity of the manipulator in the inertial coordinate system. Substituting into the yields the expression for the hydrodynamic force:

$$F_y(z,t) = \pi \rho_w b \sum_{s=1}^{+\infty} \left(K_1(mb) \sin(mz + mh) \left[D_1 \int_{z=0}^{l} \frac{\partial^2 w}{\partial t^2} \sin(mz + mh)dz + D_2 \frac{d^2 y}{dt^2} \right] \right)$$

$$-2\rho_w \left(\frac{\partial w}{\partial t} + \frac{dy}{dt} \right) \cos(\varepsilon) \sum_{s=1}^{+\infty} \left(D_3 \sin(mz + mh) \left[D_1 \int_{z=0}^{l} \frac{\partial w}{\partial t} \sin(mz + mh)dz + D_2 \frac{dy}{dt} \right] \right) \tag{8}$$

where ε denotes the modified flow separation point, which can be determined experimentally. The expression for coefficient D_3 is:

$$D_3 = mb \frac{\partial K_1(mb)}{\partial r} + K_1(mb) \tag{9}$$

Based on the aforementioned assignments, students can understand that hydrodynamics is induced by the motion of the vehicle and the deflection of the manipulator. They can also derive expressions for hydrodynamics, preparing for the subsequent modeling of the flexible manipulator.

Manipulator Dynamics. Using **Newton's second law**, students can derive the force and moment equations for a small element on the manipulator:

$$\rho_L A_L dz \frac{\partial^2 w}{\partial t^2} + dV + \rho_L A_L dz \frac{d^2 y}{dt^2} - F_y dz = 0 \tag{10}$$

$$dM - Vdz - dVdz + 0.5\rho_L A_L \left(\frac{d^2 y}{dt^2} + \frac{\partial^2 w}{\partial t^2} \right) dz^2 - 0.5 F_y dz^2 = 0 \tag{11}$$

where ρ_L represents the linear mass density of the manipulator, A_L indicates the cross-sectional area of the manipulator, and V and M denote the shearing force and the bending moment on the manipulator. The deflection of the manipulator can also be derived using the modal superposition method:

$$w(z,t) = \sum_{k=1}^{+\infty} \left[W_k(z) q_k(t) \right] \tag{12}$$

where $q_k(t)$ denotes the kth vibrational mode's time-dependent function, while $W_k(z)$ represents the corresponding mode shape at a distance z. By considering the boundary conditions of the manipulator and the relationship between bending moment and deflection, students can derive the natural frequencies and mode shapes of the manipulator:

$$\omega_k = \lambda_k^2 \sqrt{\frac{EI}{\rho_L A_L}} \tag{13}$$

$$W_k(z) = C \left(\begin{array}{c} \sin(\lambda_k z) + \alpha_1 \cos(\lambda_k z) + \alpha_2 \sinh(\lambda_k z) + \alpha_3 \cosh(\lambda_k z) \\ + \sum_{s=1}^{+\infty} \frac{G_s \sin(mz+mh)[I_{s,0}+\alpha_1 I_{s,1}+\alpha_2 I_{s,2}+\alpha_3 I_{s,3}]}{\left[1-G_s \int_{z=0}^{l} \sin^2(mz+mh)dz\right]} \end{array} \right) \tag{14}$$

XX where ω_k denotes natural frequency of the kth vibrational mode, E is Young's modulus, I denotes the moment of inertia of the manipulator cross section, and λ_k, α_1, α_2, α_3 are constant coefficients. The expressions for $G_s, I_{s,0}, I_{s,1}, I_{s,2}, I_{s,3}$ in the vibration mode function are respectively:

$$G_s = \frac{-\pi \rho_w b \lambda_k^4 D_1 K_1(mb)}{\rho_L A_L (m^4 - \lambda_k^4)} \tag{15}$$

$$I_{s,0} = \int_{z=0}^{l} \sin(\lambda_k z) \sin(mz+mh) dz \tag{16}$$

$$I_{s,1} = \int_{z=0}^{l} \cos(\lambda_k z) \sin(mz+mh) dz \tag{17}$$

$$I_{s,2} = \int_{z=0}^{l} \sinh(\lambda_k z) \sin(mz+mh) dz \tag{18}$$

$$I_{s,3} = \int_{z=0}^{l} \cosh(\lambda_k z) \sin(mz+mh) dz \tag{19}$$

For a given set of system parameters, numerical solutions for the aforementioned parameters can be obtained, thereby determining the vibration frequency and vibration mode shape of the manipulator. Assuming no coupling effect between the vibration modes of the manipulator, students can derive nonlinear equations for time-domain functions with proportional damping:

$$\frac{\partial^2 q_k}{\partial t^2} + 2\zeta_k \omega_k \frac{\partial q_k}{\partial t} + \omega_k^2 q_k + \gamma_k \frac{d^2 y}{dt^2} + e_{k,1} \frac{\partial q_k}{\partial t} \frac{dy}{dt} + e_{k,2} \left(\frac{\partial q_k}{\partial t}\right)^2 + e_{k,3} \left(\frac{dy}{dt}\right)^2 + e_{k,4} \omega_k^2 q_k^3 = 0 \tag{20}$$

where:

$$\gamma_k = \frac{\rho_L b \int_{z=0}^{l} W_k dz - \rho_w \sum_{s=1}^{+\infty} [K_1(mb) D_2 D_4]}{\rho_L b \int_{z=0}^{l} W_k^2 dz - \rho_w \sum_{s=1}^{+\infty} \left(K_1(mb) D_1 D_4^2\right)} \tag{21}$$

$$e_{k,1} = \frac{2\rho_w \cos(\varepsilon) \sum\limits_{s=1}^{+\infty} \left[D_1 D_3 D_4^2 + D_2 D_3 \int_{z=0}^{l} W_k^2 \sin(mz + mh) dz \right]}{\rho_L A_L \int_{z=0}^{l} W_k^2 dz - \pi \rho_w b \sum\limits_{s=1}^{+\infty} \left[K_1(mb) D_1 D_4^2 \right]} \quad (22)$$

$$e_{k,2} = \frac{2\rho_w \cos(\varepsilon) \sum\limits_{s=1}^{+\infty} \left[D_1 D_3 D_4 \int_{z=0}^{l} W_k^2 \sin(mz + mh) dz \right]}{\rho_L A_L \int_{z=0}^{l} W_k^2 dz - \pi \rho_w b \sum\limits_{s=1}^{+\infty} \left(K_1(mb) D_1 D_4^2 \right)} \quad (23)$$

$$e_{k,3} = \frac{2\rho_w \cos(\varepsilon) \sum\limits_{s=1}^{+\infty} (D_2 D_3 D_4)}{\rho_L A_L \int_{z=0}^{l} W_k^2 dz - \pi \rho_w b \sum\limits_{s=1}^{+\infty} \left(K_1(mb) D_1 D_4^2 \right)} \quad (24)$$

$$e_{k,4} = \frac{-1.5 \int_{z=0}^{l} W_k \frac{\partial^2 \left(\frac{\partial^2 W_k}{\partial z^2} \left(\frac{\partial W_k}{\partial z} \right)^2 \right)}{\partial z^2} dz}{\int_{z=0}^{l} W_k \frac{\partial^4 W_k}{\partial z^4} dz} \quad (25)$$

XX Through these tasks, the students complete the dynamic modeling of the fluid and the manipulator systems. They have derived the system's dynamic equations as shown in Eq., as well as the system's frequencies and mode shapes.

2.2 Analysis of Frequency

After completing the fluid dynamics and manipulator dynamics modeling, students obtain complex nonlinear dynamic equations. The natural frequency is one of the key characteristics of the system, which can be determined using the given system parameters. As indicated by Eq. (12), the response of the underwater flexible manipulator is a sum of infinite modes. The time functions of the first few modes are more significant, contributing more to the deflection. Furthermore, high-mode frequencies are difficult to detect with sensors; thus, higher-order modes are neglected.

Students use MATLAB software to observe the frequency variation trends of the first and second modes by changing the system parameters, and plot Fig. 2. Then, they need to write conclusions based on the calculated frequencies and mode shapes.

When the water depth is fixed and the manipulator length varies, both the first-mode frequency and the second-mode frequency decrease as the manipulator length increases. Additionally, it is evident from the figure that the second-mode frequency is significantly higher than the first-mode frequency.

When the manipulator length is kept constant and the liquid depth is varied, the students obtain the plot of the first- and second-mode frequencies as a function of liquid depth, as shown in Fig. 3. It can be observed that the first- and the second-mode frequencies decrease as the liquid depth increases. Consistent with the previous observations, the second-mode frequency is significantly higher than the first-mode frequency.

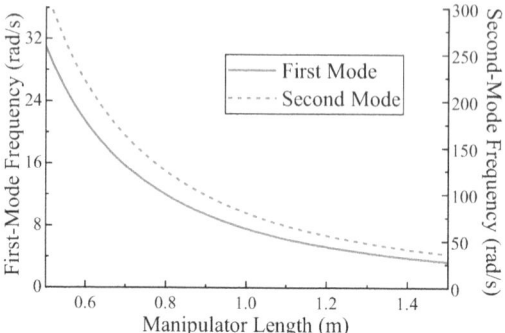

Fig. 2. First two frequencies vary with the manipulator length.

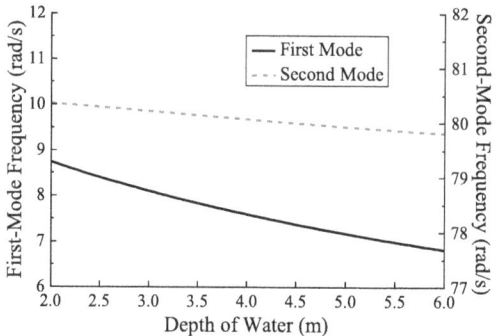

Fig. 3. First two frequencies vary with the liquid depth.

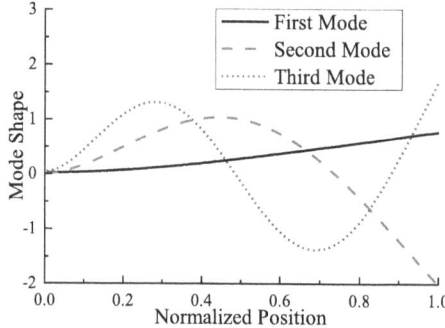

Fig. 4. First three mode shapes vary with the normalized position.

2.3 Analysis of Mode Shape

Students also need to observe the mode shapes of the first three modes by varying the normalized position and plot these mode shapes as shown in Fig. 4. Regarding the mode shapes, the mode shape of the first mode of the manipulator vibration increases with increasing normalized position. The mode shape of the second mode increases and then.

decreases with increasing normalized position. Meanwhile, the mode shape of the third mode increases, then decreases, and increases again with the increase of the normalized position. This homework provides them with a deeper understanding of the subject and the course.

3 Tests

The previous assignment addresses system modeling and dynamics analysis. In the tests, students are required to write simulation code based on the system model. First, in MATLAB, they need to develop the mathematical model to solve for the system's key parameters. Then, using the ode45 function, they compute various physical quantities of the system at each moment in time. Finally, they obtain the time-domain response of the system, plot the results, and perform an analysis.

3.1 Time-Domain Response Simulation

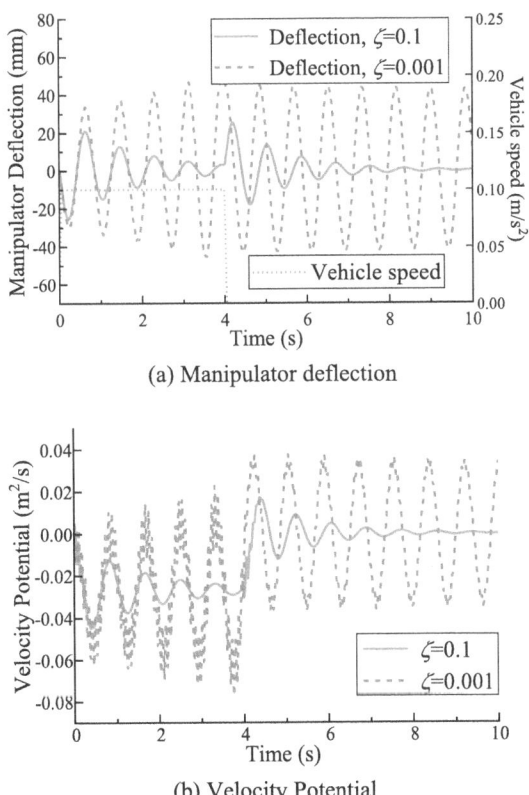

(a) Manipulator deflection

(b) Velocity Potential

Fig. 5. Time responses of manipulator deflection and velocity potential.

Students select the simulation model parameters: the Young's modulus, density, length, and radius of the manipulator are 291 MPa, 1200 kg/m3, 1 m, and 5 cm, respectively. The depth and fluid density of the flow field are 4 m and 1000 kg/m3, respectively. Under the condition where the trapezoidal velocity command drives the vehicle to move 40 cm, the students plot the system response graph, as shown in Fig. 5. The curves reflect the deflection at the end of the manipulator and the changes in the velocity potential.

From Fig. 5, it can be observed that due to the movement of the vehicle, the deflection of the flexible manipulator and the velocity potential exhibit periodic changes, and their change frequencies are identical. When the damping ratio is relatively small, the vibration frequency matches the first- and second-mode frequencies obtained in the previous chapter. However, when the damping is relatively large, the higher-order modal vibrations rapidly decay due to the damping effect, making them difficult to detect. Additionally, the presence of damping causes the amplitude of the deflection and velocity potential vibrations to decrease over time. Figure 5 also presents the speed curve of the vehicle.

Students can also determine that the transient amplitude of the deflection is 46.6 mm and the residual amplitude is 43.5 mm. The transient amplitude of the velocity potential is 4.96×10^{-2} m^2/s, and the residual amplitude is 3.89×10^{-2} m^2/s.

3.2 Structural Parameter Variations

Based on the time-domain response simulation, students obtain the transient and residual amplitudes of deflection and velocity potential for a given set of structural parameters. By altering the structural parameters, they generate response graphs of the system under different configurations. This exercise aids students in understanding robustness as they learn about controller-related concepts.

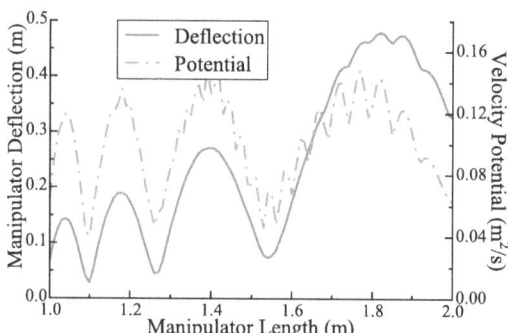

Fig. 6. The residual amplitude when the length of the manipulator arm changes.

Figure 6 shows the residual amplitudes of manipulator deflection and velocity potential when the depth of the flow field remains constant while the manipulator length varies. From Fig. 6, it can be observed that both the deflection and velocity potential residual amplitudes exhibit peaks and troughs as the manipulator length increases. Students need to explain this phenomenon: the residual amplitude reaches a peak when the vibrations caused by the vehicle's acceleration and deceleration commands are in phase.

Conversely, when the vibrations caused by the vehicle's acceleration and deceleration commands are out of phase, the residual amplitude reaches a trough.

3.3 Driving Distance Variations

The previous simulations are conducted under the condition where the vehicle is driven with a trapezoidal velocity to move 40 cm. However, for different driving distances, the system response might exhibit unique results. Therefore, students need to vary the driving distance in the simulations and plot the graphs of manipulator deflection and velocity potential for different driving distances, as shown in Fig. 7.

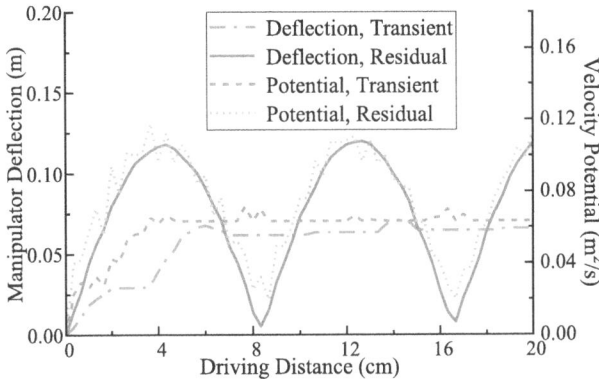

Fig. 7. The transient and residual amplitude when the driving distance changes.

It can be observed that when the driving distance varies from 0 to 20 cm, both transient and residual vibrations exhibit regular patterns. As the driving distance increases, the transient amplitude of the deflection initially increases, then briefly remains stable, and increases again, eventually reaching a steady value with small peaks. This phenomenon occurs because, as the driving distance increases, the system's vibration reaches a peak during the driving process, causing the transient amplitude to increase; when the vibration transitions from the peak back to the initial value, the transient amplitude remains unchanged; as it goes from the initial value to a trough, the transient amplitude continues to increase. Subsequently, when the vibrations induced by the vehicle's acceleration and deceleration commands are in phase, the transient vibration exhibits peaks. The trend of the transient amplitude of the velocity potential is similar to that of the deflection, but it includes more high-mode frequency vibrations. This is consistent with the high-mode vibrations in the time-domain response curve of the velocity potential.

The residual amplitude of the deflection exhibits periodic changes. In addition, the distance between peaks corresponds to the time required for the vehicle to move, matching the first-mode vibration period of the manipulator. Similarly, the residual amplitude of the velocity potential corresponds to the first- and second-mode periods. This further validates the dynamic characteristics of the model.

3.4 Controller

In industrial systems, control methods are commonly employed to minimize vibrations. Among these, the input shaper is one of the most prevalent techniques [11]. This method designs the driving signal based on the system's calculated frequency. The vehicle, driven by the newly designed signal, can effectively suppress the vibrations. The time function of the DP controller is as follows:

$$c(\tau) = \begin{cases} \dfrac{\zeta_m \omega_m e^{-\zeta_m \omega_m \tau}}{[1-e^{-h\pi\zeta_m/\sqrt{1-\zeta_m^2}}+e^{-\pi\zeta_m/\sqrt{1-\zeta_m^2}}-e^{-(1+h)\pi\zeta_m/\sqrt{1-\zeta_m^2}}]}, & 0 \leq \tau \leq \dfrac{\pi h}{\omega_m\sqrt{1-\zeta_m^2}} \\ 0, & \dfrac{\pi h}{\omega_m\sqrt{1-\zeta_m^2}} < \tau < \dfrac{\pi}{\omega_m\sqrt{1-\zeta_m^2}} \\ \dfrac{\zeta_m \omega_m e^{-\zeta_m \omega_m \tau}}{[1-e^{-h\pi\zeta_m/\sqrt{1-\zeta_m^2}}+e^{-\pi\zeta_m/\sqrt{1-\zeta_m^2}}-e^{-(1+h)\pi\zeta_m/\sqrt{1-\zeta_m^2}}]}, & \dfrac{\pi}{\omega_m\sqrt{1-\zeta_m^2}} \leq \tau \leq \dfrac{\pi(1+h)}{\omega_m\sqrt{1-\zeta_m^2}} \\ 0, & else \end{cases} \quad (26)$$

where ω_m represents the designed frequency, which is calculated previously, ζ_m is the designed damping ratio, and h is the control gain, with $0 < h < 1$. By setting $h = 0.6560$, we convolve the DP input shaper with the driving signal, generating the new command input to the system specified in Sect. 3.1. The system's time-domain response is shown in Fig. 8. From the figure, students can observe that the controller effectively suppresses the vibrations in both the velocity potential and the deflection of the manipulator.

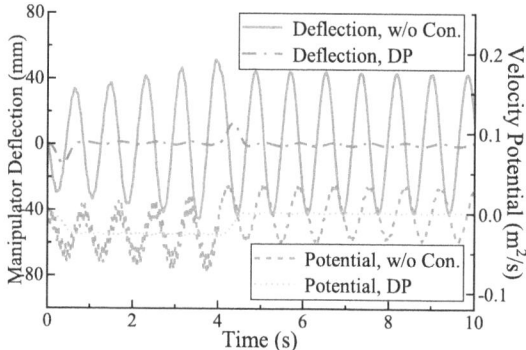

Fig. 8. Time responses with and without control.

4 Advanced Study

After analyzing the dynamic characteristics of the system, an advanced study section is provided for students to expand their extracurricular research. In this section, students can gain a deeper and more comprehensive understanding of the system's properties and learn about control methods used in industrial applications.

Robustness is a crucial performance metric for controllers. Due to inaccuracies in system parameter estimation, there may be discrepancies between the designed frequency and the actual frequency. Therefore, the controller must exhibit good robustness. This ensures that the controller can still achieve good control performance even when there are significant modeling errors or the system parameters are difficult to obtain. To validate the robustness of the controller, it is necessary to alter the system model parameters in the simulations. By keeping the designed frequency of the DP shaper and altering the structural parameters, students generate response graphs of the system under different configurations, with and without the controller.

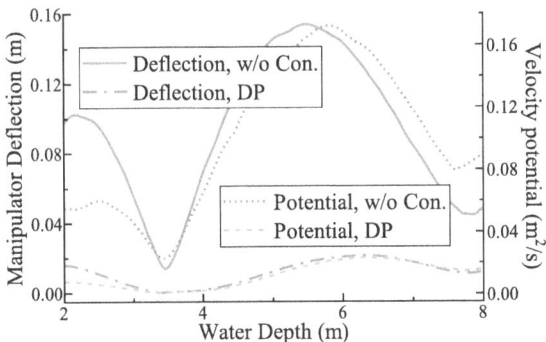

Fig. 9. The residual amplitude with and without control when the water depth changes.

Figure 9 indicates that the residual amplitude reaches a minimum when the water depth matches the controller's designed frequency of 4 m. At other water depths, the residual amplitude remains relatively low. Calculations show that the suppression rates of manipulator deflection and velocity potential at different water depths are 87.73% and 88.34%, respectively. This validates the effectiveness and robustness of the controller.

Advanced study can also be conducted to investigate the three-dimensional problem of underwater flexible manipulator. In real-world scenarios, the underwater flexible manipulator experiences not only vibration along the driving direction, but also vibration perpendicular to the driving direction [12]. This is caused by vortex-excited vibration. This will make the dynamics model of the system more complicated and more demanding on the controller. In addition, closed-loop controllers are also one of the research directions, as they can effectively resist external perturbations and do not require an accurate system model.

5 Conclusion

This paper presents a customized teaching approach for the nonlinear systems and intelligent control course based on underwater flexible manipulators. This approach provides mechanical engineering students with modeling and simulation projects. The course includes homework, tests, and advanced study. The structured curriculum, which does not require any specialized equipment, can be conducted both online and offline. While ensuring that the majority of students can easily grasp the course content, this customized teaching approach also allows outstanding students to engage in in-depth exploration.

References

1. Singhose, W., Vaughan, J., Danielson, J., Lawrence, J.: The use of tele-operated cranes for advanced controls education. In: ASME International Mechanical Engineering Congress and Exposition, vol. 43017, pp. 497–502 (2007)
2. Singhose, W., Vaughan, J., Danielson, J., Lawrence, J.: Use of cranes in system dynamics and controls education. IFAC Proc. **41**(2), 9099–9104 (2008)
3. Ding, M., Huang, J., Ye, J.: Application of flexible link manipulators in control engineering courses. In: International Conference on Education Technology and Computers, pp. 177–181. Association for Computing Machinery (2020)
4. Xue, H., Huang, J.: Use of tower cranes in dynamics and control education for mechanical-engineering students. In: Control, Instrumentation and Mechatronics: Theory and Practice, pp. 93–104. Springer, Singapore (2022)
5. Abdulghani, Z., Darus, I.Z.M., Nasir, H.: Trajectory prediction model for the angle hub of a two link flexible manipulator utilizing long short term memory. In: 2024 IEEE 10th International Conference on Smart Instrumentation, Measurement and Applications (ICSIMA), pp. 179–184. IEEE (2024)
6. Liu, X., Wang, M., Zheng, Y., Wang, X.: Fuzzy PI vibration suppression control strategy for space double flexible telescopic manipulator with fractional disturbance observer. Aerosp. Sci. Technol. **155**, 109579 (2024)
7. Xue, H., Huang, J.: Dynamic modeling and vibration control of underwater soft-link manipulators undergoing planar motions. Mech. Syst. Signal Process. **181**, 109540 (2022)
8. Qi, C., Huang, J.: Control of coupled interaction between the flexible cantilever link and airflow. Mechatronics **96**, 103088 (2023)
9. Miller, A.S., Singhose, W., Glauser, U.: Integrating PLC theory and programming into advanced controls courses. In: 2016 American Control Conference (ACC), pp. 7302–7307. IEEE (2016)
10. Singhose, W., et al.: Use of cranes in education and international collaborations. J. Robot. Mechatron. **23**(5), 881–892 (2011)
11. Li, L., Hu, X., Zou, Y.: Implementation of input shaping for the vibration control of flexible beams manipulated by industrial robot. J. Vib. Shock **38**(20), 12–17 (2019)
12. Xue, H., Huang, J.: Dynamics and control of underwater manipulators considering interaction between soft-body link and fluid flow. J. Sound Vib.Vib. **570**, 118022 (2024)

A Spring-Pendulum Educational Device for Teaching Dynamics and Control Courses

Chenglei Yang and Jie Huang(✉)

School of Mechanical Engineering, Beijing Institute of Technology, Beijing 100081, China
bit_huangjie@bit.edu.cn

Abstract. This article proposes a spring-pendulum device for teaching dynamics and control courses. This article also provides a teaching syllabus that introduces the course content, which mainly includes dynamic modeling, numerical simulation, experimental verification, and advanced learning. Students could use Lagrange equation to derive the dynamic model of forced vibration of the spring pendulum, and predict dynamic behaviors of the spring pendulum through numerical simulation. Students could also use the spring-pendulum device to verify the effectiveness of the dynamic model. In addition, students could choose the advanced learning to derive approximate solutions for the vibration of the spring pendulum by using multi-scale methods, and understand the conditions under which internal resonance occurs. The advanced learning also includes designing a smoother to suppress the coupled vibration of the spring pendulum, and comparing the robustness of the smoother with other controllers through experiments and simulations. By using the spring-pendulum educational device to study dynamics and control courses, students could also gain a reference understanding of other coupled-vibration systems similar to the spring pendulum.

Keywords: Spring Pendulum · Educational Device · Teaching Courses

1 Introduction

Students could learn theoretical and experimental knowledge to analyze dynamic characteristics and design controllers in dynamics and control courses [1–3]. The courses will teach students knowledge such as dynamic modeling, computational dynamics, controller design, and experimental operations. Cranes and cherry pickers are used as teaching devices for dynamics and control courses [4–6]. However, the experimental operations of those devices are relatively complex.

The purpose of this article is to propose a simple instructional device for teaching dynamics and control courses. The main contribution of this article is proposing a spring-pendulum educational device for better teaching dynamics and control courses. A spring pendulum is a physical system, which is a pendulum made of a spring with a mass at the tip. Spring pendulum is also called swinging spring or elastic pendulum. It is a classical model of coupled nonlinear oscillation. The planar motion model of a spring-pendulum system has two degrees of freedom and two natural frequencies. The oscillations of spring pendulum include spring oscillations and pendulum oscillations. When the structural parameters of the spring-pendulum system meet certain conditions, internal resonance will occur and it will cause relatively large transient and residual amplitudes.

By using a spring-pendulum educational device to study dynamics and control courses, students could explore the internal resonance phenomenon of spring pendulum, gain a deeper understanding of the coupling characteristics between spring-oscillation and payload-swing, and comprehend the exchange of coupling energy. In addition, students could have a reference for learning the dynamics and vibration control of coupled oscillators similar to spring-pendulum system [7–9]. Therefore, it is meaningful to use a spring-pendulum educational device to teach dynamics and control courses.

2 Teaching Contents

2.1 Syllabus

The syllabus for using the spring-pendulum device in dynamics and control courses mainly includes dynamic modeling, numerical simulation, experimental verification, and advanced learning.

Dynamic Modeling. Students will derive the kinetic and potential energy of the spring-pendulum system, and derive the dynamic equation of forced vibration in the spring-pendulum system by applying the Lagrange equation. Then, the students will solve the equilibrium point of the spring-pendulum system.

Numerical Simulation. Students will learn how to obtain time-domain plots of swing angle θ and spring deflection x through computational dynamics, and predict coupling frequency. Moreover, students will also calculate the simulation results of the transient deflection and residual amplitude of swing angle θ and spring deflection x versus various driving distance. Students will also learn to predict the dynamic behaviors of the spring-pendulum system through numerical simulations.

Experimental Verification. To verify the effectiveness of the dynamic model, students will learn to build a testbed including a motion control system, an image detection system, and the spring-pendulum system. Students will obtain the experimental results of the time-domain responses of swing angle θ and spring deflection x. Besides, when the driving distance changes, the experimental results of the transient deflection and residual amplitude of the swing angle θ and spring deflection x will also be obtained to be compared with the simulation results.

Advanced Learning. In addition to the basic learning contents mentioned above, the students who are highly interested in dynamics and control courses and have extra time could engage in advanced learning. Students could try using the multiscale method to derive an approximate solution to the dynamic equation of the spring-pendulum system. Moreover, students will learn to design a smoother to suppress the coupled vibration of the spring pendulum. Additionally, students will compare the robustness of the smoother with another input shaper.

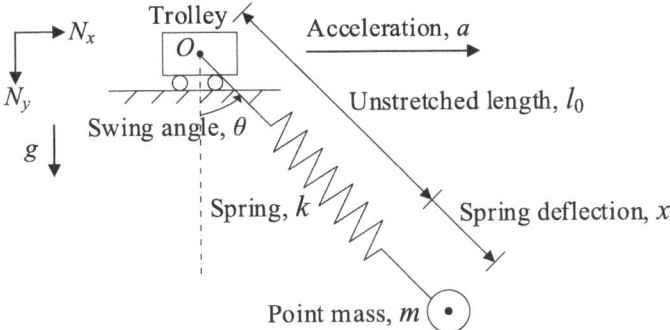

Fig. 1. Planar model of a spring pendulum.

2.2 Dynamic Modeling

Students could learn about the internal resonance through dynamic modeling and analysis of the spring pendulum. Figure 1 shows a planar model of the spring pendulum. The length of the spring without tension is l_0, and the deflection of the spring within the elastic limit is x. The stiffness coefficient of the spring is k. The swing angle of the spring pendulum is θ. The load is assumed to be a point mass with mass of m. The acceleration of gravity is g. In addition, it is assumed that the mass of spring and any damping are ignored. The spring pendulum moving in the vertical plane can be regarded as a vibration system with two degrees of freedom and two natural frequencies. The input is a horizontal acceleration a of the trolley. The output of the system is the deflection x and swing angle θ of the spring pendulum.

The kinetic energy of the system is:

$$T_1 = \frac{1}{2}m\left[\dot{x} + \left(\int adt\right)\sin\theta\right]^2 + \frac{1}{2}m\left[\dot{\theta}(l_0 + x) + \left(\int adt\right)\cos\theta\right]^2 \quad (1)$$

where t is time, $\dot{x}$ is the derivative of x with respect to time, $\dot{\theta}$ is the derivative of θ with respect to time, dt is time differentiation.

Based on the assumption that point O is the point of zero potential energy, the potential energy of the system is:

$$V_1 = -mg(l_0 + x)\cos\theta + \frac{1}{2}kx^2 \quad (2)$$

The Lagrangian is the difference between kinetic energy and potential energy:

$$L_1 = \frac{m}{2}\left\{\left[\dot{x} + \left(\int adt\right)\sin\theta\right]^2 + \left[\dot{\theta}(l_0 + x) + \left(\int adt\right)\cos\theta\right]^2\right\} + mg(l_0 + x)\cos\theta - \frac{1}{2}kx^2 \quad (3)$$

The first-order partial derivative of L_1 with respect to x is:

$$\frac{\partial L_1}{\partial x} = m\dot{\theta}^2(l_0 + x) + m\dot{\theta}\left(\int adt\right)\cos\theta + mg\cos\theta - kx \quad (4)$$

Solving the first-order partial derivative of L_1 with respect to $\dot{x}$ yields:

$$\frac{\partial L_1}{\partial \dot{x}} = m\dot{x} + m\left(\int adt\right)\sin\theta \qquad (5)$$

Taking the first-order partial derivative of L_1 with respect to θ gets:

$$\frac{\partial L_1}{\partial \theta} = m\dot{x}\left(\int adt\right)\cos\theta - m\dot{\theta}(l_0+x)\left(\int adt\right)\sin\theta - mg(l_0+x)\sin\theta \qquad (6)$$

The first-order partial derivative of L_1 with respect to $\dot{\theta}$ is:

$$\frac{\partial L_1}{\partial \dot{\theta}} = m\dot{\theta}(l_0+x)^2 + m(l_0+x)\left(\int adt\right)\cos\theta \qquad (7)$$

Solving the first-order derivative of $\frac{\partial L_1}{\partial \dot{x}}$ with respect to t yields:

$$\frac{d}{dt}\frac{\partial L_1}{\partial \dot{x}} = m\ddot{x} + ma\sin\theta + m\left(\int adt\right)\dot{\theta}\cos\theta \qquad (8)$$

Taking the first-order derivative of $\frac{\partial L_1}{\partial \dot{\theta}}$ with respect to t gets:

$$\frac{d}{dt}\frac{\partial L_1}{\partial \dot{\theta}} = m(l_0+x)^2\ddot{\theta} + 2m\dot{\theta}\dot{x}(l_0+x) + m\dot{x}\left(\int adt\right)\cos\theta$$
$$+ m(l_0+x)a\cos\theta - m(l_0+x)\left(\int adt\right)\dot{\theta}\sin\theta \qquad (9)$$

The Lagrange equation is:

$$\begin{cases} \dfrac{d}{dt}\dfrac{\partial L_1}{\partial \dot{\theta}} - \dfrac{\partial L_1}{\partial \theta} = 0 \\ \dfrac{d}{dt}\dfrac{\partial L_1}{\partial \dot{x}} - \dfrac{\partial L_1}{\partial x} = 0 \end{cases} \qquad (10)$$

Substituting Eqs. (4), (6), (8), (9) into Eq. (10) results in the dynamical model describing the swing oscillation and spring deflection of the spring pendulum:

$$\begin{cases} (l_0+x)\ddot{\theta} + 2\dot{x}\dot{\theta} + \omega_1^2 l_0\sin\theta + a\cos\theta = 0 \\ \ddot{x} + (\omega_2^2 - \dot{\theta}^2)x - l_0\dot{\theta}^2 - g\cos\theta + a\sin\theta = 0 \end{cases} \qquad (11)$$

where

$$\omega_1^2 = \frac{g}{l_0}, \quad \omega_2^2 = \frac{k}{m} \qquad (12)$$

In the absence of input in the system, assume that the approximate solutions of swing angle θ and spring deflection x are:

$$\begin{cases} \theta(t,\varepsilon) = \varepsilon\theta_1(T_0,T_1) + \varepsilon^2\theta_2(T_0,T_1) + O(\varepsilon^3) \\ x(t,\varepsilon) = \varepsilon x_1(T_0,T_1) + \varepsilon^2 x_2(T_0,T_1) + O(\varepsilon^3) \end{cases} \qquad (13)$$

where $0 < \varepsilon << 1$, and:

$$T_0 = t, \ T_1 = \varepsilon t \tag{14}$$

Define the partial derivative operators to represent derivative operators:

$$\frac{d}{dt} = D_0 + \varepsilon D_1, \ \frac{d^2}{dt^2} = D_0^2 + 2\varepsilon D_0 D_1 + \varepsilon^2 D_1^2 \tag{15}$$

Substituting Eq. (13) into Eq. (11) when $a = 0$, results in:

$$\begin{cases} (l_0 + \varepsilon x_1 + \varepsilon^2 x_2)\left[\varepsilon(D_0^2\theta_1 + 2\varepsilon D_0 D_1\theta_1 + \varepsilon^2 D_1^2\theta_1) + \varepsilon^2(D_0^2\theta_2 - 2\varepsilon D_0 D_1\theta_2 + \varepsilon^2 D_1^2\theta_2)\right] \\ +2(\varepsilon D_0 x_1 + \varepsilon^3 D_1 x_2)(\varepsilon D_0\theta_1 + \varepsilon^3 D_1\theta_2) + \omega_1^2 l_0 \sin(\varepsilon\theta_1 + \varepsilon^2\theta_2) = 0 \\ \left[\varepsilon(D_0^2 x_1 + 2\varepsilon D_0 D_1 x_1 + \varepsilon^2 D_1^2 x_1) + \varepsilon^2(D_0^2 x_2 - 2\varepsilon D_0 D_1 x_2 + \varepsilon^2 D_1^2 x_2)\right] \\ +\left[\omega_2^2 - (\varepsilon D_0\theta_1 + \varepsilon^3 D_1\theta_2)^2\right](\varepsilon x_1 + \varepsilon^2 x_2) - l_0(\varepsilon D_0\theta_1 + \varepsilon^3 D_1\theta_2)^2 - g\cos(\varepsilon\theta_1 + \varepsilon^2\theta_2) = 0 \end{cases} \tag{16}$$

Applying Taylor expansion to trigonometric function results in:

$$\sin(\varepsilon\theta_1 + \varepsilon^2\theta_2) \approx (\varepsilon\theta_1 + \varepsilon^2\theta_2) - \frac{(\varepsilon\theta_1 + \varepsilon^2\theta_2)^3}{3!} + o[(\varepsilon\theta_1 + \varepsilon^2\theta_2)^3] \tag{17}$$

$$\cos(\varepsilon\theta_1 + \varepsilon^2\theta_2) \approx 1 - \frac{(\varepsilon\theta_1 + \varepsilon^2\theta_2)^2}{2!} + o[(\varepsilon\theta_1 + \varepsilon^2\theta_2)^2] \tag{18}$$

Substituting Eq. 17, 18, into Eq. (16) obtains:

$$\begin{cases} (l_0 + \varepsilon x_1 + \varepsilon^2 x_2)\left[\varepsilon(D_0^2\theta_1 + 2\varepsilon D_0 D_1\theta_1 + \varepsilon^2 D_1^2\theta_1) + \varepsilon^2(D_0^2\theta_2 - 2\varepsilon D_0 D_1\theta_2 + \varepsilon^2 D_1^2\theta_2)\right] \\ +2(\varepsilon D_0 x_1 + \varepsilon^3 D_1 x_2)(\varepsilon D_0\theta_1 + \varepsilon^3 D_1\theta_2) \\ +\omega_1^2 l_0 \left\{(\varepsilon\theta_1 + \varepsilon^2\theta_2) - \frac{(\varepsilon\theta_1+\varepsilon^2\theta_2)^3}{3!} + o[(\varepsilon\theta_1+\varepsilon^2\theta_2)^3]\right\} = 0 \\ \left[\varepsilon(D_0^2 x_1 + 2\varepsilon D_0 D_1 x_1 + \varepsilon^2 D_1^2 x_1) + \varepsilon^2(D_0^2 x_2 - 2\varepsilon D_0 D_1 x_2 + \varepsilon^2 D_1^2 x_2)\right] \\ +\left[\omega_2^2 - (\varepsilon D_0\theta_1 + \varepsilon^3 D_1\theta_2)^2\right](\varepsilon x_1 + \varepsilon^2 x_2) - l_0(\varepsilon D_0\theta_1 + \varepsilon^3 D_1\theta_2)^2 \\ -g\left\{1 - \frac{(\varepsilon\theta_1+\varepsilon^2\theta_2)^2}{2!} + o[(\varepsilon\theta_1+\varepsilon^2\theta_2)^2]\right\} = 0 \end{cases} \tag{19}$$

The solution of ε^1 corresponding to Eq. (19) is:

$$\begin{cases} D_0^2\theta_1 + \omega_1^2\theta_1 = 0 \\ D_0^2 x_1 + \omega_2^2 x_1 = 0 \end{cases} \tag{20}$$

The solution of ε^2 corresponding to Eq. (19) is:

$$\begin{cases} l_0(2D_0 D_1\theta_1 + D_0^2\theta_2) + x_1 D_0^2\theta_1 + 2D_0 x_1 D_0\theta_1 + \omega_1^2 l_0\theta_2 = 0 \\ 2D_0 D_1 x_1 + D_0^2 x_2 + \omega_2^2 x_2 - l_0(D_0\theta_1)^2 + \dfrac{g\theta_1^2}{2} = 0 \end{cases} \tag{21}$$

The general solution of Eq. (20) is:

$$\begin{cases} \theta_1 = A_1(T_1)e^{j\omega_1 T_0} + \overline{A_1(T_1)}e^{-j\omega_1 T_0} \\ x_1 = A_2(T_1)e^{j\omega_2 T_0} + \overline{A_2(T_1)}e^{-j\omega_2 T_0} \end{cases} \quad (22)$$

Assuming the complex forms of A_1 and A_2 are:

$$A_1(T_1) = C_1(t)e^{j\varphi_1(t)}, \ A_2(T_1) = C_2(t)e^{j\varphi_2(t)} \quad (23)$$

where $C_i(t)$ and $\varphi_i(t)$ ($i = 1, 2$) are real functions.

Substituting Eq. (23) and Eq. (14) into Eq. (22) obtains:

$$\begin{cases} \theta_1 = 2C_1(t)\cos[\omega_1 t + \varphi_1(t)] \\ x_1 = 2C_2(t)\cos[\omega_2 t + \varphi_2(t)] \end{cases} \quad (24)$$

Substituting Eq. (22) into Eq. (21), results in:

$$\begin{cases} D_0^2\theta_2 + \omega_1^2\theta_2 = -2j\omega_1\frac{\partial A_1}{\partial T_1}e^{j\omega_1 T_0} + 2j\omega_1\frac{\partial \overline{A_1}}{\partial T_1}e^{-j\omega_1 T_0} + \frac{1}{l_0}(A_1\overline{A_2}\omega_1^2 - 2A_1\overline{A_2}\omega_1\omega_2)e^{j(\omega_1-\omega_2)T_0} \\ + \frac{1}{l_0}(\overline{A_1}A_2\omega_1^2 - 2\overline{A_1}A_2\omega_1\omega_2)e^{j(\omega_2-\omega_1)T_0} + \frac{1}{l_0}(A_1A_2\omega_1^2 + 2A_1A_2\omega_1\omega_2)e^{j(\omega_1+\omega_2)T_0} \\ + \frac{1}{l_0}(\overline{A_1A_2}\omega_1^2 + 2\overline{A_1A_2}\omega_1\omega_2)e^{-j(\omega_1+\omega_2)T_0} \\ D_0^2 x_2 + \omega_2^2 x_2 = -2j\omega_2\frac{\partial A_2}{\partial T_1}e^{j\omega_2 T_0} + 2j\omega_2\frac{\partial \overline{A_2}}{\partial T_1}e^{-j\omega_2 T_0} - (l_0\omega_1^2 A_1^2 + \frac{g}{2}A_1^2)e^{2j\omega_1 T_0} \\ - (l_0\omega_1^2\overline{A_1}^2 + \frac{g}{2}\overline{A_1}^2)e^{-2j\omega_1 T_0} + (2l_0\omega_1^2 - g)A_1\overline{A_1} \end{cases} \quad (25)$$

In order to eliminate the secular terms, we obtain:

$$-2j\omega_1\frac{\partial A_1}{\partial T_1} = 0, \ -2j\omega_2\frac{\partial A_2}{\partial T_1} = 0, \ \frac{dA_1}{dt} = 0, \ \frac{dA_2}{dt} = 0 \quad (26)$$

Substituting Eq. (23) into Eq. (26) results in:

$$\begin{cases} \dot{C}_1(t)e^{j\varphi_1(t)} + C_1(t)e^{j\varphi_1(t)}j\dot{\varphi}_1(t) = 0 \\ \dot{C}_2(t)e^{j\varphi_2(t)} + C_2(t)e^{j\varphi_2(t)}j\dot{\varphi}_2(t) = 0 \end{cases} \quad (27)$$

The real part and imaginary part of the left and right parts of Eq. (27) are corresponding equivalent, we obtain:

$$\dot{C}_1 = 0, \ \dot{C}_2 = 0, \ \dot{\varphi}_1 = 0, \ \dot{\varphi}_2 = 0 \quad (28)$$

Solving Eq. (28) yields:

$$C_1 = const, \ C_2 = const, \ \varphi_1 = const, \ \varphi_2 = const \quad (29)$$

where $C_1, C_2, \varphi_1, \varphi_2$ are determined by boundary conditions.

After deleting the secular terms, Eq. (25) becomes:

$$\begin{cases} D_0^2\theta_2 + \omega_1^2\theta_2 = \dfrac{1}{l_0}(A_1\overline{A_2}\omega_1^2 - 2A_1\overline{A_2}\omega_1\omega_2)e^{j(\omega_1-\omega_2)T_0} + \dfrac{1}{l_0}(\overline{A_1}A_2\omega_1^2 - 2\overline{A_1}A_2\omega_1\omega_2)e^{j(\omega_2-\omega_1)T_0} \\ +\dfrac{1}{l_0}(A_1A_2\omega_1^2 + 2A_1A_2\omega_1\omega_2)e^{j(\omega_1+\omega_2)T_0} + \dfrac{1}{l_0}(\overline{A_1A_2}\omega_1^2 + 2\overline{A_1A_2}\omega_1\omega_2)e^{-j(\omega_1+\omega_2)T_0} \\ D_0^2 x_2 + \omega_2^2 x_2 = -(l_0\omega_1^2 A_1^2 + \dfrac{g}{2}A_1^2)e^{2j\omega_1 T_0} - (l_0\omega_1^2\overline{A_1}^2 + \dfrac{g}{2}\overline{A_1}^2)e^{-2j\omega_1 T_0} + (2l_0\omega_1^2 - g)A_1\overline{A_1} \end{cases} \quad (30)$$

Solving Eq. (30) results in:

$$\begin{cases} \theta_2 = \dfrac{\frac{1}{l_0}(A_1\overline{A_2}\omega_1^2 - 2A_1\overline{A_2}\omega_1\omega_2)}{\omega_1^2 - (\omega_1-\omega_2)^2}e^{j(\omega_1-\omega_2)T_0} + \dfrac{\frac{1}{l_0}(\overline{A_1}A_2\omega_1^2 - 2\overline{A_1}A_2\omega_1\omega_2)}{\omega_1^2 - (\omega_2-\omega_1)^2}e^{j(\omega_2-\omega_1)T_0} \\ + \dfrac{\frac{1}{l_0}(A_1A_2\omega_1^2 + 2A_1A_2\omega_1\omega_2)}{\omega_1^2-(\omega_1+\omega_2)^2}e^{j(\omega_1+\omega_2)T_0} + \dfrac{\frac{1}{l_0}(\overline{A_1A_2}\omega_1^2 + 2\overline{A_1A_2}\omega_1\omega_2)}{\omega_1^2-(\omega_1+\omega_2)^2}e^{-j(\omega_1+\omega_2)T_0} \\ x_2 = \dfrac{-(l_0\omega_1^2 A_1^2 + \frac{g}{2}A_1^2)}{\omega_2^2-(2\omega_1)^2}e^{j(2\omega_1)T_0} + \dfrac{-(l_0\omega_1^2\overline{A_1}^2 + \frac{g}{2}\overline{A_1}^2)}{\omega_2^2-(2\omega_1)^2}e^{-j(2\omega_1)T_0} + \dfrac{(2l_0\omega_1^2-g)A_1\overline{A_1}}{\omega_2^2} \end{cases} \quad (31)$$

Substituting Eq. (23) and Eq. (14) into Eq. (31) obtains:

$$\begin{cases} \theta_2 = \dfrac{\frac{2}{l_0}C_1 C_2(\omega_1^2 - 2\omega_1\omega_2)\cos\left[\begin{array}{c}(\omega_2-\omega_1)t \\ +\varphi_2-\varphi_1\end{array}\right]}{\omega_1^2-(\omega_1-\omega_2)^2} + \dfrac{\frac{2}{l_0}C_1 C_2(\omega_1^2 + 2\omega_1\omega_2)\cos\left[\begin{array}{c}(\omega_1+\omega_2)t \\ +\varphi_1+\varphi_2\end{array}\right]}{\omega_1^2-(\omega_1+\omega_2)^2} \\ x_2 = \dfrac{-(2l_0\omega_1^2 + g)C_1^2}{\omega_2^2-(2\omega_1)^2}\cos(2\omega_1 t + 2\varphi_1) + \dfrac{(2l_0\omega_1^2 - g)C_1^2}{\omega_2^2} \end{cases} \quad (32)$$

Substituting Eq. (32) and Eq. (24) into Eq. (13), and ignoring $O(\varepsilon^3)$, results in the approximate solution for the free vibration of the spring pendulum:

$$\begin{cases} \theta(t,\varepsilon) = 2\varepsilon C_1\cos(\omega_1 t + \varphi_1) + \dfrac{\frac{2}{l_0}\varepsilon^2 C_1 C_2(\omega_1^2 - 2\omega_1\omega_2)}{(2\omega_1 - \omega_2)\omega_2}\cos[(\omega_2-\omega_1)t + \varphi_2 - \varphi_1] \\ + \dfrac{\frac{2}{l_0}\varepsilon^2 C_1 C_2(\omega_1^2 + 2\omega_1\omega_2)}{-(2\omega_1+\omega_2)\omega_2}\cos[(\omega_1+\omega_2)t + \varphi_1+\varphi_2] \\ x(t,\varepsilon) = 2\varepsilon C_2\cos(\omega_2 t+\varphi_2) + \dfrac{-\varepsilon^2(2l_0\omega_1^2 + g)C_1^2\cos(2\omega_1 t + 2\varphi_1)}{(\omega_2+2\omega_1)(\omega_2-2\omega_1)} + \dfrac{\varepsilon^2(2l_0\omega_1^2-g)C_1^2}{\omega_2^2} \end{cases} \quad (33)$$

where the amplitudes of the second term on the right side of the first equation and the second term on the right side of the second equation in (33) tend to infinity if $\omega_2 = 2\omega_1$, because their denominators tend to zero. Therefore, the students could understand the conditions under which internal resonance occurs.

2.3 Controller Design

In order to suppress the coupled oscillation of the spring pendulum, the students could learn to design a non-smooth piecewise smoother with fifteen parts (NS15P smoother)

[10] in the course of dynamics and control. The time function of the NS15P smoother was represented by:

$$c(\tau) = \begin{cases} \mu(\delta^{p\tau} - \delta^{q\tau}), 0 \leq \tau < hpT_m \\ \mu\Theta\delta^{p\tau}, hpT_m \leq \tau < pT_m \\ \mu(\delta^{p\tau} + H), pT_m \leq \tau < hqT_m \\ \mu(H + \vartheta^h\delta^{q\tau}), hqT_m \leq \tau < (1+h)pT_m \\ \mu(\xi + \Lambda\delta^{p\tau}), (1+h)pT_m \leq \tau < qT_m \\ \mu(\delta^{p\tau} + \Lambda\delta^{p\tau} + \chi), qT_m \leq \tau < (hp+hq)T_m \\ \mu([1+\sigma\Theta]\delta^{p\tau} - \vartheta\delta^{q\tau}), (hp+hq)T_m \leq \tau < (p+hq)T_m \\ \mu(\delta^{p\tau} + \chi), (p+hq)T_m \leq \tau < (q+hp)T_m \\ \mu(\xi + \Theta\delta^{p\tau}), (q+hp)T_m \leq \tau < (p+q)T_m \\ \mu(\delta^{p\tau} + \Lambda\delta^{p\tau} + \chi), (p+q)T_m \leq \tau < (p+hp+hq)T_m \\ \mu(\delta^{p\tau} + \Gamma), (p+hp+hq)T_m \leq \tau < (1+h)qT_m \\ \mu(\Gamma + \vartheta^{(1+h)}\delta^{q\tau}), (1+h)qT_m \leq \tau < (p+q+hp)T_m \\ \mu([\Lambda - \sigma^{(1+h)}]\delta^{p\tau} + \vartheta^{(1+h)}\delta^{q\tau}), (p+q+hp)T_m \leq \tau < (q+hp+hq)T_m \\ \mu\sigma\Theta\delta^{p\tau}, (q+hp+hq)T_m \leq \tau < (p+q+hq)T_m \\ \mu(\vartheta^{(1+h)}\delta^{q\tau} - \sigma^{(1+h)}\delta^{p\tau}), (p+q+hq)T_m \leq \tau < (p+q+hp+hq)T_m \end{cases} \quad (34)$$

where

$$\mu = \frac{pq\zeta_m\Omega_m}{(q-p)[1 - e^{-hK} + e^{-K} - e^{-(1+h)K}]^2}, \quad T_m = \frac{\pi}{pq\Omega_m\sqrt{1-\zeta_m^2}}, \quad K = \frac{\pi\zeta_m}{\sqrt{1-\zeta_m^2}} \quad (35)$$

$$\delta = e^{-\zeta_m\Omega_m}, \quad \sigma = e^{\frac{(p-q)}{q}K}, \quad \vartheta = e^{\frac{(q-p)}{p}K}, \quad \Lambda = \sigma - \sigma^h, \quad \Psi = \vartheta^h - \vartheta, \quad \Theta = 1 - \sigma^h \quad (36)$$

$$\xi = \vartheta^h\delta^{q\tau} - \sigma^{(1+h)}\delta^{p\tau}, \quad \Gamma = \Lambda\delta^{p\tau} - \vartheta\delta^{q\tau}, \quad H = \Lambda\delta^{p\tau} - \delta^{q\tau}, \quad \chi = \Psi\delta^{q\tau} - \sigma^{(1+h)}\delta^{p\tau} \quad (37)$$

2.4 Simulation and Experimental Results

In order to verify the effectiveness of the dynamic model and the NS15P smoother, students could build a testbed under the guidance of the professor to conduct relevant experiments. Figure 2 shows the testbed for the motion of the spring-pendulum system, which consists of three parts: an image detection system, a motion control system, and the spring-pendulum system. A camera was used to detect deflections of a black marker at the point-mass load. The deflections of the marker are related to the swing angle θ and the spring deflection x. In this experiment, the stiffness coefficient of the spring was 35 N/m, the length of the spring without tension was 0.305 m, and the mass of the load was 0.135 kg.

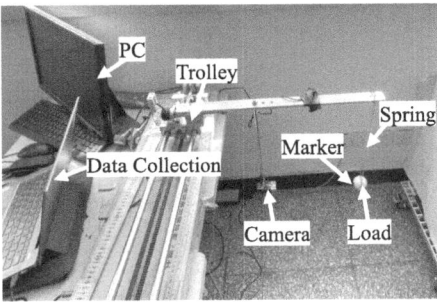

Fig. 2. Testbed of a spring-pendulum system.

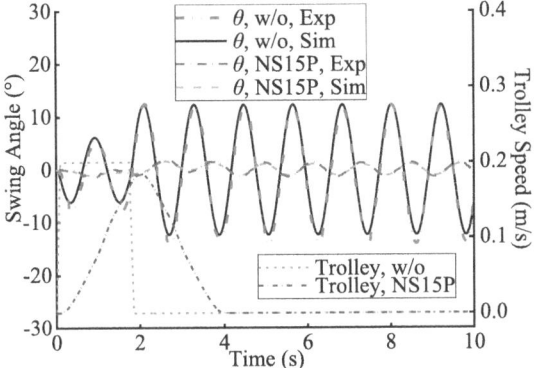

Fig. 3. Responses of swing angle and trolley speed.

Figure 3 displays the experimental and simulated responses of swing angle θ under control of the NS15P smoother and without control. In the simulated curves, the uncontrolled transient deflection and residual amplitude of the swing angle θ were 12.33° and 24.69°, respectively. Under the control of the NS15P smoother, the transient deflection and residual amplitude of the swing angle θ were 2.63° and 2.13° in simulation, respectively. Additionally, the uncontrolled transient deflection and residual amplitude of the swing angle θ in the experimental results were 13.96° and 26.69°, respectively. However, under the control of the NS15P smoother, the transient deflection and residual amplitude of the swing angle θ were 3.63° and 2.76° in experiment, respectively. In the simulation results, the suppression rates of the NS15P smoother on the transient deflection and residual amplitude of the swing angle θ were 78.67% and 91.37%, respectively. According to the experimental results, the NS15P smoother had suppression rates of 74.00% and 89.66% on the transient deflection and residual amplitude of the swing angle θ, respectively. Besides, the maximum speed of the trapezoidal speed was 20 Cm/s, and the Non-Zero Speed of the Trolley Lasted Longer Under the Control of the NS15P Smoother.

Figure 4 shows the experimental and simulated responses of spring deflection x under control of the NS15P smoother and without control. In the case of no control of the NS15P

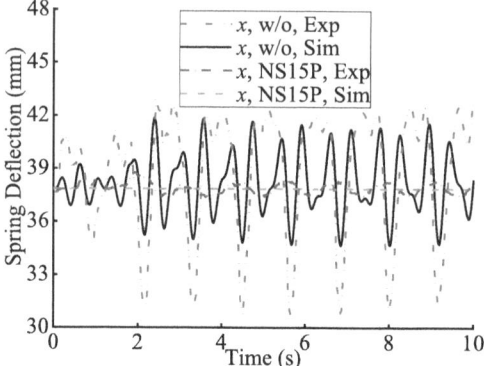

Fig. 4. Response of spring deflection.

smoother, the transient deflection and residual amplitude of the spring deflection x in the simulated curves were 2.47 mm and 7.23 mm, respectively. But when the NS15P smoother applied control, the transient deflection and residual amplitude of the spring deflection x were 0.04 mm and 0.04 mm in simulation, respectively. Furthermore, when there was no control, the transient deflection and residual amplitude of spring deflection x were 6.34 mm and 12.48 mm in experiment, respectively. However, when the control of the NS15P smoother was applied in the experiment, the transient deflection and residual amplitude of the spring deflection x were 1.41 mm and 1.03 mm, respectively. According to the simulation findings, the NS15P smoother had suppression rates of 98.38% and 99.45% on the transient deflection and residual amplitude of the spring deflection x, respectively. The experimental findings showed that the NS15P smoother had suppression rates of 77.76% and 91.75% on the transient deflection and residual amplitude of the spring deflection x.

The frequency of 5.39 rad/s could be found from the experimental uncontrolled swing angle. And this was similar to the frequency of 5.32 rad/s, which could be obtained according to the simulated uncontrolled swing angle. Besides, the frequency of 14.78 rad/s could be found from the experimental uncontrolled spring deflection. And this frequency was similar to the frequency of 13.96 rad/s, which could be found from the simulated uncontrolled spring deflection. Considering the frequency errors in experiments and simulations were small, and the NS15P smoother exhibited high suppression rates, the effectiveness of the dynamic model and controller suppression was verified.

Figure 5 shows the variation of the transient deflection of swing angle θ with driving distance under control of the NS15P smoother and without control in simulation and experiment. The displacement of the trolley was the driving distance. The maximum speed of the trolley was 20 cm/s. Changing the duration of the maximum speed of the trolley could alter its displacement. The uncontrolled transient deflection of swing angle θ increased with the increase of the driving distance from 0 to 16.5 cm in simulation. It was because the transient deflection was related to the duration of the transient stage and the acceleration of the trolley. After the driving distance of 20 cm, the change of the uncontrolled transient deflection of swing angle θ was very small. Within the range of 0 to 50 cm of the driving distance, the trends of change of peaks and troughs in the experimental and simulation curves were similar. Under the control of the NS15P smoother, the average suppression rates of the transient deflection of the swing angle θ were 67.47% and 80.48% in experimental and simulation results, respectively.

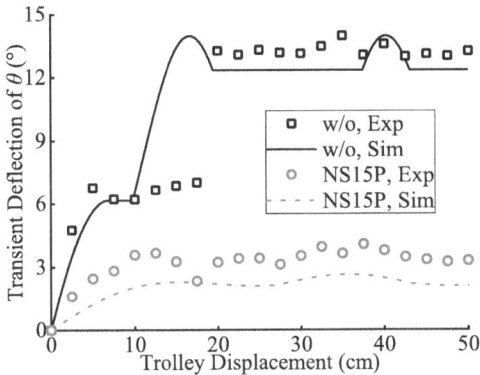

Fig. 5. Transient deflection of swing angle for various trolley displacement.

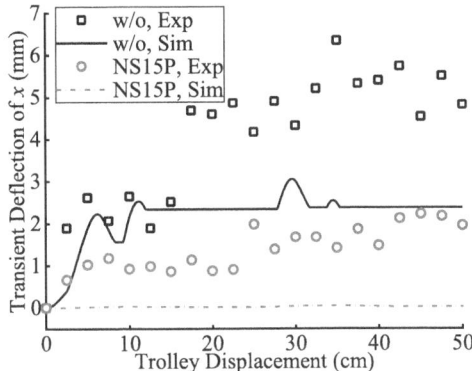

Fig. 6. Transient deflection of spring deflection for various trolley displacement.

Figure 6 displays the transient deflection of spring deflection x versus various driving distances under control of the NS15P smoother and without control in simulation and experiment. The dynamic characteristics of the uncontrolled transient deflection of spring deflection x were similar to those of swing angle θ. The results of the experiment and simulation showed that the average suppression rates of the transient deflection of spring deflection x under the control of the NS15P smoother were 64.54% and 99.05%, respectively.

Figure 7 shows the residual amplitude of swing angle θ versus changing trolley displacement under control of the NS15P smoother and without control in simulation and experiment. The trends of change of peaks and troughs in the experimental and simulation curves were similar within the range of 0 to 50 cm of the trolley displacement. The driving distance between the two peaks of 11.8 cm and 35.2 cm corresponds to the frequency of the swing angle. According to the findings in experiment and simulation, under the control of the NS15P smoother, the average suppression rates of the residual amplitude of the swing angle θ were 81.67% and 91.25%, respectively.

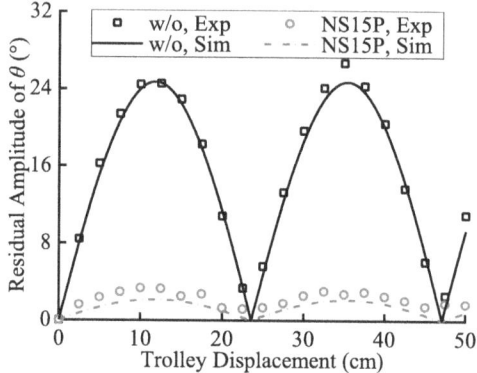

Fig. 7. Residual amplitude of swing angle for various trolley displacement.

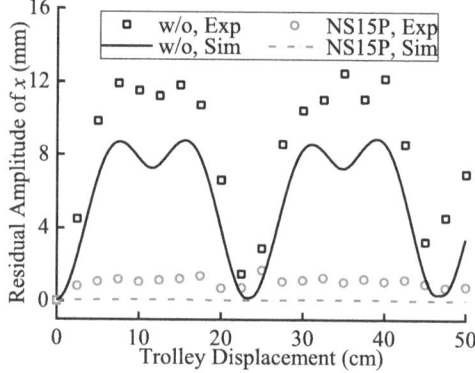

Fig. 8. Residual amplitude of spring deflection for various trolley displacement.

Figure 8 shows the variation of the residual amplitude of spring deflection x with driving distance under control of the NS15P smoother and without control in simulation and experiment. The driving distance between the two peaks of 31.1 cm and 39.1 cm corresponds to the frequency of spring vibration. The dynamic behaviors of the uncontrolled residual amplitude of spring deflection x were similar to those of swing angle θ. From the experiment and simulation results of Fig. 8, the residual amplitude of spring deflection x under the control of the NS15P smoother was suppressed by an average of 83.43% and 99.31%, respectively.

Therefore, Fig. 5 to Fig. 8 show that the NS15P smoother exhibited high suppression rates for transient deflections and residual amplitudes of swing angle θ and spring deflection x under different driving distances. Additionally, the dynamic characteristics have also been effectively predicted.

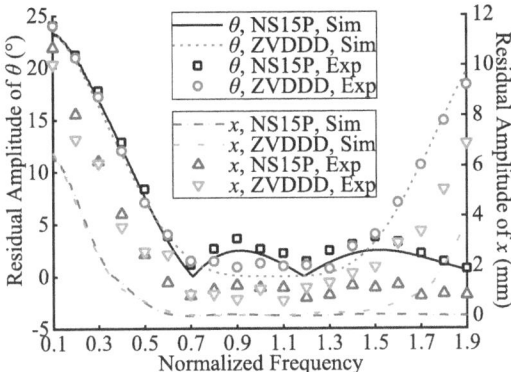

Fig. 9. Residual amplitudes of swing angle and spring deflection under different design errors in the frequency.

Considering the inevitable errors of frequency in practical applications, the controller needs to have good robustness. Because the rise time of the ZVDDD shaper is similar to that of the NS15P smoother, the ZVDDD shaper was used to compare robustness with the NS15P smoother.

Figure 9 shows the residual amplitudes of swing angle θ and spring deflection x changing with the normalized frequency under the control of the NS15P smoother and the ZVDDD shaper in simulations and experiments. The normalized frequency was defined as the ratio of the actual frequency to the design frequency. When the NS15P smoother applied control, the residual amplitude of the swing angle θ decreased when the normalized frequency increased from 0 to 0.7, while that had peaks at normalized frequencies of 0.9 and 1.5. Under the control of the ZVDDD shaper, the residual amplitude of the swing angle θ decreased when the normalized frequency increased from 0 to 1.1, while that increased when the normalized frequency ranged from 1.1 to 1.9. Additionally, there were similarities between the variation trends of the simulated curves and those of the experimental curves. When the normalized frequency increased from 0 to 1.1, the residual amplitude of the spring deflection x controlled by the two controllers decreased. Then, within the range of the normalized frequency increasing from 1.1 to 1.9, the NS15P smoothed residual amplitude of the spring deflection x remained unchanged, while that shaped by the ZVDDD shaper began to increase.

Therefore, the residual amplitudes of swing angle θ and spring deflection x suppressed by the NS15P smoother and the ZVDDD shaper were similar at low frequencies and near the normalized frequency of 1. However, the ZVDDD shaper had a much worse effect at high frequencies than the NS15P smoother in attenuating the residual amplitudes of swing angle θ and spring deflection x.

3 Conclusions

This article mainly introduced a spring-pendulum educational device for teaching dynamics and control courses. The course contents involved dynamic modeling, numerical simulation, and experimental verification. Students could learn to use Lagrange equation to derive the dynamic equation of forced vibration of the spring pendulum. Besides, students could observe the dynamic behaviors of the spring-pendulum vibration through simulations and experiments. Additionally, the advanced learning included deriving approximate solutions for the vibration of the spring-pendulum system by using multi-scale methods. The advanced learning also included designing a smoother to suppress coupled vibrations and comparing the robustness of controllers. Students could deepen their understanding of internal resonance and other similar mechanical systems with coupled vibrations by using the spring-pendulum device to study this course.

References

1. Pridgen, B., Maleki, E., Singhose, W., Seering, W., Glauser, U., Kaufmann, L.: A small-scale cherrypicker for experimental and educational use. In: Proceedings of the 2011 American Control Conference, pp. 681–686. IEEE (2011)
2. Xue, H., Huang, J.: Use of tower cranes in dynamics and control education for mechanical-engineering students. In: Control, Instrumentation and Mechatronics: Theory and Practice, pp. 93–104. Springer, Singapore (2022)
3. Ding, M., Huang, J., Ye, J.: Application of flexible link manipulators in control engineering courses. In: Proceedings of the 12th International Conference on Education Technology and Computers, pp. 177–181. Association for Computing Machinery (2020)
4. Miller, A., Singhose, W., Glauser, U.: Integrating PLC theory and programming into advanced controls courses. In: American Control Conference, pp. 7302–7307. IEEE (2016)
5. Maleki, E., Pridgen, B., Singhose, W., Glauser, U., Seering, W.: Educational use of a small-scale cherrypicker. Int. J. Mech. Eng. Educ. **40**(2), 104–120 (2012)
6. Singhose, W., et al.: Use of cranes in education and international collaborations. J. Robot. Mechatron. **23**(5), 881–892 (2011)
7. Bek, M.A., Amer, T.S., Sirwah, M.A., Awrejcewicz, J., Arab, A.A.: The vibrational motion of a spring pendulum in a fluid flow. Results Phys. **19**, 103465 (2020)
8. Amiri-Hezaveh, A., Ostoja-Starzewski, M.: A convolutional-iterative solver for nonlinear dynamical systems. Appl. Math. Lett. **130**, 107990 (2022)
9. Sypniewska-Kamińska, G., Starosta, R., Awrejcewicz, J.: Quantifying nonlinear dynamics of a spring pendulum with two springs in series: an analytical approach. Nonlinear Dyn. **110**(1), 1–36 (2022)
10. Yang, C., Huang, J., Singhose, W.: Dynamic modeling and oscillation control of industrial cranes transporting upright slender flexible payloads. Mech. Syst. Signal Process. **220**, 111676 (2024)

3D Scene Reconstruction Using Lidar Point Clouds and Images

Sulaiman Sabikan[1], Kenneth Edmond Ntende[2], Sophan Wahyudi Nawawi[2(✉)], and Shahrudin Zakaria[1]

[1] Faculty of Electrical Technology and Engineering, Universiti Teknikal Malaysia Melaka, Hang Tuah Jaya, 76100 Durian Tunggal, Melaka, Malaysia
[2] Faculty of Electrical Engineering, Universiti Teknologi Malaysia, 81310 UTM Skudai, Johor, Malaysia
e-sophan@utm.my

Abstract. LiDAR technology has been a key component in the field of 3D mapping in recent years due to its relative advantages, such as high accuracy regardless of the environmental setting. Consequently, it has been utilized in numerous outdoor and indoor applications, including city planning, autonomous driving, and more . However, the adoption of LiDAR for 3D mapping is limited due to challenges such as the high cost of 3D LiDAR sensors, low resolution and refresh rate and speed shift of targets among others. To enhance and achieve a comprehensive 3D understanding of an environment, the use of multiple sensors in conjunction with LiDAR has been shown to yield significant results. The integration of LiDAR and cameras has demonstrated the ability to provide a rich context for 3D scene construction. Notable performance in various 3D vision tasks has been observed with these two common sensor types: LiDAR and cameras. Cameras provide scene context and semantic information, while LiDAR sensors deliver precise 3D geometry. However, due to the difference in data types—namely, image data and point clouds—a challenge exists in determining the best method to combine such data for the reconstruction task. A data processing pipeline using the convolutional occupancy networks framework has been developed, incorporating a PointNet encoder and a ResNet model for the image data to achieve scene reconstruction.

Keywords: LiDAR · Scene Reconstruction · 3D reconstruction · Point clouds

1 Introduction

LiDAR technology has been a key component in the field of 3D mapping in recent years due to its significant advantages, such as high accuracy in various environmental settings. Consequently, it has been widely utilized in numerous outdoor and indoor applications, including city planning and autonomous driving [1–3, 13]. However, the adoption of LiDAR for 3D mapping is constrained by several factors, including the high cost of 3D LiDAR sensors, low resolution and refresh rates, and target speed shifts, among others.

To enhance and achieve a comprehensive 3D understanding of an environment, the use of multiple sensors together with lidar has been shown to achieve significant results [4–6]. Specifically, the combination of LiDAR and cameras has been shown to provide a richer context for 3D scene construction [6]. Recent studies highlight notable performance in various 3D vision tasks when using these two common sensor types: cameras, which provide scene context and semantic information, and LiDAR sensors, which produce precise 3D geometry [7, 8]. However, the differences in the data generated by these sensors pose a challenge. Cameras produce projections of the real world in the form of image data (ordered, discrete, and regular), while LiDAR generates point clouds (sparse and unordered), storing spatial geometric information of the target [9].

Additionally, image processing and sensor calibration are essential to mitigate noise and distortions in LiDAR data, which could otherwise hinder the quality of the final output. To address these challenges, a data fusion technique combining LiDAR and camera data has been proposed to enhance the understanding and reconstruction of an environment.

This project employs an implicit 3D representation based on the convolutional occupancy networks approach to achieve 3D scene reconstruction using two inputs: point clouds and color images of the scene.

1.1 LiDAR Technology

LiDAR (Light Detection and Ranging) technology is a remote sensing technique widely used for three-dimensional mapping and reconstruction purposes. This approach involves sending laser pulses to a specific location and measuring time taken for the reflected light from the sensor to return to the sensor [10].

LiDAR Sensors

A LiDAR sensor consists of two major components: a transmitter and a receiver. The transmitter performs the primary task of generating the laser beam that is emitted towards the target, while the receiver is responsible for capturing and analyzing the reflected signal from the target object. Furthermore, the receiver unit comprises a telescope for photon collection, an optical analyzer for filtering signals based on wavelength, and a data acquisition module for calculating the elapsed time [11].

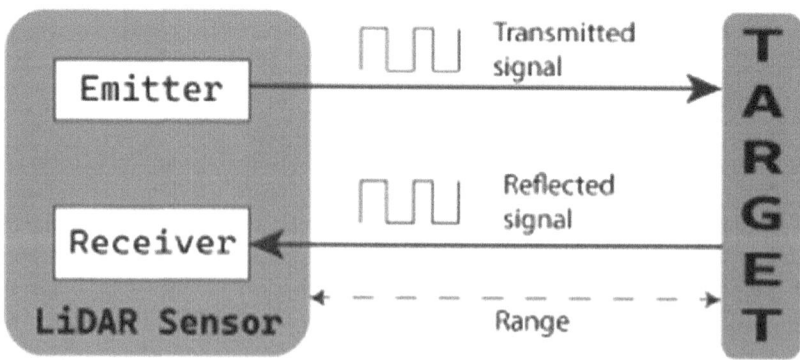

Fig. 1. A Simplified model of a LiDAR sensor

A simplified sketch for the operation of a LiDAR sensor is illustrated in Fig. 1. LiDAR sensors determine distance or range by calculating the round-trip time taken by a transmitted signal to return to the emitter. LiDAR sensing systems can be categorized into three types: rotor-based mechanical LiDAR, scanning solid-state LiDAR, and full solid-state LiDAR [12]. The most popular type of mechanical LiDAR, rotor-based, uses a motor-controlled rotating structure to generate a wide field of view. It has a high signal-to-noise ratio but heavy components and consumes a lot of power. There are two varieties of scanning solid-state LiDAR that don't have moving parts: optical phased arrays, which use phase modulators for accurate beam steering and quicker scanning, and Mirror MEMS, which uses tiny mirrors for beam control. Flash LiDAR, also known as full solid-state LiDAR, works similarly to digital cameras using photodetectors to capture backscattered light. It has a higher power consumption and a shorter range, but it can collect data more quickly.

Challenges to LiDAR Technology

One major issue is sensor calibration, as LiDAR sensors often require adjustments to eliminate errors resulting from component tolerances during manufacturing. Additionally, environmental signal interference can degrade the sensor's performance, causing a low signal-to-noise ratio (SNR). Data compression is another challenge, as LiDAR sensors generate large volumes of points quickly, creating difficulties for data handling, especially in real-time processing. Weather conditions also affect LiDAR performance, with point clouds being significantly impacted by adverse weather such as rain, snow, and fog. Finally, the segmentation of points on surfaces and those above surfaces can be difficult, complicating data interpretation.

1.2 3D Data Representations

3D structure representations can be split into either explicit representations (based on a predefined shape) such as voxel-based representations, point cloud representations, and mesh-based or implicit representations based on neural networks [14].

Voxel Based Representations

A voxel is a volumetric unit in three dimensions containing data about the characteristics of the associated volume, such as density, color, and material [15, 16]. Voxelization refers to a method of converting a continuous 3D representation for instance point clouds, or a 3D mesh into discrete volumes/cubes and it may be split into binary voxelization and non-binary voxelization [17]. Voxel-based 3D representations face several challenges, including high memory consumption, particularly when applied to detailed objects, as voxels require significant memory. These representations also suffer from low resolution, as memory requirements grow cubically with increased resolution, leading to higher computational costs during rendering. Additionally, modeling complex shapes or objects is difficult, limiting voxels to simpler forms. Due to their resolution constraints, voxels are unsuitable for reconstruction tasks involving smooth surfaces or complex, curved geometries.[18].

Point Clouds Representations

A point cloud refers to a set of points describing an object shape, or scene in a 3D space hence each point in the set has values for the x, y, and z coordinates [19]. A point cloud can be used to capture much more information concerning an object or scene such as underlying geometries which provide the accurate shape representation hence have been applied in various sectors e.g. scene reconstruction [1, 20, 21], depth map creation [22, 23], autonomous driving [24, 25], among others. Point cloud data may be obtained through the use of structured light sensors such as LiDAR sensors [26], stereo cameras [27] and RGBD cameras [28]. To obtain a point cloud for an entire environment, multiple point clouds of smaller environments are obtained considering different viewpoints and these are integrated together through a process of point cloud registration [19].

The formation of point clouds for a scene through point cloud registration presents several challenges, including variant overlap from obtaining point clouds from different viewpoints, which complicates the merging process. Point clouds are typically unordered and scattered, which makes them prone to noise and outliers, ultimately degrading the quality of the output data [19]. Additionally, the irregular structure of point cloud data leads to high computation costs, as it cannot be directly used for the required application and often requires further processing [29].

Mesh Representation

A mesh representation is a set of 3D points or vertices united with edges to create surfaces or faces of the 3D object from which a face is made of many polygons adjacent to each other [30]. Mesh representations will require much less memory constraints than voxel representations for the same resolution since they only represent a boundary surface rather than the entire volume [31]. Non-uniform polygonal meshes are a usually applied 3D data representation especially in computer graphics due to their ability to clearly illustrate and distinguish the structure from its surrounding [32].

Although mesh-based representations are widely used in computer graphics, reconstruction, and computer vision tasks, they face several limitations. One challenge is their high complexity and large file size due to the numerous polygons in detailed meshes, leading to significant computational and memory demands. Additionally, meshes are rigid and difficult to modify because they are based on polygonal structures, which can be problematic when changes are necessary. They are also highly topology-dependent, meaning that a good template shape is required to create an effective mesh representation. Furthermore, meshes can struggle with achieving detailed reconstructions, particularly for complex objects. Lastly, applying mesh representations to large scenes can be challenging due to the vast number of polygons needed, making the task computationally intensive and complex [33].

Implicit Representations

Unlike explicit functions such as meshes, voxels and point clouds, implicit functions approach the representation of objects by the use of neural networks. The implicit representations can be categorized into two major categories;

Neural Parametric Surfaces. This approach proposes learning the shape representations directly from raw input data such as point clouds through the use of a parameter, θ

such that the 3D object surface can be obtained from a signed function $f(x; \theta)$ being approximated to represent the surface for point data, x [34, 35]. From which each surface is presented as a local parameterized chart, however producing the required overlapping charts for the surface is challenging and not straight forward [36].

Neural Implicit Surfaces. This approach represents a 3D surface as a continuous classifier function (such as a neural network classifier) in 3D space, which predicts whether a given input point lies on the surface of an object. Examples of such methods include signed distance functions [37], occupancy networks [38], neural radiance fields [39] among others.

Implicit 3D data representations offer several advantages over voxel-based, mesh-based, and point cloud representations. These include high flexibility, as they are not tied to a specific surface discretization, allowing for smoother and more well-defined topologies [35, 38, 40]. They also provide high-resolution surface reconstructions, making them better suited for modeling complex shapes [39, 41]. Implicit representations have a lower memory footprint compared to mesh-based models and are more efficient for scene reconstruction [41], and they do not require a predefined template shape, making them more versatile for a wide range of 3D reconstruction tasks.

2 Methodology

The proposed pipeline consists of the following major elements namely; volume encoder, volume decoder, reconstruction block, and the texture rendering block. By utilizing these major components an input scene point cloud obtained from a LiDAR sensor and a reference image maybe used to generate a 3D model representation for a given scene. The general data processing pipeline for 3D scene reconstruction is shown in Fig. 2.

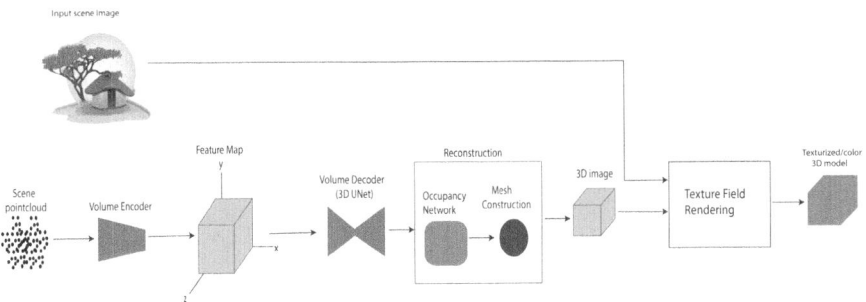

Fig. 2. The proposed data processing pipeline

2.1 Volumetric Encoding

The volumetric encoding constitutes the volume encoder and the volume decoder sections.

Volume Encoder

For an input point cloud, the volume encoder obtains feature encodings corresponding to the point cloud. These are converted into feature maps which can then be stored onto one or more selected planes. These features are then projected using into a 3D volume by use of an average pooling process [41]. For the volume encoder a PointNet architecture is considered and is shown in the Fig. 3. It consists of an input layer which takes a set of 3D points representing the coordinates (x, y, z), the feature transformation network and this learns linear tranformation and normalizes the input point cloud. The segmentation network is part of the classification network and it generates point scores coresponding to local and global features and the fully connected layers which generate feature vectors corresponding to classification and segementation tasks.

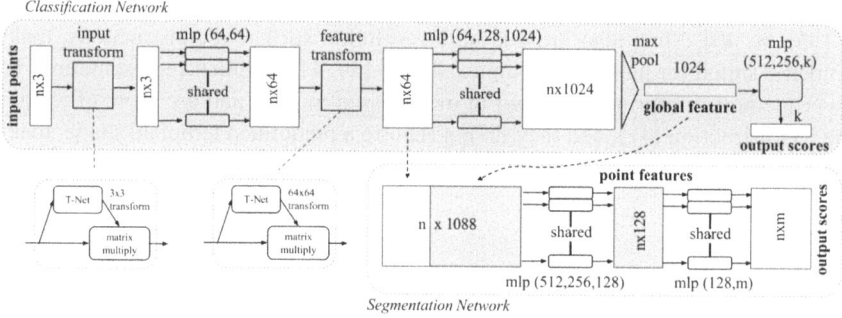

Fig. 3. PointNet architecture adopted from [42]

Volume Decoder

This makes use of a 3D U-Net architecture to process the feature planes and feature volume derived from the volume encoder. The U-Net architecture makes use of up-sampling and down sampling convolutions which can be used to extract global and local feature information obtained from the point cloud as shown in Fig. 4. The U-Net consists of an encoder, bottleneck, decoder block, skip connections, and the final output layer [43].

2.2 Reconstruction Block

The reconstruction block consists of two components namely an occupancy network and mesh construction.

Occupancy Network

An occupancy network is an implicit 3D representation which models the 3D surface as a conditional boundary based on a nonlinear classifier function. The nonlinear classifier in this case is a fully connected neural network whose task is to determine whether a point lies on the 3D surface or not [38]. The architecture of the occupancy network is shown in the Fig. 4. An occupancy network architecture consists of the following elements;

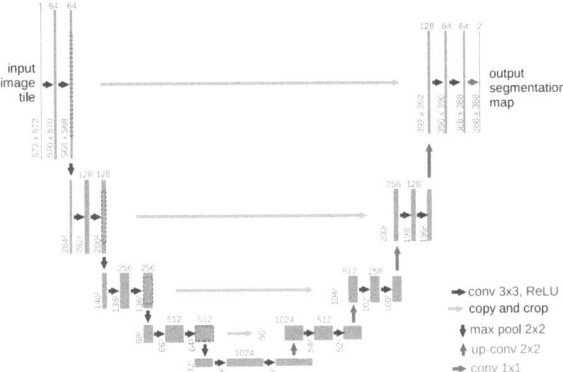

Fig. 4. The U-Net architecture adopted from[43].

Encoder block

The purpose of this block is to extract the key features from the input modality which could be a point cloud, or an image. For our case, the input being a point cloud, the Point-Net is applied as an encoder. It therefore follows, that the PointNet encoder processes the input point cloud and extracts a latent representation [18, 38].

Decoder block

This takes as an input the extracted latent representation from the encoder and uses it to understand the 3D shape to be reconstructed. For this case the decoder block utilized is constructed using a fully connected neural network consisting of five ResNet blocks [38] shown in the Fig. 5. Conditional Batch Normalization is applied to the decoder input at different stages in the decoder layers.

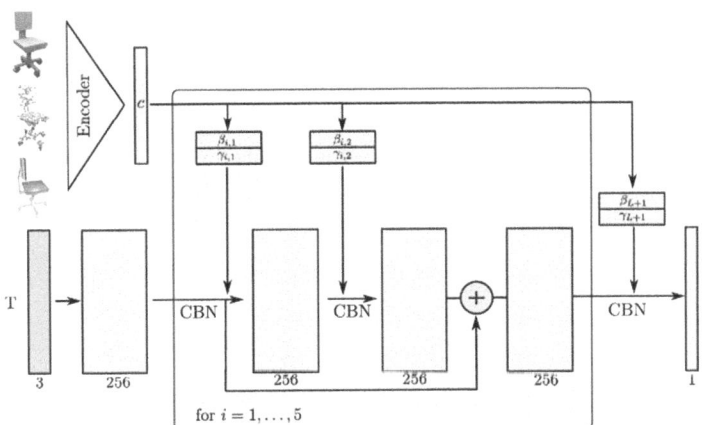

Fig. 5. The architecture of an occupancy network

Mesh construction

Due to the fact that an implicit 3D representation such as an occupancy network cannot directly provide a tangible 3D structure, a method is thus required to generate a tangible structure. In this case, the occupancy networks make use of the Multiresolution Iso Surface Extraction (MISE) algorithm to create a mesh structure as shown in Fig. 6. The MISE algorithm involves building an octree datastructure, marking off grid points at a given resolution, marking voxels for adjacent points for which are occupied or unoccupied, subdividing the voxels into 8 vertices and obtaining new vertices [38]. These steps are repeated till the required mesh resolution is achieved and then the marching cubes algorithm is applied to construct the mesh.

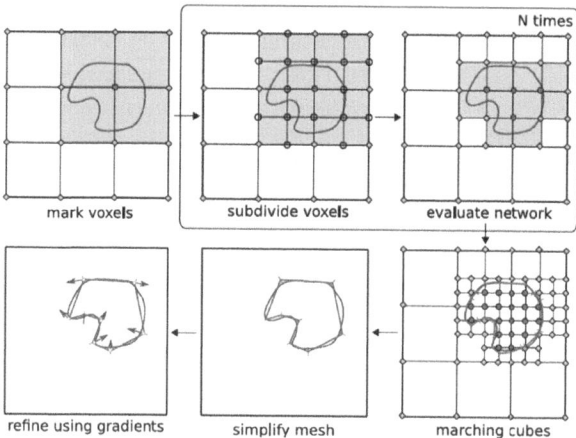

Fig. 6. Mesh construction using the MISE algorithm

Reconstruction Training

During the training phase of the network, a uniform sample of the points is selected from the feature volume of interest and the occupancy probability is determined. Since the problem is considered as a binary classification, a binary cross entropy loss is applied between the predicted occupancy value $\hat{O}_p$ and the true occupancy value O_p given by [41] and shown in Eq. (1);

$$\mathcal{L}(\hat{o}_p, o_p) = -[o_p \cdot \log(\hat{o}_p) + (1 - o_p)\log(1 - \hat{o}_p)] \qquad (1)$$

2.3 Reconstruction Block

The texture rendering phase consists of an image encoder, a shape encoder and a texture fields implicit function t_θ whose purpose is to learn the color/texture representation of an image from the input 2D image which is then applied to a 3D object. A simplified processing pipeline for the color information rendering is illustrated in the Fig. 7.

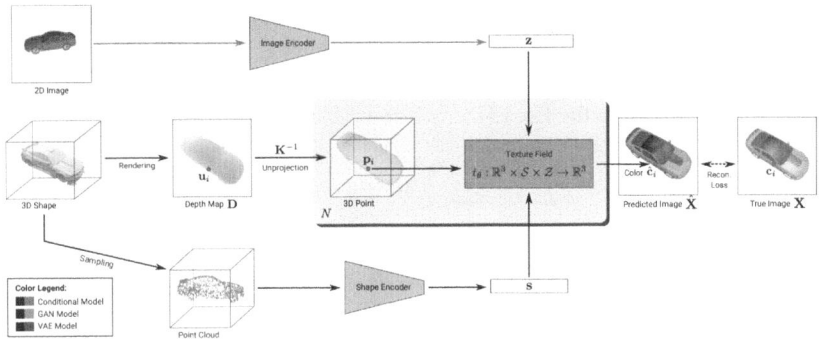

Fig. 7. Texture rendering approach adopted from [44]

The function t provides a relationship for an input point $\mathbb{R}^3$ in a 3D point cloud to a corresponding color value (RGB 3D vector $\mathfrak{R}$) i.e. $t : \mathbb{R}^3 \to \mathfrak{R}^3$. For the texture fields function t, it makes use of a shape embed S obtained from the shape encoder to learn the geometric information that is edges, lines of the 3D object and a color feature embedding Z obtained from an image encoder [44]. Therefore, the function t parameterized with θ, becomes as Eq. (2),

$$t_\theta : \mathbb{R}_i^3 \times S \times Z \to \mathbb{R}_o^3 \qquad (2)$$

From which t_θ is a neural network, S is a shape embed from the 3D object, Z is an image embed from the input image, $\mathbb{R}_i$ is the input 3D point cloud and $\mathbb{R}_o$ is the resulting texturized 3D model.

Image Encoder

The image encoder was employed to extract local features from the input image, particularly color information, and convert it into the required latent size Z. In this work, a ResNet-18 model was utilized as part of the convolutional occupancy network framework, serving as the backbone for the image encoder.

Shape Encoder

The input 3D object is sampled uniformly to obtain a well spread point cloud. From the resulting point cloud, a fixed dimension vector with the corresponding geometric information can be obtained using a PointNet encoder architecture as shown in Fig. 2.

Texture Field

This makes use of an input shape s, and colour embed z to predict the corresponding colour value c_i at each input point p_i. To predict the colour value, it makes use of a depth value from the image which may be obtained using depth completion.

2.4 Texture Field Rendering

In the texture rendering scenario an image is introduced, from which an embedding z can be extracted representing the input image. In order to train our network effectively, we adopt a supervised setting wherein we minimize the L_1 loss between the predicted image $\hat{X}$ and the rendered image X. This ensures that the predicted image closely resembles the actual rendered image, thereby enhancing the accuracy and reliability. The loss function used can be obtained from the Eq. (3).

$$\mathcal{L}_c = \frac{1}{B}\sum_{b=1}^{B}\sum_{i=1}^{N_b} \|t_\theta(p_{bi}, s_b, z_b) - c_{bi}\| \qquad (3)$$

From which B is the mini-batch size, p_{bi} is the input point cloud, N_b is the foreground pixels, s_b is the shape encoding, z_b is the input encoding from the image encoder and c_{bi} is the color value of the input image. The neural network t_θ is a fully connected ResNet architecture similar to that utilized in the occupancy network.

3 Results and Discussion

A summary of the results obtained using a similar method to generate 3D geometry scenes from synthetic indoor datasets, featuring objects from ShapeNet [41] is presented in Table 1. The results demonstrated the effectiveness of convolutional occupancy networks (CONs) in scene reconstruction. The dataset consisted of 10,000 points with Gaussian noise (standard deviation of 0.05), and the reconstruction was evaluated using both plane-based and volumetric methods. Plane-based reconstruction (map features onto a 2D plane with a resolution 128^2) and volumetric reconstruction methods (map input to a 3D feature volume) is applied. The input scenes were created with increasing complexity, and results were assessed at different voxel resolutions, specifically 323 and 643, to examine the impact of resolution on reconstruction quality. The performance was evaluated using metrics such as Intersection over Union (IoU), Chamfer-L1 distance, and F-Score. The IoU metric quantifies the overlap between the predicted and ground truth volumes, with higher values indicating better performance. The Chamfer-L1 distance measures the average distance between corresponding points in the predicted and ground truth volumes, with lower values reflecting better accuracy. The F-Score combines precision and recall into a single performance metric, where higher values indicate better overall performance.

As indicated in Table 1, at higher resolutions (e.g., 128^3 voxel resolution), the model successfully captured finer details of the scene but struggled with noise, which impacted the accuracy of the reconstruction. On the other hand, lower-resolution volumetric features (e.g., 32^3) demonstrated more resilience to noise and resulted in smoother surfaces, though at the cost of finer details. The best overall performance was achieved at a voxel resolution of 64^3, as it provided a balanced trade-off between detail retention and noise tolerance. The 64^3 voxel resolution yielded the highest IoU (0.849), the lowest Chamfer-L1 distance (0.042), and the highest F1-Score (0.964), suggesting that this resolution struck the optimal balance for scene reconstruction in this experiment. Additionally, combining 2D and 3D feature maps, such as using a 3×128^2 grid along with a 32^3 voxel resolution, further improved the model's performance with an IoU of 0.816, Chamfer-L1 of 0.044, and F-Score of 0.952. This approach demonstrated that the integration of 2D plane-based features with 3D volumetric data could enhance scene reconstruction, especially in environments with varying levels of complexity.

The proposed method achieves an F-Score of 0.96 at a voxel resolution of 64^3, outperforming key competitors such as PointNet++ (F-Score: 0.88) in both accuracy and computational efficiency [45]. By combining LiDAR point clouds and image-based features, the method produces smooth, detailed 3D reconstructions while maintaining high precision and recall, demonstrating its superiority and significant contribution to 3D reconstruction tasks.

Table 1. Results from the reconstruction section

Voxels resolution	IoU	Chamfer-L1	F-Score
3D, 32^3	0.782	0.047	0.941
3D, 64^3	0.849	0.042	0.964
2D-3D, $(3 \times 128^2 + 32^3)$	0.816	0.044	0.952

3.1 Qualitative Results

The illustrations in Fig. 8 indicate the qualitative results using a similar method, obtained at 32^3 voxel resolution with consideration of 3 planes at plane resolution of 128^2. These results highlight the model's capability to reconstruct 3D scenes with smooth and accurate surfaces. The original created indoor scene also referred to as the ground truth is shown in the Fig. 9. The model was supplied with an input point cloud from their synthetic indoor dataset from objects from the ShapeNet dataset.

The input for the model was a point cloud derived from this synthetic dataset, which contained geometric and spatial data about the objects. Upon analyzing the results, it is evident that the proposed method performs effectively in generating scene reconstructions that preserve the essential details and maintain smoothness across surfaces. This demonstrates the model's robustness in handling noisy input data and its ability to recover fine-grained details of the 3D geometry, even at relatively low voxel resolutions.

The use of plane-based features further contributes to improving the quality of the reconstructions, allowing the method to balance computational efficiency with visual fidelity. Overall, the proposed approach shows significant promise for practical applications in scenarios requiring detailed 3D reconstruction from sparse input data.

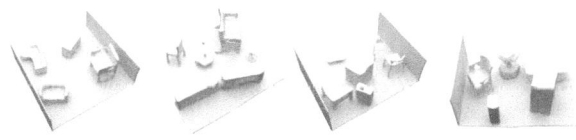

Fig. 8. Illustration of the reconstructed scene 3×128^2, 32^3

Fig. 9. An input point cloud generated from the synthetic dataset

3.2 Qualitative Results

The texture fields approach was used for texturization of a 3D model while using color information from a 2D input image [44]. The method takes an input 3D object plus a single image input to generate a texture reconstruction for the 3D object. The model uses images/shapes such as chairs, cars, lamps, benches etc. obtained from the ShapeNet dataset as the actual/ground truth images and transfers the predicted texture to the 3D object. The approach makes use of the Fréchet Inception Distance (FID), and Structural Similarity index (SSIM) functions to evaluate the performance.

Table 2 presents the results obtained from the texture fields approach for individual objects. The results highlight the versatility of the texture fields approach across a range of object categories, with certain categories (e.g., Chairs, Airplanes) consistently outperforming others. Categories like Cars and Cabinets show higher FID scores, likely due to their complex geometries or highly varied textures, which the model might struggle to capture fully. The SSIM values, however, remain relatively high for all categories, indicating that the method is generally successful in preserving structural integrity, even when the texture realism (as reflected in FID) varies. Overall, the texture fields approach demonstrates strong potential for 3D texture reconstruction but may require further optimization for challenging categories.

Table 2. Results obtained from the texture fields approach for individual objects

Object	FID	SSIM
Airplanes	9.236	0.968
Cars	24.271	0.885
Chairs	5.791	0.941
Tables	8.846	0.943
Benches	12.965	0.937
Cabinets	24.251	0.932
Sofas	9.618	0.938
Lamps	16.747	0.960
vessels	21.483	0.929
Average	14.801	0.939

From the Fig. 10, the texture fields approach is tested on input images obtained from the ShapeNet dataset. The input shapes act as the ground truth and it can be observed that the model is capable of successfully extracting texture from a single image.

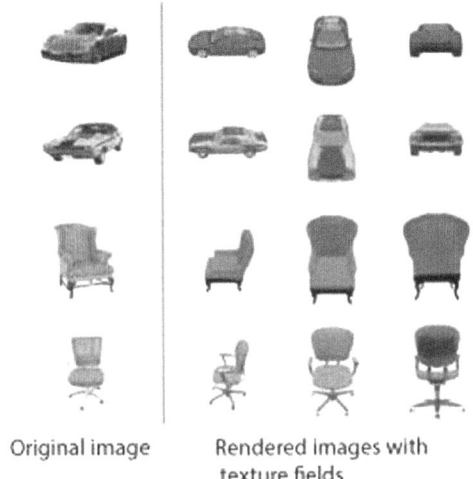

Original image   Rendered images with texture fields

Fig. 10: Illustration of the reconstructed scene 3 x 128^2, 32^3, adopted from [44].

4 Conclusion

The conclusion emphasizes the effectiveness of the proposed methods in achieving high-quality 3D reconstructions in terms of both structural detail and color representation. The findings suggest that the integration of LiDAR point cloud data with corresponding image information results in smooth, precise, and detailed 3D models. This combination enhances the fidelity of the reconstructed scenes, as the point cloud provides accurate spatial information, while the image contributes essential texture and color details.

The conclusion postulates that applying this integrated approach broadly could lead to consistent, desirable 3D reconstructions across a wide range of applications. This insight underscores the potential of the method for industries such as urban planning, robotics, and simulation, where accurate 3D models are critical. Future work could explore optimizing this integration for faster processing, improved handling of noisy data, or enhanced scalability to handle larger or more complex datasets.

References

1. Caminal, I., Casas, J.R., Royo, S.: SLAM-based 3D outdoor reconstructions from lidar data. In: 2018 International Conference on 3D Immersion, pp. 1–8 (2018)
2. Shan, T., Englot, B.: LeGO-LOAM: lightweight and ground-optimized LiDAR odometry and mapping on variable terrain. In: 2018 IEEE/RSJ International Conference on Intelligent Robots and Systems (2018). https://doi.org/10.1109/IROS.2018.8594299.

3. Song, Z., Lu, J., Yao, Y., Zhang, J.: Self-supervised depth completion from direct visual-LiDAR odometry in autonomous driving. IEEE Trans. Intell. Transp. Syst. **23**(8), 11654–11665 (2022). https://doi.org/10.1109/TITS.2021.3106055
4. Lin, Y.C., Fei, Y., Gao, Y., Shi, H., Xie, Y.: A LiDAR-camera calibration and sensor fusion method with edge effect elimination. In: International Conference Control Automation Robotics and Vision (2022). https://doi.org/10.1109/ICARCV57592.2022.10004361.
5. Wu, X., et al.: Sparse fuse dense: towards high quality 3d detection with depth completion. In: Proceedings of the IEEE Comput. Soc. Conference Computer Vision Pattern Recognition, pp. 5408–5417 (2022). https://doi.org/10.1109/CVPR52688.2022.00534
6. Li, Z., Gogia, P.C., Kaess, M.: Dense surface reconstruction from monocular vision and LiDAR. In: 2019 International Conference on Robotics and Automation (ICRA), pp. 6905–6911 (2019). https://doi.org/10.1109/ICRA.2019.8793729
7. Wang, T.H., Hu, H.N., Lin, C.H., Tsai, Y.H., Chiu, W.C., Sun, M.: 3D LiDAR and stereo fusion using stereo matching network with conditional cost volume normalization. In: 2019 International Conference Intelligent and Robotic Systems, vol. 2, pp. 5895–5902, IEEE (2019). https://doi.org/10.1109/IROS40897.2019.8968170
8. Zhong, H., Wang, H., Wu, Z., Zhang, C., Zheng, Y., Tang, T.: A survey of LiDAR and camera fusion enhancement. Procedia Comput. Sci. **183**, 579–588 (2021). https://doi.org/10.1016/j.procs.2021.02.100
9. Farahnakian, F., Heikkonen, J.: Fusing LiDAR and color imagery for object detection using convolutional neural networks. In: 2020 Proceedings 23rd International Conference Information Fusion, FUSION (2020) https://doi.org/10.23919/FUSION45008.2020.9190620
10. Raj, T., Hashim, F.H., Huddin, A.B., Ibrahim, M.F., Hussain, A.: A survey on LiDAR scanning mechanisms. Electron. **9**(5), 741 (2020). https://doi.org/10.3390/electronics9050741
11. Roriz, R., Cabral, J., Gomes, T.: Automotive LiDAR technology: a survey. IEEE Trans. Intell. Transp. Syst. **23**(7), 6282–6297 (2022). https://doi.org/10.1109/TITS.2021.3086804
12. Li, Y., Ibanez-Guzman, J.: Lidar for autonomous driving: the principles, challenges, and trends for automotive LiDAR and perception systems. IEEE Signal Process. Mag. **37**(4), 50–61 (2020). https://doi.org/10.1109/MSP.2020.2973615
13. Zhang, J., Singh, S.: LOAM: LiDAR odometry and mapping in real-time. Robot. Sci. Syst. **2**(9), 1-9 (2014). https://doi.org/10.15607/RSS.2014.X.007
14. Zheng, Z., Yu, T., Dai, Q., Liu, Y.: Deep implicit templates for 3D shape representation. In: Proceedings IEEE Comput. Soc. Conference Computer Vision and Pattern Recognition, pp. 1429–1439 (2021). https://doi.org/10.1109/CVPR46437.2021.00148
15. C.B.C.B, Xu, D., Gwak, J., Chen, K., Savarese, S.: 3D-R2N2: a unified approach for single and multi-view 3D object reconstruction deep residual GRU / LSTM network indicates equal contribution computer vision – ECCV 2016, Eccv, vol. 9905, pp. 628–644 (2016). https://doi.org/10.1007/978-3-319-46484-8
16. Sitzmann, V., Thies, J., Heide, F., Niebner, M., Wetzstein, G., Zollhofer, M.: DeepVoxels: learning persistent 3D feature embeddings. In: Proceedings of the IEEE Comput. Soc. Conf. Computer Vision Pattern Recognition, pp. 2432–2441 (2019). https://doi.org/10.1109/CVPR.2019.00254
17. Aleksandrov, M., Zlatanova, S., Heslop, D.J.: Voxelisation algorithms and data structures: a review. Sensors **21**(24), 1–22 (2021). https://doi.org/10.3390/s21248241
18. Geiger, A.: Learning 3D Reconstruction in Function Space. Autonomous Vision Group, pp. 1–57 (2020)
19. Li, L., Wang, R., Zhang, X.: A tutorial review on point cloud registrations: principle, classification, comparison, and technology challenges. Hindawi Math. Probl. Eng. **2021**(1), 9953910 (2021). https://doi.org/10.1155/2021/9953910

20. Tao, Y., Popovic, M., Wang, Y., Digumarti, S.T., Chebrolu, N., Fallon, M.: 3D LiDAR reconstruction with probabilistic depth completion for robotic navigation. In: IEEE International Conference on Intelligent Robots and Systems, pp. 5339–5346 (2022). https://doi.org/10.1109/IROS47612.2022.9981531
21. Tian, F., Gao, Y., Fang, Z., Gu, J., Yang, S.: 3D reconstruction with auto-selected keyframes based on depth completion correction and pose fusion. J. Vis. Commun. Image Represent. **79**, 103199 (2021). https://doi.org/10.1016/j.jvcir.2021.103199
22. Popescu, C.R., Lungu, A.: Real-time 3D reconstruction using a kinect sensor. Comput. Sci. Inf. Technol. **2**(2), 95–99 (2014). https://doi.org/10.13189/csit.2014.020206
23. Kuhner, T., Kummerle, J.: Large-scale volumetric scene reconstruction using LiDAR. In: Proceeding-IEEE International Conference on Robotics and Automation, pp. 6261–6267 (2020). https://doi.org/10.1109/ICRA40945.2020.9197388
24. Fu, C., Dong, C., Mertz, C., Dolan, J.M.: Depth completion via inductive fusion of planar LiDAR and monocular camera. In: IEEE International Conference Intelligent Robots and Systems (IROS), pp. 10843–10848, (2020). https://doi.org/10.1109/IROS45743.2020.9341385
25. Guan, L., Chen, Y., Wang, G., Lei, X.: Real-time vehicle detection framework based on the fusion of LiDAR and camera. Electron. **9**(3), 451 (2020). https://doi.org/10.3390/electronics9030451
26. Moosmann, F., Stiller, C.: Velodyne SLAM. In: 2011 IEEE Intelligent Vehicles Sympposium (2011). https://doi.org/10.1109/IVS.2011.5940396
27. Kirsten, E., Inocencio, L.C., Veronez, M.R., Da Silveira, L.G., Bordin, F., Marson, F.P.: 3D data acquisition using stereo camera. In: International Geoscience and Remote Sensing Symposium, pp. 9214–9217 (2018). https://doi.org/10.1109/IGARSS.2018.8519568
28. Billings, S.D., et al.: 3-D Mapping with an RGB-D camera. IEEE Trans. Robot. **30**(1), 177–187 (2015). https://doi.org/10.1109/TRO.2013.2279412
29. Han, X., Jin, J.S., Wang, M., Jiang, W., Gao, L., Xiao, L.: A review of algorithms for filtering the 3D point cloud. Signal Process. Image Commun. **57**(February), 103–112 (2017). https://doi.org/10.1016/j.image.2017.05.009
30. Khan, M.S.U., Pagani, A., Liwicki, M., Stricker, D., Afzal, M.Z.: Three-dimensional reconstruction from a single RGB image using deep learning: a review. J. Imaging **8**(9), 225 (2022)
31. Xu, Q., Mu, T., Yong-Liang, Y.: A survey of deep learning-based 3D shape generation. Comput. Vis. Media **9**(3), 407–442 (2023). https://doi.org/10.1007/s41095-022-0321-5
32. Hanocka, R., Hertz, A., Fish, N., Giryes, R., Fleishman, S., Cohen-or, D.: MeshCNN: a network with an edge. ACM Trans. Graph. **38**(4), 1–12 (2019). https://doi.org/10.1145/3306346.3322959
33. Xiao, Y., Lai, Y., Zhang, F., Li, C., Gao, L.: A survey on deep geometry learning : from a representation perspective **6**(2), 113–133 (2020)
34. Gropp, A., Yariv, L., Haim, N., Atzmon, M., Lipman, Y.: Implicit geometric regularization for learning shapes. In: 37th International Conference Mach. Learn. ICML, vol. PartF16814, pp. 3747–3757 (2020)
35. Palafox, P., Božič, A., Thies, J., Nießner, M., Dai, A.: NPMs: neural parametric models for 3d deformable shapes. In: Proceedings IEEE International Conference on Computer Vision, no. Iccv, pp. 12675–12685 (2021). https://doi.org/10.1109/ICCV48922.2021.01246
36. Schirmer, L., et al.: Neural networks for implicit representations of 3D scenes. In: Proceedings-2021 34th SIBGRAPI Conference on Graphics Patterns and Images, SIBGRAPI, pp. 17–24 (2021). https://doi.org/10.1109/SIBGRAPI54419.2021.00012
37. Park, J.J., Florence, P., Straub, J., Newcombe, R., Lovegrove, S.: DeepSDF : Learning Continuous Signed Distance Functions for Shape Representation (2019)

38. Mescheder, L., Oechsle, M., Niemeyer, M., Nowozin, S., Geiger, A.: Occupancy networks : learning 3D reconstruction in function space. In: Proceedings of the IEEE/CVF Conference on Computer Vision and Pattern Recognition (CVPR), pp. 4460–4470 (2019)
39. Yu, A., Ye, V., Tancik, M., Kanazawa, A.: PixelNeRF: neural radiance fields from one or few images. In: Proceedings IEEE Comput. Soc. Conference on Computer Vision Pattern Recognition, pp. 4576–4585 (2021). https://doi.org/10.1109/CVPR46437.2021.00455
40. Jiang, C., Sud, A., Makadia, A., Huang, J., Niebner, M., Funkhouser, T.: Local implicit grid representations for 3D scenes. In: Proceedings of the IEEE Computer Soc. Conference Computer Vision Pattern Recognition, pp. 6000–6009 (2020). https://doi.org/10.1109/CVPR42600.2020.00604
41. Peng, S., Niemeyer, M., Mescheder, L., Pollefeys, M., Geiger, A.: Convolutional occupancy networks. In: European Conference Computuer Vision, vol. 12348, pp 523–540 (2020)
42. Qi, C.R., Su, H., Mo, K., Guibas, L.J.: PointNet : deep learning on point sets for 3D classification and segmentation. In: Proceedings of the IEEE Conference on Computer Vision and Pattern Recognition (CVPR), pp. 652–660 (2017)
43. Olaf, R., Philipp, F., Thomas, B.: U-net: convolutional networks for biomedical image segmentation. In: Medical Image Computing and Computer-Assisted Intervention - MICCAI, pp. 12–20 (2015). https://doi.org/10.1007/978-3-319-24574-4
44. Oechsle, M., Mescheder, L., Niemeyer, M., Strauss, T., Geiger, A.: Texture fields : learning texture representations in function space. In: Proceedings of the IEEE/CVF International Conference on Computer Vision (ICCV), pp. 4531–4540 (2019)
45. Qi, C.R., Yi, L., Su, H., Guibas, L.J.: PointNet++: deep hierarchical feature learning on point sets in a metric space. In: Proceedings of the IEEE/CVF Conference on Computer Vision and Pattern Recognition (CVPR) (2017)

Dissolved Oxygen Concentration Modelling Based on Nanobubble Technology in Recirculating Aquaculture System

Indra Sakti[1,2], Nadiatulhuda Zulkifli[1(✉)], Sofia[3], Hilman Syaeful Alam[2], Mohd Fua'ad Rahmat[1], Sevia Mahdaliza Idrus Nameh[1], Hanif Fakhrurroja[2], Anto Tri Sugiarto[2], and Ahmad Aminudin[3]

[1] Faculty of Electrical Engineering, Universiti Teknologi Malaysia, 81310 Johor Bahru, Malaysia
nadiatulhuda@utm.my

[2] Research Centre for Smart Mechatronics, National Research and Innovation Agency, Bandung 40135, Indonesia

[3] Faculty of Mathematics and Natural Sciences, Universitas Pendidikan Indonesia, Bandung 40154, Indonesia

Abstract. Recirculating Aquaculture Systems (RAS) represent a sophisticated aquaculture methodology that optimises water utilisation while sustaining environmental integrity. A critical parameter within RAS is the concentration of dissolved oxygen (DO), which significantly influences the growth and overall health of aquatic organisms. Nanobubble (NB) technology is expected to increase the effectiveness of oxygenation in aquaculture systems because NB can produce nanometre-sized oxygen bubbles with longer residence time. This study aims to model DO concentration in RAS by utilising nanobubble technology, to improve system stability and efficiency. The DO dynamics model developed uses a numerical approach to analyse the dynamic balance of DO between the increase and consumption of DO in a water body by considering factors such as mechanical aeration, water circulation, and the oxygen demand of organisms. In mechanical aeration using NB, in terms of volumetric mass transfer coefficient ($k_L a$), it has an advantage over conventional aeration methods. The $k_L a$ value indicates how effectively oxygen can be transferred from the gas phase (air bubbles) to the liquid phase (water) in the system. One of the things that affects the value of $k_L a$ is the size of the bubbles formed, therefore the DO dynamic model equation is directly affected by the bubble size of the NB. In this study, it has been successfully proven that mechanical aeration using NB with a bubble size of 50–130 nm has a $k_L a$ value of $0.0649\ s^{-1}$ compared to MB which reaches $0.00024\ s^{-1}$ and $0.00026\ s^{-1}$ with bubble sizes of 40 and 500 μm.

Keywords: model dynamic dissolved oxygen · recirculating aquaculture system · nanobubble technology · mass transfer

1 Introduction

Nanobubbles are very small gas bubbles ranging from 1 to 200 nm. The main characteristics of nanobubbles are high stability and large surface area. NB can stay in solution for longer periods and can disperse well. Applications in various industries, including agriculture (such as irrigation and water purification), wastewater treatment, the food and beverage industry, and other research and development fields [1].

The recirculating aquaculture system is a highly efficient and water-saving method for factory aquaculture, making it a significant production method for the future of aquaculture. The system allows for recycling water, which is first filtered, sterilised, and aerated before being fed back into the aquaculture tank. This makes it an ideal system for studying the control of dissolved oxygen (DO) concentration, which is an important factor for the growth and development of aquatic products [2].

System dynamics modelling is used to create a mathematical model of the system dynamics of DO in aquaculture systems. The model serves as the primary function for examining and formulating the DO concentration level control system. A mechanism modelling approach is used to develop the model and seeks to investigate the interrelationships between system behaviour and the underlying mechanisms, thereby establishing a mathematical representation to analyse the principles governing DO variations in response to changes in external variables. This approach allows the model to comprehensively summarise the quantitative impact of diverse physical or chemical factors on DO, resulting in a model characterised by high-precision [2].

2 Related Work

2.1 Mechanism of Nanobubble Formation

Tsuge [3] found a phenomenon in microbubbles (MB) with size < 50 µm shrank and after a while disappeared or were assumed to be broken in water, later, very small particles with size < 2 µm were detected, and appeared stable within a few days. Therefore, it is assumed that one of the mechanisms for the formation of stable NB in water is residual or resulting from the shrinkage of MB based on assumptions from several previous studies [4, 5].

The assumption of NB formation due to the shrinkage of MB has been proven using improved image processing techniques. Jin et al. [6] used the light scattering method in dark-field microscopy (DFM) to investigate this assumption. Based on the results, there are two different evolutionary mechanisms of MB related to their size. In air MB exceeding 50 µm in diameter, aggregation tendencies are observed, which then lead to their rupture within seconds. MB between 10–50 µm in size tends to shrink and form NB. NB with bright dots appears stable in the water during observation until they disappear on the observation area's other side. Although the motion of the nanobubbles is not visible due to the limitation of the image magnification resolution, it is enough to show the stability of the NB.

2.2 Mass Transfer Characteristics of Gas to Liquid

Micro-Nano bubbles (M-NB) have higher gas-to-liquid mass transfer efficiency compared to MB. The gas-liquid mass transfer rate is 125 times faster and the DO concentration is 16 times higher than that of MB, so they have broad potential to be utilised in the field of water treatment [7].

In the gas-liquid mass transfer using air, the mass transfer rate from the gas phase to the liquid phase can be approximated by the following equation [8]:

$$\frac{dC}{dt} = k_L a (C_s - C_t) \tag{1}$$

Where $\frac{dC}{dt}$: rate of change of DO concentration in water concerning time; $k_L a$: volumetric mass transfer coefficient; C_s: DO concentration at saturation; C_t: DO concentration was measured during the experiment.

Levitsky et al. [9] conducted research on the effectiveness of DO transfer in a bubble-based aeration system by manipulating the dimensions of the MB, and then evaluating the impact on $k_L a$ through computational simulations. The variation in bubble size is influenced by the airflow used. The research findings are presented as shown in Table 1.

Table 1. Volumetric mass transfer coefficients in air for 44- and 500- μm MB [9]

Condition of Experiment	Diameter of Bubbles (μm)	$k_L a$ (/sec)
Stable	44	0.00026
Stable	500	0.00024

The research findings showed that the reduction in bubble diameter correlated with an increase in the resulting $k_L a$ value.

2.3 Development of Rotating Flow Nozzle Type Micro And Nanobubble Generator

This research was conducted through the development of a hydrodynamic cavity-based nano bubble generator utilising a two-chamber rotating flow nozzle [1]. The nozzle was constructed from Plexiglas, having an external diameter of 80 mm and a height of 160 mm. The external chamber of the nozzle is fitted with 16 guide holes that facilitate the directed flow of liquid towards the bubble column, which has a diameter of 7 mm. The bubble column is also constructed from Plexiglas, with a diameter of 145 mm and a total volume of 7.4 L, and is equipped with an outlet pipe connected to the cycle drainage system.

Research shows that nanobubbles with dimensions starting at 200 nm can be effectively produced in pure, negatively charged aqueous solutions. The stability of the nanobubbles has been demonstrated for 5 to 10 months, showing no substantial change in size or zeta potential.

2.4 Dissolved Oxygen Modelling in Recirculating Aquaculture Systems Based on Mechanism Analysis

Referring to [2], dissolved oxygen (DO) modelling in recirculating aquaculture systems (RAS) based on mechanism analysis aims to accurately reflect dynamic changes in DO levels by considering various influencing factors. This method is preferred over experimental modelling due to its ability to incorporate multiple variables affecting DO, such as respiration consumption, atmospheric reaeration, water circulation dynamics, aeration processes, and photosynthetic activity.

Mechanism analysis enables a comprehensive understanding of the quantitative influence of physical and chemical factors on DO. This results in a high-accuracy model, which is essential for precise DO control in aquaculture systems.

This study uses the principle of mass conservation, where the increase in DO is balanced by its consumption. This balance is critical to maintaining optimal DO levels for aquatic life. The model incorporates four key determinants: water circulation, mechanical aeration processes, surface reaeration phenomena, and the respiratory activity of aquatic organisms, including species such as shrimp.

3 Methodology

3.1 Data Collection on Micro and Nano Bubble Testing

In the experimental data acquisition of micro and nanobubbles, the test facility used in this investigation was carefully engineered to generate micro and nanobubbles by recirculation, using centrifugal pumps and vortex flow nozzles systematically configured in a bubble column reactor [1].

The design for testing micro and nanobubbles in this study involved several key components to ensure effective generation and characterisation. Here are the main elements:

- Swirling Flow Nozzle: This nozzle is essential for generating nanobubbles through hydrodynamic cavitation, which is effective for continuous production at high flow rates.
- Recirculation System: The experimental configuration includes a recirculation apparatus featuring a stainless steel centrifugal pump in conjunction with a vortex flow nozzle connected to the bubble column. This system enables the formation of nanobubbles under liquid saturation or supersaturation conditions, thereby promoting uninterrupted testing and production.
- Bubble Column: This bubble column has a waste line connected to the sewage system, which reintroduces water to the feed reservoir, thus facilitating the cyclical operation of the system for experimental purposes.
- Quantification of dissolved oxygen concentration in nanobubble water was performed using a WA-2017SD DO meter manufactured by Lutron Instruments, Taiwan.
- Deionised (DI) water: Deionised water was obtained from a duo of water purification equipment configured in a sequential arrangement, thus guaranteeing the high purity standards required for the experimental procedure. This choice of solvent is crucial for the proper evaluation of the properties and stability of the nanobubbles, as it significantly reduces the chance of contaminants potentially affecting the results.

3.2 Dimensional Characterisation of Microbubbles Using Particle Image Velocimetry (PIV) Method

The main mechanism underlying the formation of nanobubbles is attributed to the contraction of microbubbles with a diameter of less than 50 μm. Therefore, the study and characterisation of bubbles at the micro-scale is included in the early stages of research and further developed at the nanoscale. Characterisation of microbubble dimensions using the Particle Image Velocimetry (PIV) method is an important aspect of understanding the dynamics and efficiency of microbubble generators. PIV is an advanced optical method used to visualise and measure fluid flow by tracking the movement of particles, in this case, microbubbles, within the flow. This method is particularly advantageous due to its ability to provide reliable measurements in multiphase flows, which are common in microbubble applications [1, 10].

3.3 Dimensional Characterisation of Nanobubbles Using Dynamic Light Scattering (DLS) Method

Dynamic Light Scattering (DLS) is an advanced analytical methodology used to quantify the size distribution of small particulates, such as nanobubbles, in suspensions. In this investigation, the ZetaSizer Nano ZS particle size analyser was used to ascertain the size distribution of nanobubbles. This equipment functions based on the principles of the DLS technique, which can characterise particles or molecules dispersed in an aqueous environment. In the context of this study, the resulting nanobubbles were successfully synthesised, showing dimensions less than 200 nm [1].

The basic principle of DLS is the analysis of light scattered by particles undergoing Brownian motion. Smaller particles move faster, causing faster fluctuations in the intensity of the scattered light, while larger particles move slower, resulting in slower fluctuations. These fluctuations are detected and converted to particle size using the Stokes-Einstein equation [11].

The magnitude of the dispersed light correlates with the particle dimensions according to the Rayleigh light scattering principle, where emissions from more substantial particles exhibit greater strength than small particles [1].

This method provides a reliable way to assess the size distribution of nano-bubbles, which is crucial for understanding their stability and behaviour [1].

3.4 Testing the Zeta Potential and Stability of Nanobubbles

Measurement of zeta potential in nanobubbles is an important aspect of understanding their stability and surface charge characteristics. In the study of bulk nanobubbles in NaCl solution, the zeta potential was measured using the electrophoretic light scattering (ELS) technique. This method involves assessing the electrophoretic mobility of nanobubbles, which is directly related to their zeta potential [12].

The zeta potential generated in this study was −15.91 mV, immediately after generation.

3.5 Testing the Gas Flow Rate

The ratio of fluid flow rate to gas flow rate, represented as Q_l/Q_g, is an important factor in influencing the dimensions of the generated nanobubbles. Research [1] showed that an increase in the Q_l/Q_g ratio is associated with reduced nanobubble dimensions. This phenomenon is explained by the observation that an increase in the Q_l/Q_g value is associated with an increase in operational pressure, which consequently results in a decrease in the average diameter of the nanobubbles.

Investigations revealed that an increase in the Q_l/Q_g ratio corresponds to a reduction in the average diameter of the bubbles. This relationship is consistent with Henry's law, which states that the solubility of a gas in a solvent is directly proportional to the partial pressure exerted by the gas. Consequently, an increased operating pressure, as evidenced by an increase in the Q_l/Q_g ratio, increases the solubility of the gas and consequently leads to a small bubble size.

In addition, this study observed that the size distribution of nanobubbles, when measured by number, showed that most of the bubbles were less than 200 nm.

In summary, the Q_l/Q_g ratio is an important parameter in controlling the size of nanobubbles, with higher ratios leading to smaller bubble sizes due to increased operating pressure and increased gas solubility.

3.6 Testing the Mass Transfer Characteristics of Gas to Liquid

The mass transfer rate of gaseous substances to the liquid medium is a significant variable affecting the efficacy of the aeration process, which can be determined through the quantification of dissolved gas concentration in the liquid medium. The level of dissolved oxygen in an aquatic environment increases according to the increase in the rate of oxygen transfer from atmospheric air to water. In cases where the mass transfer occurs with optimal efficiency, the water will show an increase in DO concentration.

In this investigation, the DO concentration in the nanobubble solution was assessed in real-time using a WA-2017 SD DO meter that uses polarographic methodology and has a measurement range that extends from 0 to 50 mg/L [1].

3.7 Determination Volumetric Mass Transfer Coefficient ($k_L a$)

Data collection of DO concentration was carried out for 61 min, with the amount of data presented totalling 732, with a step of 5 s to see the dynamic changes in DO concentration. The gas used in the mechanical aeration of the nanobubble generator is free air.

The value of the volumetric mass transfer coefficient ($k_L a$) was determined using the logarithmic method based on the data of DO concentration versus time. The exponential equation of DO concentration was used to calculate $k_L a$ by performing regression analysis on the collected data using Eq. (1) [8].

Equation (1) is integrated into Eq. (2),

$$\ln\left(\frac{C_s - C_0}{C_s - C}\right) = k_L a t \qquad (2)$$

where C_s: DO concentration at saturation; C_0: initial dissolved oxygen concentration; C: DO concentration during measurement; k_L: Mass transfer coefficient; a: Contact surface area per unit volume of liquid; t: time; ln: natural logarithm.

The value of $k_L a$ is obtained from the gradient of the curve $\ln\left(\frac{C_s - C_0}{C_s - C}\right)$ at time t.

Next, the value of R-squared (R^2) will be calculated, which is used as an indicator to evaluate the fit of the regression model. A model with an R^2 value close to 1 means that the model can provide very accurate predictions.

3.8 Dynamic Modelling of DO

In this study, the architecture of the RAS is shown in Fig. 1. Which consists of:

a. Aquaculture tank: this is where aquatic organisms are raised. The tank is cylindrical with a volume of about 3 m³.
b. Waste filtering device: this equipment is used to purify effluent coming from aquaculture systems. The resulting filtered effluent undergoes a sterilisation and aeration process before reintegration into the aquaculture system, thus facilitating the recycling of water resources.

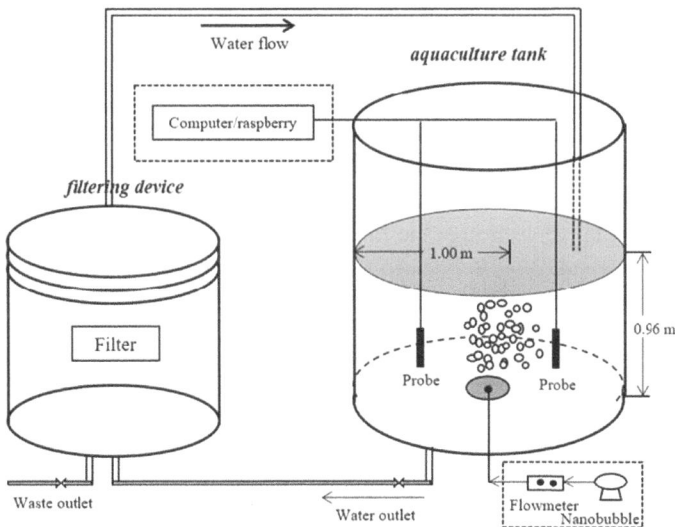

Fig. 1. The architecture of RAS.

In the context of recirculating aquaculture systems, the concept of dynamic balance of DO is essential to maintain optimal conditions for aquatic life. The balance between the increase and consumption of DO in a water body is governed by the principle of mass conservation. Where at any given time, the increase of DO should be equal to its consumption which ensures the dynamic balance in the system [2].

By the characteristics of the recirculating aquaculture system, this paper comprehensively considers the influence of three main factors on the dynamic change of DO, such as circulating water flow (E_{wf}), mechanical aeration (A_{er}), and shrimp respiration (R_{shrimp}) shown in Fig. 2, and developed a differential equation model capable of describing variations in DO levels in aquatic systems, as in Eq. (3).

$$\frac{dC}{dt} = E_{wf} + A_{er} - R_{shrimp} \tag{3}$$

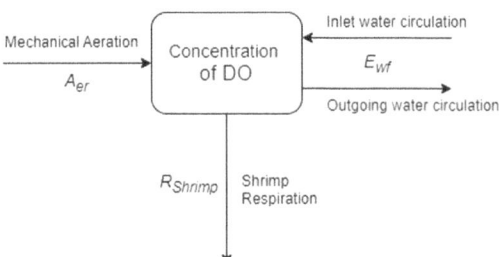

Fig. 2. Diagram of DO dynamic balance.

a. Circulating water flow E_{wf}

In RAS, the wastewater exiting the tank undergoes a filtration process, where it undergoes sterilisation, and aeration and is then reintroduced into the aquaculture tank to facilitate the recycling mechanism. [13]. As a result, the hydrodynamic properties of water will significantly affect the DO level in aquaculture tanks, which can be expressed quantitatively through the following equation:

$$E_{wf} = \frac{1}{V}(Q_{in}C_{in} - Q_{out}C_{out}) \tag{4}$$

Assuming that the volumetric flow rate of wastewater is equivalent to the volumetric flow rate of inlet water and the DO concentration in wastewater corresponds to the DO concentration in the reservoir, the above equation can be written as follows:

$$E_{wf} = \frac{Q_{in}}{V}(C_{in} - C) \tag{5}$$

b. Mechanical aeration A_{er}

Mechanical aeration serves as the primary means of oxygenation in aquatic environments used for industrial aquaculture, especially those generated by nanobubble technology using the concept of bubble mass transfer coefficient Eqs. (1) and the parameter constant used based on the determination of k_La.

c. Shrimp respiration R_{shrimp}

In this study, referring to research [2] using shrimp, shrimp need oxygen and energy to move, grow, and digest food. As a result, the aerobic respiration process will reduce the dissolved oxygen concentration in the aquatic environment. The shrimp respiration equation is written as follows:

$$R_{shrimp} = \frac{MR}{V} \quad (6)$$

where, R: respiration rate per unit mass (mg/h.kg); M: total mass of cultured organisms (kg); V: volume of water in the cultivation tank (L).

By combining the R and M values divided by V, this equation calculates the total oxygen consumption rate by all shrimp per volume of water. This gives an idea of how much oxygen is being used in the system per litre per hour.

4 Results Dan Discussions

4.1 Determination $k_L a$

In laboratory experiments conducted in obtaining DO concentration values as mechanical aeration using nanobubbles obtained DO values as shown in the graph in Fig. 3. The DO concentration showed a marked increase during the initial five minutes of production, then reached a state of stabilisation after ten minutes, and reached a peak DO concentration of 12.40 mg/L, with an average value of 10.67 mg/L.

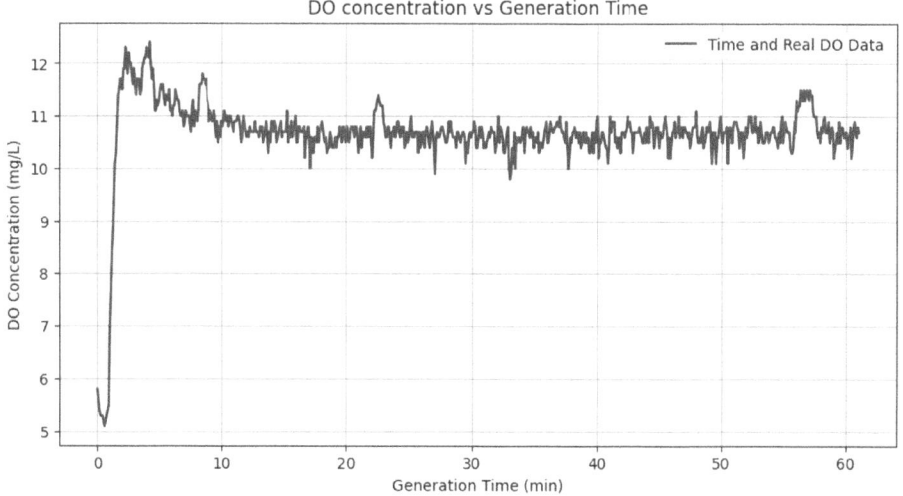

Fig. 3. Graph of the experiment.

By Eq. (2), the results of the $k_L a$ graph can be seen in Fig. 4. Judging from the $k_L a$ data pattern on the graph, the experimental data is blue with a dot shape, and the data distribution shows an exponential pattern. To analyse the data, the following exponential model was used:

$$y = \beta_0 \cdot e^{\beta_1 \cdot t} \quad (7)$$

where y $=\ln\left(\frac{C_s-C_0}{C_s-C}\right)$; t: time in seconds; β_0: initial constant or coefficient (value y at t = 0); β_1: a constant that determines the rate of growth or decline; t: time.

With the exponential model, a graph is produced in Fig. 4 with exponential line fitting data in red colour with a dotted line shape.

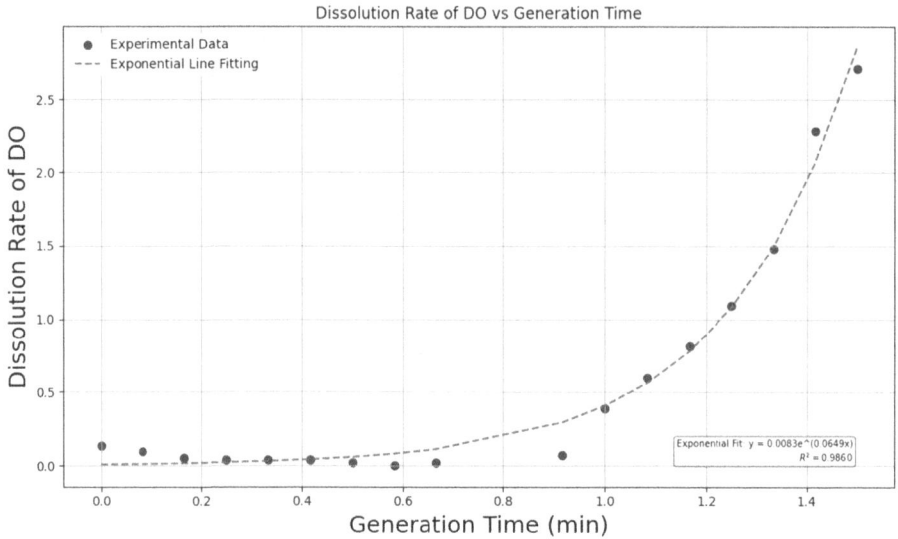

Fig. 4. Graph function $k_L a$ with exponential regression.

From the parameter β_1, the value of $k_L a$ is 0.0649 s^{-1} with a high R^2 of 0.986.

The increase in DO concentration during the aeration process indicates a diffusion gas transfer process. Diffusion occurs between the air in the nanobubble and DI water. Based on research by [14] that the value of $k_L a$ is also influenced by the size of the bubbles formed.

Furthermore, by comparing the $k_L a$ value of experimental data results with previous research by [9], aeration using nanobubble produces a greater $k_L a$ value compared to microbubbles because the current work with rotating flow instrumentation conditions can produce smaller bubbles in the range of 50 to 130 nm.

4.2 Dynamic Model DO

Dynamic modelling is very useful for applications to improve plant performance, especially to optimise plant design and processes [15]. Based on the variables that affect the DO concentration equilibrium in RAS (Eq. 3), the final equation for DO concentration in RAS versus time is as follows:

$$\frac{dC}{dt} = \frac{Q_{in}}{V}(C_{in} - C) + k_L a(C_{sat} - C) - \frac{MR}{V} \tag{8}$$

The characteristics of dynamic models related to DO systems are shown in Table 2.

Table 2. Dynamic modelling parameters of the DO system is RAS

Parameter	Value	SI	Ref
Q_{in}	0.34 m³/h	9.4 ×10–5 m³/s	[2]
C_{sat}	7.7 mg/L	7.7 × 10–3 kg/m³	[2]
V	3 m³	3 m³	Experiment
C_{in}	7.5 mg/L	7.5 × 10⁻³ kg/m³	[2]
M	3.25 kg	3.25 kg	[16]
R	110.2 mg/h/kg	3.06 × 10–8 kg/sec/kg	[16]
$k_L a$	3.894/min	0.0649/sec	Experiment Data

In order to analyse the system, an input transfer function written in Laplace transform (in s-domain) is required.

The differential Eq. (8) is rearranged to facilitate the application of Laplace transform by grouping C.

$$\frac{dC}{dt} + \left(\frac{Q_{in}}{V} + k_L a\right) C = \frac{Q_{in}}{V} C_{in} + k_L a C_{sat} - \frac{MR}{V} \quad (9)$$

$$SC(s) + \left(\frac{Q_{in}}{V} + k_L a\right) C(s) = \frac{Q_{in}}{V} \frac{C_{in}}{S} + k_L a \frac{C_{sat}}{S} - \frac{MR}{VS} \quad (10)$$

$$\left(S + \frac{Q_{in}}{V} + k_L a\right) C(s) = \frac{Q_{in}}{V} \frac{C_{in}}{S} + k_L a \frac{C_{sat}}{S} - \frac{MR}{VS} \quad (11)$$

Hence $C(s)$ is

$$C(s) = \frac{\frac{Q_{in}}{V} \frac{C_{in}}{S} + k_L a \frac{C_{sat}}{S} - \frac{MR}{VS}}{S + \frac{Q_{in}}{V} + k_L a} \quad (12)$$

$$C(s) = \frac{\frac{Q_{in} C_{in} + k_L a C_{sat} - MR}{VS}}{S + \frac{Q_{in}}{V} + k_L a} \quad (13)$$

where is

$$G(s) = \frac{C(s)}{C_{in}(s)} \quad (14)$$

Assuming $C_{in(s)}$ is constant, then

$$G(s) = \frac{\frac{Q_{in}}{V} C_{in}(s) + k_L a V C_{sat}(s) - MR}{VS(S + \frac{Q_{in}}{V} + k_L a)} \quad (15)$$

$$G(s) = \frac{Q_{in} C_{in} + k_L a V C_{sat} - MR}{VS(S + \frac{Q_{in}}{V} + k_L a)} \quad (16)$$

By substituting the parameter values in Table 2 with SI units, the transfer function is obtained as the following equation:

$$G(s) = \frac{1.5 \times 10^{-3}}{3s^2 + 0.06s} \tag{17}$$

Equation (17) is a dynamic model of DO in RAS using nanobubble technology as a DO generator.

4.3 Validation of the Dynamic Model DO

The transfer function of Eq. (17) is an open-loop equation; it can be analysed using the Zero-Pole Map.

Numerator: 1.5×10^{-3} has no factor s, so there is no zero in the system.

Denominator: $s(3s + 0.06)$ has two poles, the first pole $s = 0$ (origin) which indicates that the system has an integration component and $s = -0.02$ where this Pole is to the left of the $j\omega$ axis which indicates the stable part of the system.

From the poles above, it can be concluded that in the time domain, the system is unstable because there is a pole at the origin ($s = 0$), while in the frequency domain, the system is only partially stable because the pole at $s = -0.02$, which shows exponential decay for the component. The dominance of the pole at the origin causes the overall system response to not reach a steady state as the output continues to increase indefinitely. Overall, the system is unstable and requires feedback or a controller to stabilise the response.

From the open loop transfer function, a simple feedback control system can be embedded by converting the open loop into a closed loop transfer function given by:

$$T(s) = \frac{G(s)}{1 + G(s)H(s)} \tag{18}$$

where: $G(s)$: open-loop transfer function; $H(s)$: feedback path transfer function (usually $H(s) = 1$ for unity feedback).

If $H(s) = 1$, then the closed-loop transfer function becomes:

$$T(s) = \frac{G(s)}{1 + G(s)} = \frac{\frac{1.5 \times 10^{-3}}{3s^2 + 0.06s}}{1 + \frac{1.5 \times 10^{-3}}{3s^2 + 0.06s}} \tag{19}$$

simplify to:

$$T(s) = \frac{1.5 \times 10^{-3}}{3s^2 + 0.06s + 1.5 \times 10^{-3}} \tag{20}$$

Give a step input to Eq. 20, the graph will display the response of the closed-loop system to the step input as shown in Fig. 5.

From Fig. 5, the system reaches the steady-state value (setpoint) after 600 s, indicating that the system is stable, although there is an initial overshoot and oscillation.

Overshoot occurs at the beginning with a value greater than 1.2 (or 20% overshoot).

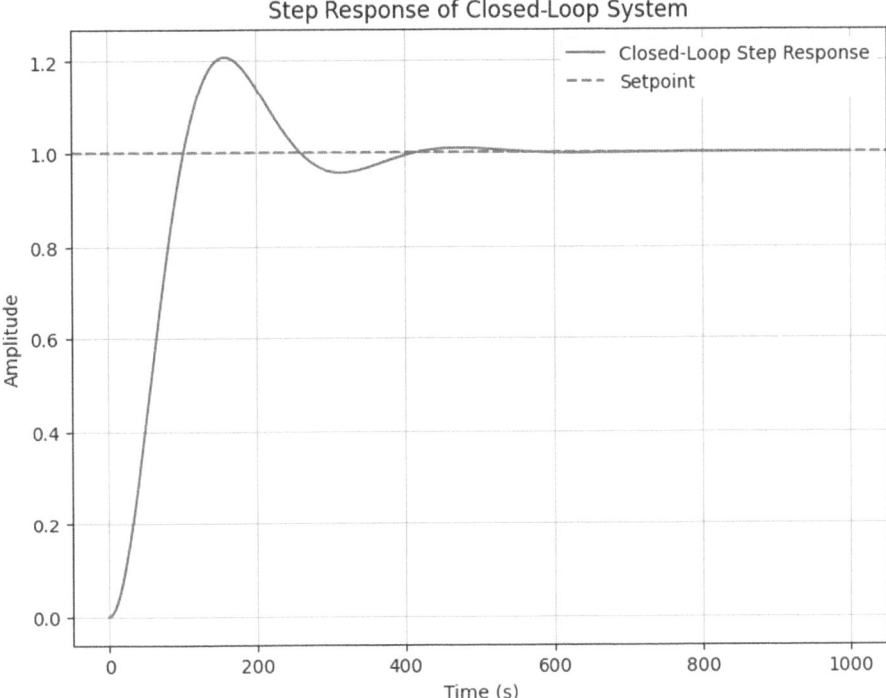

Fig. 5. Graphical representation of the step response for a closed-loop system T(s).

The settling time occurs at about 500–600 s, indicating that the system has a slow response and the oscillation frequency is relatively low.

In conclusion, by changing the system from open-loop to closed-loop, the system becomes stable, although it requires further refinement to improve its dynamic performance.

5 Conclusions

In this study, Some conclusions can be drawn, as outlined below:

1. The use of nanobubbles as mechanical aeration was successful in increasing the DO concentration. The DO concentration showed a marked increase during the initial five minutes of production, then reached a state of stabilisation after ten minutes, and reached a peak DO concentration of 12.40 mg/L, with an average value of 10.67 mg/L.
2. The volumetric mass transfer value ($k_L a$) generated through the logarithmic method shows that nanobubble aeration has $k_L a$ of 0.0649 s^{-1} or 3.894 m^{-1} means that aeration using nanobubble has a much higher rate of oxygen transfer from gas to liquid than microbubble aeration (0.00026 s^{-1}). The regression model of nanobubble aeration has a coefficient of determination (R^2) value of 0.986 which means that the model has excellent predictive ability and fits the observed data.

3. This study has successfully obtained a dynamic model of DO in RAS using nanobubbles as a DO generator, wherein DO dynamic modelling describes changes in DO concentration over time influenced by the oxygen transfer rate from the air into the water. The oxygen transfer rate is influenced by the dimensions of the bubble, the smaller the bubble, the more k_La increases and accelerates the achievement of DO saturation conditions.

Acknowledgements. This research was conducted as part of a collaborative research grant from Universiti Teknologi Malaysia (UTM) and the National Research and Innovation Agency (BRIN) under grant number R.J13000.7351.4B734. This research is also a collaboration between Lightwave Communication Research Group, Faculty of Electrical Engineering, UTM and Smart Instrumentation Research Group, Research Center for Smart Mechatronics, Research Organization of Electronics and Informatics, BRIN.

References

1. Alam, H.S., Sutikno, P., Soelaiman, T.A.F., et al.: Bulk nanobubbles: generation using a two-chamber swirling flow nozzle and long-term stability in water. J. Flow Chem. **12**, 161–173 (2022)
2. Zhou, X., Wang, J., Huang, L., et al.: Modelling and controlling dissolved oxygen in recirculating aquaculture systems based on mechanism analysis and an adaptive PID controller. Comput. Electron. Agric. **192**, 106583 (2022)
3. Tsuge, H.: Micro- and Nanobubbles. CRC Press Taylor & Francis Group, Boca Raton (2015)
4. Huynh, E., Leung, B.Y.C., Helfield, B.L., et al.: In situ conversion of porphyrin microbubbles to nanoparticles for multimodality imaging. Nat. Nanotechnol. **10**, 325–332 (2015)
5. Ushikubo, F.Y., Furukawa, T., Nakagawa, R., et al.: Evidence of the existence and the stability of nano-bubbles in water. Colloids Surf. A **361**, 31–37 (2010)
6. Jin, J., Wang, R., Tang, J., et al.: Dynamic tracking of bulk nanobubbles from microbubbles shrinkage to collapse. Colloids Surf. A **589**, 124430 (2020)
7. Li, H., Hu, L., Song, D., et al.: Subsurface transport behavior of micro-nano bubbles and potential applications for groundwater remediation. Int. J. Environ. Res. Public Health **11**, 473–486 (2013)
8. Dong, X., Liu, Z., Liu, F., et al.: Effect of liquid phase rheology and gas–liquid interface property on mass transfer characteristics in bubble columns. Chem. Eng. Res. Des. **142**, 25–33 (2019)
9. Levitsky, I., Tavor, D., Gitis, V.: Microbubbles, oscillating flow, and mass transfer coefficients in air-water bubble columns. J. Water Process. Eng. **49**, 103087 (2022)
10. Yusuf, A., Asdak, C., Muhaemin, M., et al.: The implementation of micro/nanobubbles (MNBs) technology to treat basin water as the primary water source for hydroponics in greenhouse. J. Tek. Pertan. Lampung (J. Agric. Eng.) **13**, 197 (2024)
11. Nirmalkar, N., Pacek, A.W., Barigou, M.: On the existence and stability of bulk nanobubbles. Langmuir **34**, 10964–10973 (2018)
12. Alam, H.S., Sutikno, P., Soelaiman, T.A.F., et al.: A diffused double-layer model of bulk nanobubbles in aqueous NaCl solutions. Therm. Sci. Eng. Prog. **50**, 102590 (2024)
13. Yang, H., Csukás, B., Varga, M., et al.: A quick condition adaptive soft sensor model with dual scale structure for dissolved oxygen simulation of recirculation aquaculture system. Comput. Electron. Agric. **162**, 807–824 (2019)

14. Bai, M., Liu, Z., Zhang, J., et al.: Prediction and experimental study of mass transfer properties of micronanobubbles. Ind. Eng. Chem. Res. **60**, 8291–8300 (2021)
15. Džubur, A., Serdarević, A., Šuvalija, S.: Modelling Steps for Dynamic Simulation of Wastewater Treatment Processes BT-Advanced Technologies, Systems, and Applications VII. In: Ademović, N., Mujčić, E., Mulić, M., et al. (eds.) pp. 122–137. Springer, Cham (2023)
16. Yao, X., Zhang, G., Yang, S., et al.: Adaptive anti-disturbance control of dissolved oxygen in circulating water culture systems. Symmetry (Basel) **15** (2023)

Analysis of Ultrasonic Signal Arrival Time in Concrete with Multiple Defects

Farah Aina Jamal Mohamad[1], Anita Ahmad[1]($\boxtimes$), Ruzairi Abdul Rahim[1], Juliza Jamaludin[2], and Syarfa Najihah Raisin[2]

[1] Faculty of Electrical Engineering, Universiti Teknologi Malaysia, 81310 UTM Skudai, Johor, Malaysia
anita@utm.my

[2] Faculty of Engineering and Built Environment, Universiti Sains Islam Malaysia, 71800 Nilai, Negeri Sembilan, Malaysia

Abstract. Concrete is the most essential material in the construction industry. Assessing its structural integrity is crucial for ensuring safety and durability. Non-destructive ultrasonic testing is commonly employed to detect defects in concrete. However, the presence of multiple defects can significantly affect the accuracy of these methods. Hence, the aim of this study is to evaluate the effects of multiple defects in terms of their arrangements, types, and sizes, on ultrasonic signal arrival time in concrete models, by using COMSOL Multiphysics 6.0. Two groups of concrete models were evaluated: Group A consisting of close, non-linearly arranged defects, and Group B with linearly arranged defects distanced 100 mm apart. Each group comprised six models with varying combinations of air hole and rust defects. The ultrasonic signal arrival times were measured using the through-transmission method. The results indicate that distanced, linear arrangement defects cause longer delays compared to close, non-linear configurations. The findings provide insights for enhancing non-destructive ultrasonic testing techniques for detecting and assessing structural defects in concrete.

Keywords: Arrival time · Concrete · Multiple defects · Ultrasonic signal · Wave propagation

1 Introduction

Concrete is a fundamental material in modern construction, known for its structural strength and durability. However, over time, internal defects such as air voids, cracks, and rust formations can develop within the concrete, compromising its structural integrity. These defects, if left undetected, can lead to significant safety hazards, which could cause loss of life and property failure [1]. Therefore, it is essential to accurately detect these hidden defects to ensure the safety of concrete structures and extend their service life.

The development of practical and reliable non-destructive testing (NDT) methods is critical for monitoring the state of concrete structures, particularly for identifying internal damages. NDT methods are highly advantageous as they enable the evaluation of structural health without destroying the tested structure. Among various NDT methods, ultrasonic testing is widely employed due to its ability to provide immediate results. In several cases, it requires only one surface of the structure to be accessed [2]. Ultrasonic testing operates on the principle of propagating ultrasonic waves through the concrete. The presence of defects alters the wave velocity, attenuation, and arrival times [3], allowing for the detection of internal flaws.

Previous studies have demonstrated the feasibility of ultrasonic testing in various applications. For example, Hongbin and Jinying [4] applied ultrasonic-guided wave testing to detect internal defects in steel-concrete modular construction, which is commonly found in nuclear structures. Similarly, Jeongnam et al. [5] utilized the time-of-flight (TOF) technique to estimate crack depth in concrete. They combined finite element analysis (FEA) with experimental approaches to measure the TOF in both non-cracked and vertically cracked concrete blocks. In another study, Tonghao et al. [3] proposed an advanced ultrasonic imaging technique that integrates linear and nonlinear wave properties within a single measurement. Their study focused on linear ultrasonic imaging to investigate inclusions with varying sizes and ratios. Additionally, Sai Teja and Debdutta [6] used ultrasonic shear horizontal (SH) waves that are transmitted by a transducer array and utilized array full matrix capture (FMC) to detect and size internal cracks in reinforced concrete. Their study focused on the geometric characterization of defects to predict crack progression and failure patterns.

While much research has focused on the presence of single defects in concrete structures [5, 7–9], real-world concrete structures often contain multiple defects of varying types, sizes, and orientations. These multiple defects complicate the wave behavior and can affect the accuracy of defect detection. While the detection of defects, whether single or multiple, generally implies that the concrete is faulty, the detailed characterization of defect types and their interactions provides essential insights for structural integrity assessments and targeted repair strategies.

Hence, this study aims to analyze the effect of multiple defects on ultrasonic signal arrival time in concrete. Specifically, this research investigates both air-type defects (air hole) and solid-type defects (rust), arranged in linear and non-linear configurations. By using COMSOL Multiphysics simulations, this research provides significant insights into the effect of type, size, and arrangement of multiple defects on ultrasonic wave propagation and arrival time.

2 Defects in Concrete

Degradation and potential failure of concrete structures are becoming significant concerns. The condition of concrete can be influenced by various factors, including the quality of design, manufacturing processes, applied loads, load characteristics, environmental conditions, and aging [10, 11]. Some of these factors may cause adverse impacts such as concrete segregation, delamination, cracking, and spalling, resulting in the formation of air holes and structural discontinuities. Two commonly encountered defect types in concrete are air voids and rust-induced damage, both of which significantly affect the material's mechanical and acoustic properties.

Air holes or voids in concrete are small cavities that occur during the concrete casting process due to improper compaction or entrained air. These voids can vary in size and distribution. Air voids weaken the concrete matrix and alter the propagation of acoustic waves, as the wave travels faster through solid material compared to air. Meanwhile, rust forms due to the corrosion of steel reinforcement within concrete. It often occurs when moisture and oxygen penetrate the concrete cover, initiating electrochemical reactions [12]. Rust expands within the matrix, causing cracking, delamination, and spalling, thereby reducing structural capacity. Its distribution is generally linear along the reinforcing bars, but localized rust pockets can also form due to non-uniform environmental exposure. This defect not only compromises the structural integrity but also accelerates further deterioration if left unaddressed.

Defects in concrete can occur in linear or non-linear arrangements, as shown in Fig. 1. Defects in linear arrangements commonly occur in cases of reinforcement corrosion, where damage propagates along the length of steel bars. Meanwhile, defects in non-linear arrangements result from random air voids, localized stress concentrations, or irregular damage caused by environmental factors. Non-linear distributions can represent more complex damage scenarios and are essential for understanding defect interaction and acoustic wave behavior.

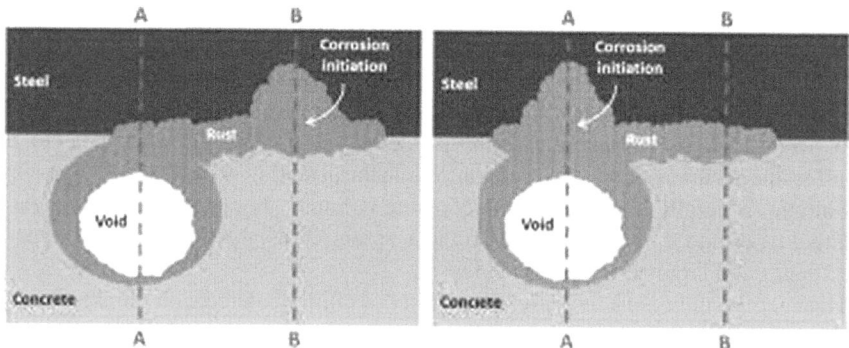

Fig. 1. Distribution of voids and corrosion in concrete structures [13].

3 Methodology

The simulation model was developed by using COMSOL Multiphysics 6.0, a versatile modeling platform due to its capability for direct multi-physics coupling. This software integrates wave propagation modeling, including the interaction between acoustic waves and structural boundaries, with signal detection via the piezoelectric effect. In this study, COMSOL Multiphysics is used to simulate and analyze the propagation of acoustic waves within concrete domains.

3.1 Geometrical Modeling

To improve computational efficiency, a simplified two-dimensional model was created to simulate the propagation of ultrasonic waves in defective concrete. The propagation characteristics of ultrasonic waves in concrete are studied, where the ultrasonic wave is excited by piezoelectric force, so the numerical simulation considered each of the piezoelectric, electrostatic, and multi-physics circuit fields in solid mechanics. In this paper, a concrete block model with cross-sectional dimensions of 300 mm × 300 mm was established, as shown in Fig. 2, under established industry standards for concrete block size [14, 15].

The ultrasonic transmitter and receiver are probes with a diameter of 20 mm, positioned on the top and bottom surfaces of the concrete model, respectively. The transducers are spaced 300 mm apart, corresponding to the size of the concrete block used in the simulation [16].

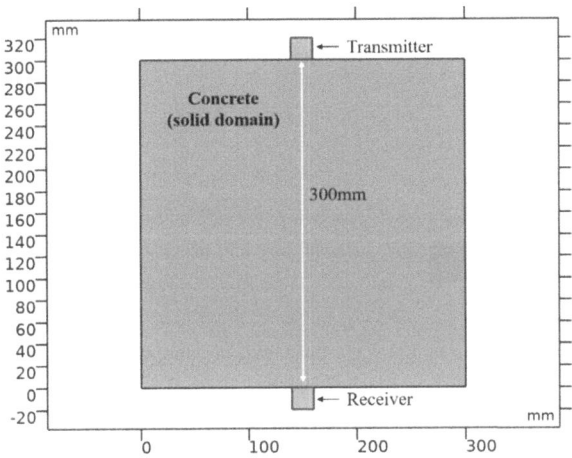

Fig. 2. A 2D schematic diagram of a concrete model.

This study utilized the ultrasonic through transmission technique due to its ability to transmit the maximum pulse energy at a right angle to the transmission line, thereby allowing for clear and easily determinable measurements of the traveled waves [17]. An ultrasonic transducer transmitted waves that propagated through the concrete models and the arrival time of the wave was recorded.

As illustrated in Fig. 3, the ultrasonic waves are represented in red and blue colors, indicating the acoustic pressure values in different directions. The intensity of the color correlates with the magnitude of the acoustic pressure. As the waves propagate, the color gradually fades, indicating a reduction in acoustic pressure.

The simulation study involved two groups of concrete models, as in Table 2, each containing six samples with varying combinations of defects. Group A consisted of defects arranged in a close, non-linear configuration, while Group B featured a linear arrangement of defects along the wave propagation path with a consistent separation distance of 100 mm apart. Each defect was modeled as either an air hole (air-type defect) or rust (solid-type defect) with diameters of 30 mm and 60 mm for various cases. These sizes were chosen based on their practical relevance to real-world scenarios, where a 30 mm defect size represents smaller voids or localized areas of corrosion that might develop in the early stages of deterioration, while a 60 mm defect size simulates more severe corrosion or larger air pockets that could significantly impact structural integrity. Rust defects were included to mimic real-world corrosion commonly found in reinforced concrete. In this study, the defects were modeled as round shapes to simplify simulations while approximating natural voids and localized rust damage.

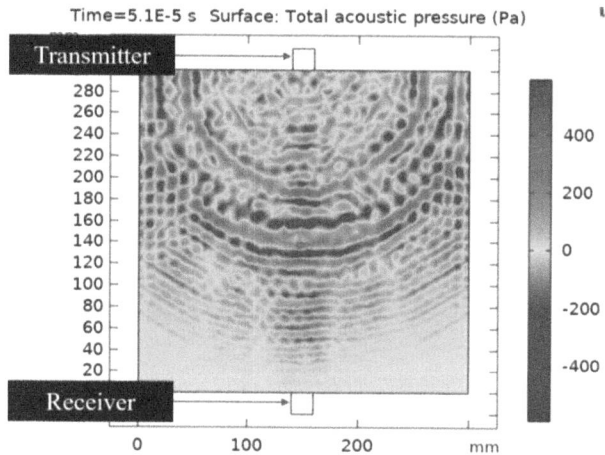

Fig. 3. Ultrasonic wave propagation in concrete model.

3.2 Properties of Materials

Table 1 lists the material properties that were used in the simulations. The concrete properties are set based on the values from the previous study [18].

Table 1. Properties of concrete, air, and rust.

Material	Density (kg/m^3)	Young's Modulus (GPa)	Poisson's ratio	Speed of sound (m/s)
Concrete	2000	3500	0.2	3400
Air	–	–	–	343
Rust	3925	102	0.15	5890

A pair of lead zirconate titanate (PZT-5H) piezoelectric transducers were used as the sensing element. These transducers are commonly used for ultrasonic monitoring

Table 2. Two groups of concrete models, Group A and Group B, each containing six samples with varying combinations of defects.

Group	Wave Propagation	
A (non-linear arrangement)	Case 1 Air = 30 mm, Air = 30 mm	Case 2 Air = 30 mm Rust = 30 mm
	Case 3 Air = 30 mm Rust = 60 mm	Case 4 Air = 60 mm Rust = 30 mm
	Case 5 Air = 60 mm Rust = 60 mm	Case 6 Rust = 30 mm Rust = 30 mm

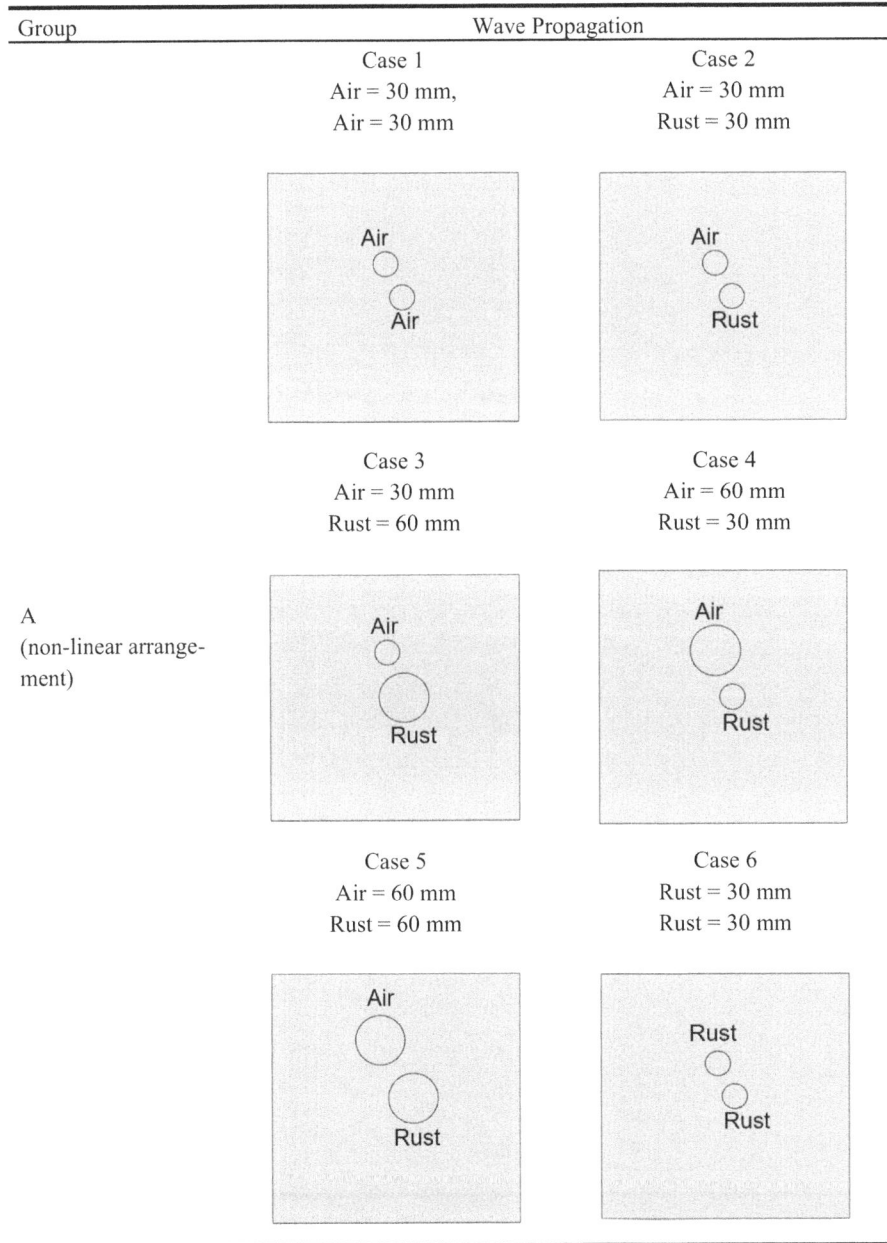

(continued)

Table 2. (*continued*)

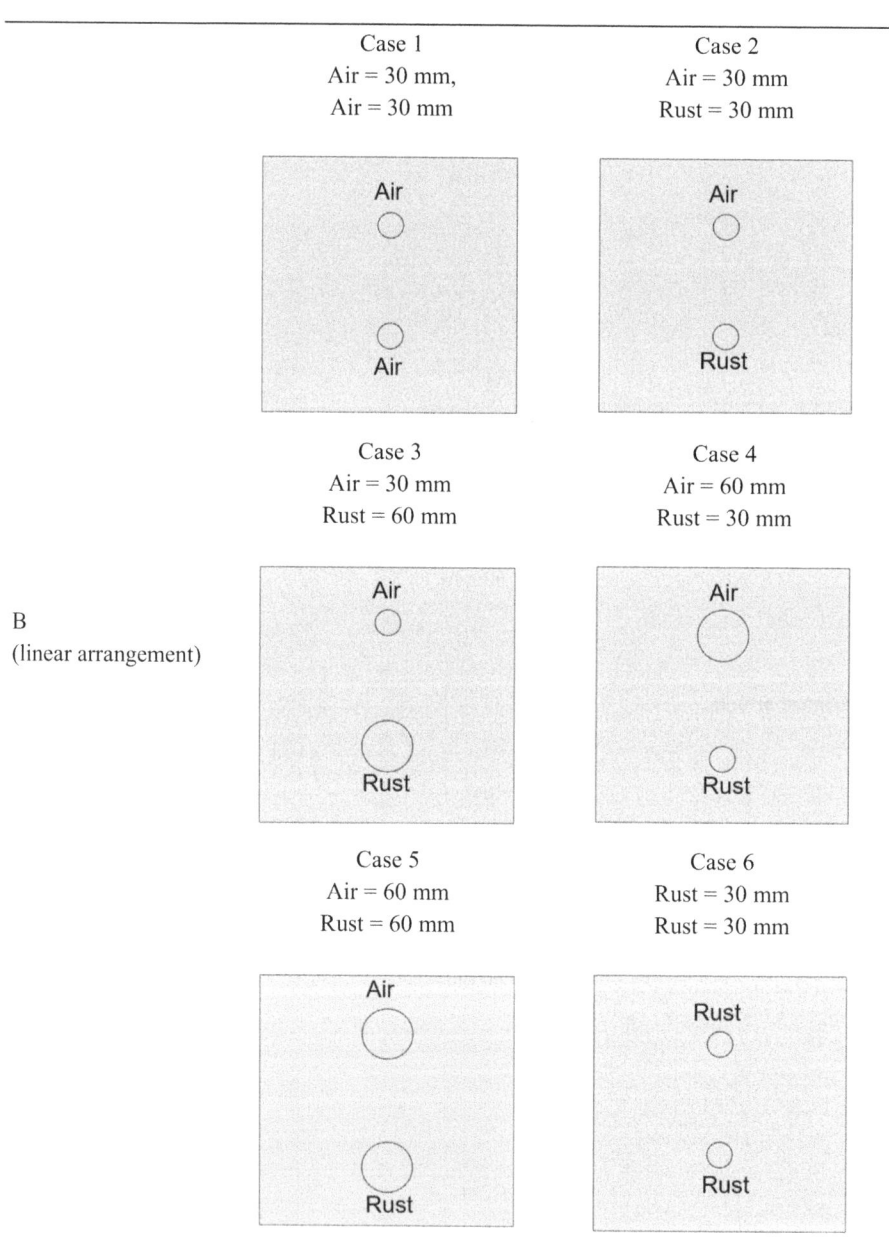

because of their compact size, affordability, and broad frequency range [19]. They operate by converting electrical energy into mechanical energy and vice versa, generating electrical output in response to structural vibrations. The piezoelectric material, under pressure, generates electricity.

3.3 Excitation Signal

The excitation signal is set to a frequency of 200 kHz, and adopts a sine function, as shown in Fig. 4. The mathematical expression for the excitation waveform is as follows:

$$V(t) = V_0 \sin(2\pi f t) \quad (1)$$

where *V(t)* represents the voltage source signal, V_0 is its amplitude, f is the source frequency, and t is the total testing time.

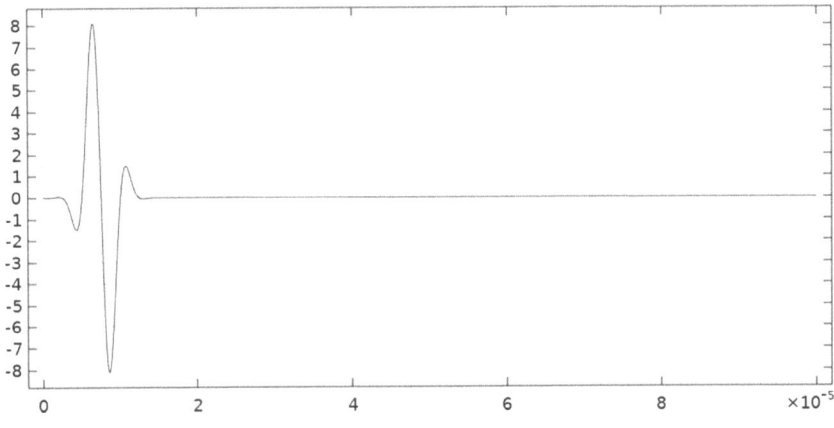

Fig. 4. Excitation signal diagram.

3.4 Mesh Generation and Study Parameter

A high-resolution mesh was generated for each concrete model to ensure an accurate wave propagation. In this study, each domain of the model was meshed into triangular finite elements. A finer mesh was applied to capture wave interactions precisely. In addition to meshing, the time-dependent solver in COMSOL was employed to simulate wave propagation through the concrete models. The integration time step is set to 1/20 of the source period.

3.5 Data Analysis

The primary data obtained from the simulation was the arrival time of the ultrasonic signal at the receiver. This arrival time was identified by detecting the first significant arriving peak in the recorded signal, which indicates the moment the ultrasonic signal reached the receiver. Additionally, the propagation of the ultrasonic wave through the concrete was visualized in the COMSOL software at different time intervals. These visualizations allowed for the analysis of the multiple defects that affected the propagation, including scattering, reflection, and attenuation patterns.

4 Result and Discussion

Table 3 presents visualizations of the ultrasonic wave propagation for each case in Group A and Group B. These visualizations illustrate how the wave interacted with the defects, showing reflections, scattering, and attenuation effects.

Table 3. Ultrasonic wave propagation in concrete with defects.

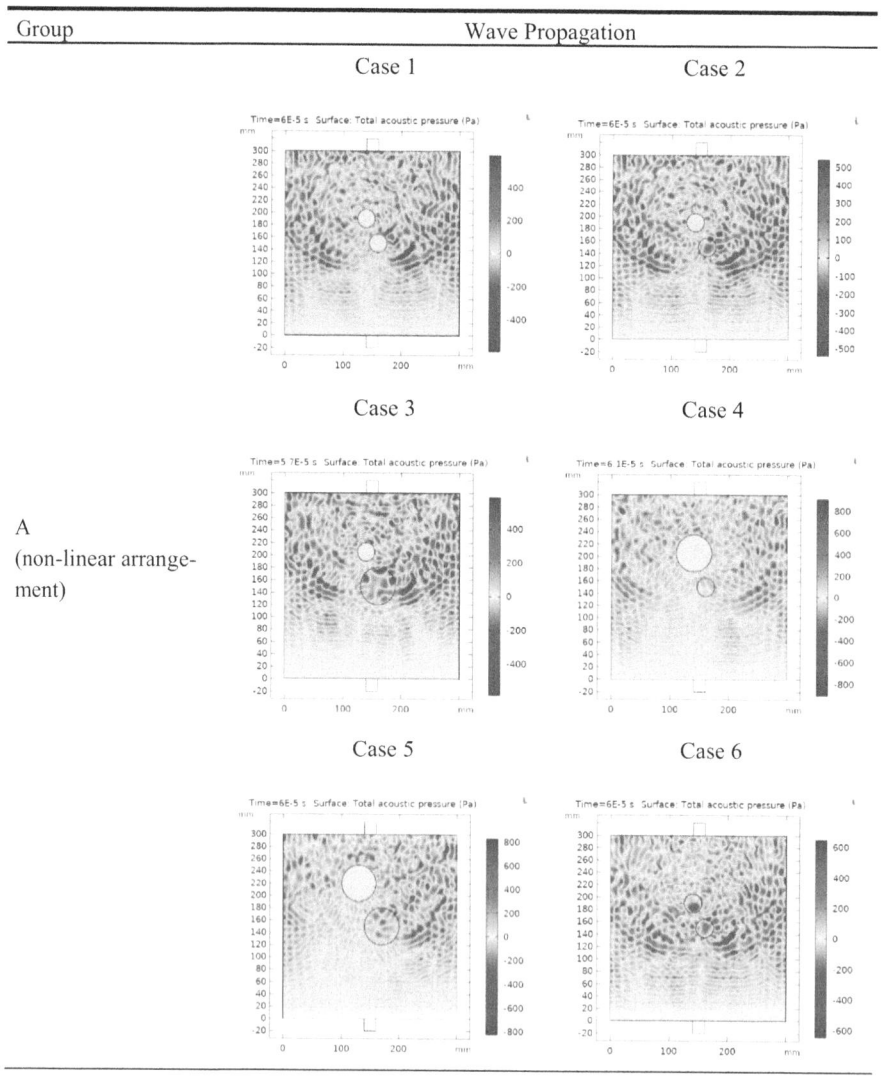

(*continued*)

Table 3. (*continued*)

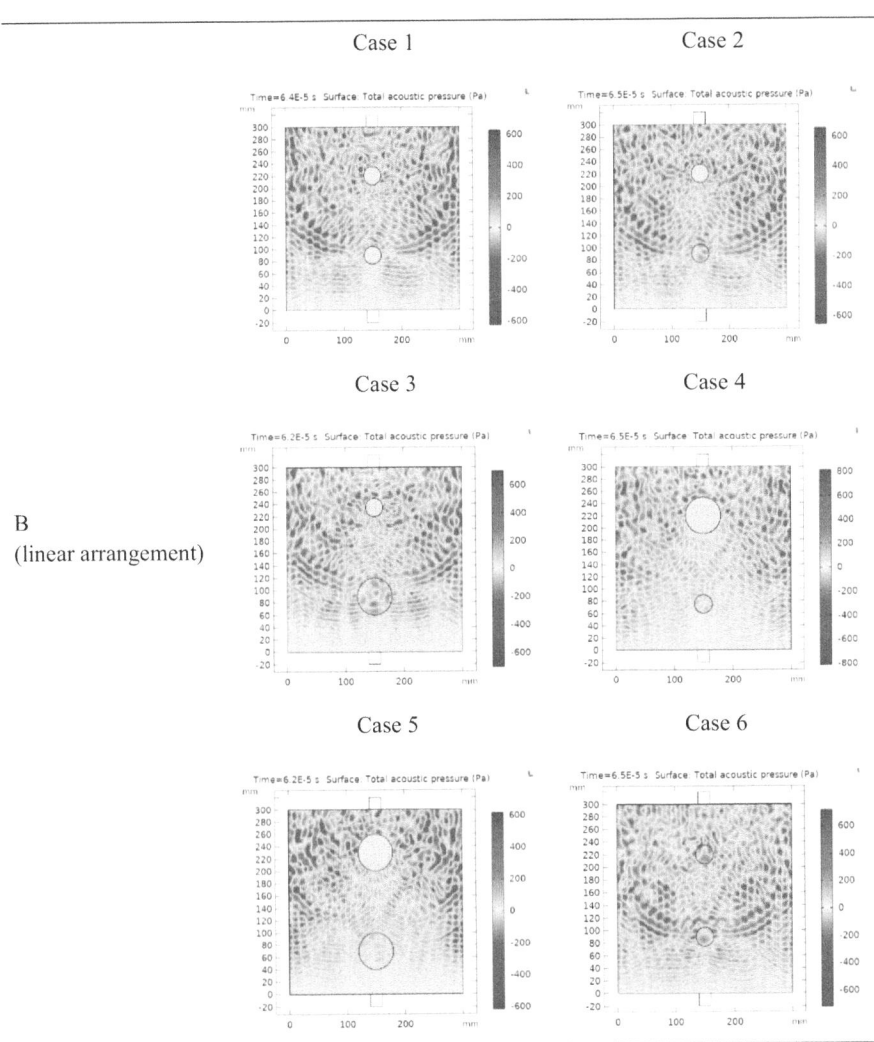

In Group A, the waves displayed more complex interaction patterns, with multiple reflections between defects occur. Meanwhile, in Group B, the waves propagated more freely between the 100 mm-separated defects, however, they encountered obstruction in different areas, resulting in longer travel times.

Table 4 presents the recorded arrival times of the ultrasonic signal for each case in Group A and Group B.

In Group A, where defects are positioned in a non-linear arrangement, the arrival time ranged from 57.43 μs to 61.42 μs. As can be seen, the ultrasonic signal was detected earliest in concrete with a 30-mm air hole and 60-mm rust, with an arrival time of

Table 4. The arrival time of ultrasonic signal for different defect configurations

Group	Case number	Defect #1	Defect #2	Arrival time (μs)
A (non-linear arrangement)	1	Air (30 mm)	Air (30 mm)	60.83
	2	Air (30 mm)	Rust (30 mm)	60.81
	3	Air (30 mm)	Rust (60 mm)	57.43
	4	Air (60 mm)	Rust (30 mm)	61.42
	5	Air (60 mm)	Rust (60 mm)	60.35
	6	Rust (30 mm)	Rust (30 mm)	60.67
B (linear arrangement)	1	Air (30 mm)	Air (30 mm)	64.90
	2	Air (30 mm)	Rust (30 mm)	65.14
	3	Air (30 mm)	Rust (60 mm)	62.54
	4	Air (60 mm)	Rust (30 mm)	65.53
	5	Air (60 mm)	Rust (60 mm)	62.44
	6	Rust (30 mm)	Rust (30 mm)	65.43

57.43 μs. This was followed by concrete containing a 60-mm air hole and 60-mm rust (60.35 μs), concrete with both 30-mm rusts (60.67 μs), concrete with a 30-mm air hole and 30-mm rust (60.81 μs), and concrete containing both 30-mm air holes (60.83 μs). The greatest delay was observed in concrete with a 60-mm air hole and 30-mm rust, where the signal was received at 61.42 μs.

In Group B, a significant variation in arrival times was observed between the concrete models with 60-mm rust and those concrete with other defect combinations. The arrival time for concrete with 60-mm air hole and 60-mm rust, and concrete with 30-mm air hole and 60-mm rust, are 62.44 μs and 62.54 μs, respectively. They showed the earliest arrival time compared to the other concretes. Conversely, concrete with 30-mm air holes recorded received signal at 64.9 μs, followed by concrete with 30-mm air hole and 30-mm rust (65.14 μs), concrete with 30-mm rusts (65.43 μs), and concrete with 60-mm air hole and 30-mm rust (65.53 μs).

To have a clear comparison between Group A and Group B, the results are graphically presented in Fig. 5.

The line plot in Fig. 5 illustrates that Group B consistently had higher arrival times, indicating that the arrangement of defects significantly impacted wave propagation compared to the closer non-linear arrangement in Group A.

4.1 Effect of Arrangement of Defects

The data clearly shows that the arrangement of defects significantly affects the ultrasonic signal arrival times. Group B featured defects arranged linearly with a fixed 100 mm separation between them, while Group A had defects arranged in closer, non-linear configurations. The arrival times in Group A were generally shorter compared to Group B. For example, in Case 1, the arrival time for 30 mm air-type defects in Group A was

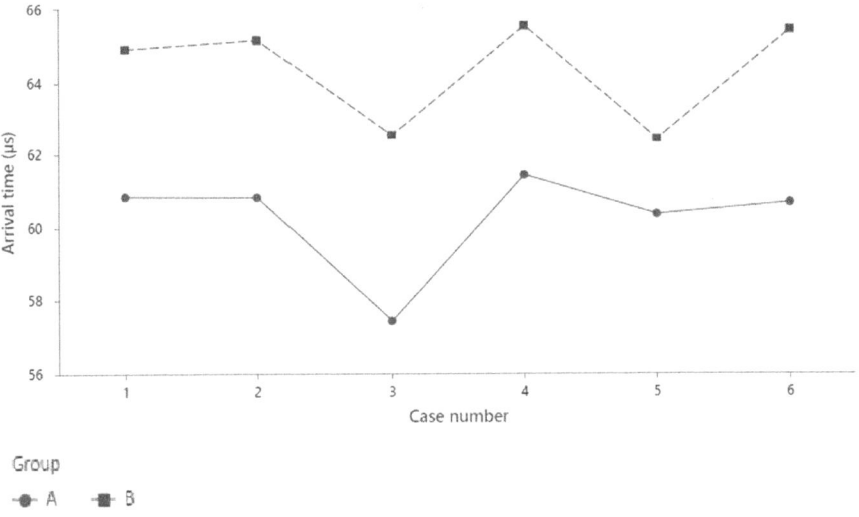

Fig. 5. Comparison of arrival time obtained from Group A and Group B.

60.83 μs, whereas in Group B, with the same defect type and size but separated by 100 mm in a linear arrangement, the arrival time increased to 64.90 μs.

This trend was observed across all cases, indicating that the greater separation distance between defects in Group B had a measurable impact on the wave propagation, causing longer delays, likely due to the more direct and multiple obstructions of the wave path. In contrast, the propagation of the ultrasonic wave in Group A, experienced more complex interactions, such as scattering, reflections, and diffractions, as they encountered multiple defects in a close, non-linear pattern. However, these one-time interactions shortened the effective distance the wave had to travel, hence the arrival time was shorter.

To address potential variations in defect arrangements, additional simulations were conducted to assess the effect of switching the positions of rust and air defects. Specifically, simulations were performed with configurations where the acoustic wave encountered the rust defect first, followed by the air defect. The results showed that the signal propagation times were consistent across all configurations. This finding indicates that the sequence defects do not significantly influence wave interactions under the parameters of the current simulation model. This consistency suggests that the acoustic wave's interaction with defects is predominantly influenced by their physical properties. However, under different conditions or with more complex defect scenarios, the results could vary.

4.2 Effects of Type of Defects

Air-type defects (air holes) caused greater delays in ultrasonic wave propagation than solid-type defects (rust). In Case 1, where both defects were 30-mm air holes, the arrival time was 60.83 μs in Group A and 64.90 μs in Group B. In comparison, in Case 6, where both defects were 30 mm-rust, the arrival time was 60.67 μs in Group A and 65.43 μs in

Group B. This highlights the significant impact of the air-type defects due to their higher acoustic impedance contrast with concrete. This is consistent with previous studies [20], where voids in concrete have been shown to reflect and scatter ultrasonic waves, causing delays in arrival times.

On the other hand, solid-type defects like rust are more similar to the surrounding concrete in terms of acoustic impedance. The ultrasonic wave can pass through these defects with less resistance and attenuation, resulting in shorter arrival times compared to air-type defects. In cases where rust was combined with air-type defects, the arrival time was influenced more by the air defect due to its higher impedance contrast. Rust-type defects caused less disruption but still had a noticeable impact, especially in linear arrangements.

4.3 Effects of Size of Defects

The size of the defects also affected arrival times. In Group A, Case 3, where the combination was 30 mm-air and 60 mm-rust, the arrival time was 57.43 μs, indicating a reduction compared to smaller defects. However, in Group B, larger defects generally resulted in longer arrival times. For example, in Case 4 (60 mm-air hole and 30 mm-rust), the arrival time increased to 65.53 μs, the longest delay recorded in Group B.

Larger defects (60 mm) consistently led to longer arrival times compared to smaller defects (30 mm). This is because larger defects present a bigger obstacle for the wave, causing more scattering, reflection, and attenuation, which increases the overall travel time. However, the arrival time for air-type defects (30 mm or 60 mm) was generally longer than for rust defects of the same size. The larger air voids caused greater disruption to the wave's path, amplifying the effect of scattering and reflection. For rust-type defects, while larger defects also caused more scattering and reflections than smaller ones, the effect was less dramatic than for air-type defects. The wave could still propagate relatively smoothly through rust, though larger rust defects did introduce more reflections than smaller ones.

5 Conclusion

The analysis of ultrasonic signal arrival times in concrete with multiple defects provides a solution for estimating the quality of the concrete structure. This study demonstrates that the type, size, and arrangement of defects significantly impact ultrasonic wave arrival times in concrete models. Separated, linear arrangements of defects caused greater delays in wave propagation compared to close, non-linear arrangements, and air-type defects had a more significant effect than solid-type defects. These findings are important for improving the accuracy of non-destructive ultrasonic testing techniques, which allows better detection and characterization of defects in concrete structures. In addition, the analysis of multiple defects, as performed in this study, highlights the variations in time-of-arrival signals and acoustic wave propagation caused by the interaction between different defect types. Such analyses are particularly relevant for understanding the severity and distribution of damage in concrete structures.

For future works, combining such analyses with advanced ultrasonic techniques could further improve the robustness and applicability of defect detection methods. Additionally, combining simulation data with real-time in-situ measurements could provide more reliable and practical results. The application of array-based approaches will also be explored to complement the current methodology, aiming to overcome the limitations of one-dimensional signal analysis and provide a more robust solution for defect identification in three-dimensional environments.

Acknowledgment. This research was supported by Universiti Teknologi Malaysia.

References

1. Ji, Y., et al.: A state-of-the-art review of concrete strength detection/monitoring methods: with special emphasis on PZT transducers. Constr. Build. Mater. **362**, 129742 (2023)
2. Felice, M.V., Fan, Z.: Sizing of flaws using ultrasonic bulk wave testing: a review. Ultrasonics **88**, 26–42 (2018)
3. Zhang, T., Zhang, L., Ozevin, D., Attard, T.: Multi-scale ultrasonic imaging of sub-surface concrete defects. Meas. Sci. Technol. **35**(3) (2024)
4. Sun, H., Zhu, J.: Nondestructive evaluation of steel-concrete composite structure using high-frequency ultrasonic guided wave. Ultrasonics **103**, 106096 (2020)
5. Kim, J., Cho, Y., Lee, J., Kim, Y.: Defect detection and characterization in concrete based on fem and ultrasonic techniques. Materials (Basel) **15**(22), 8160 (2022)
6. Kuchipudi, S.T., Ghosh, D.: An ultrasonic wave-based framework for imaging internal cracks in concrete. Struct. Control. Health Monit. **29**(12), e3108 (2022)
7. Ramamoorthy, S.K., Kane, Y., Turner, J.A.: Ultrasound diffusion for crack depth determination in concrete. J. Acoust. Soc. Am. **115**(2), 523–529 (2004)
8. Pahlavan, L., Zhang, F., Blacquière, G., Yang, Y., Hordijk, D.: Interaction of ultrasonic waves with partially-closed cracks in concrete structures. Constr. Build. Mater. **167**, 899–906 (2018)
9. Rucka, M., Witkowski, W., Chróścielewski, J., Burzyński, S., Wilde, K.: A novel formulation of 3D spectral element for wave propagation in reinforced concrete. Bull. Pol. Acad. Sci. Tech. Sci. **65**(6), 805–813 (2017)
10. Yıldırım, G., Öztürk, O., Al-Dahawi, A., Afşın Ulu, A., Şahmaran, M.: Self-sensing capability of engineered cementitious composites: effects of aging and loading conditions. Constr. Build. Mater. **231**, 117132 (2020)
11. Zhou, A., Büyüköztürk, O., Lau, D.: Debonding of concrete-epoxy interface under the coupled effect of moisture and sustained load. Cem. Concr. Compos. **80**, 287–297 (2017)
12. Perez, N.: Electrochemical corrosion. In: Materials Science: Theory and Engineering, pp. 835–898 (2024)
13. Wong, H.S., et al.: Methods for characterising the steel–concrete interface to enhance understanding of reinforcement corrosion: a critical review by RILEM TC 262-SCI. Mater. Struct. **55**(4), 124 (2022)
14. Wu, B., Li, Z.: Mechanical properties of compound concrete containing demolished concrete lumps after freeze-thaw cycles. Constr. Build. Mater. **155**, 187–199 (2017)
15. Luo, Y., Su, J., Xu, Y., Ou, T., Peng, X.: Axial compression performance of post-fire concrete columns strengthened using thin-walled steel tubes. Sustainability **11**(18), 4971 (2019)
16. Kang, L., Feeney, A., Dixon, S.: The high frequency flexural ultrasonic transducer for transmitting and receiving ultrasound in air. IEEE Sens. J. **20**(14), 7653–7660 (2020)

17. Rahiman, M.H.F., Rahim, R.A., Ayob, N.M.N.: The front-end hardware design issue in ultrasonic tomography. IEEE Sens. J. **10**(7), 1276–1281 (2010)
18. Ai, D., Du, L., Li, H., Zhu, H.: Corrosion damage identification for reinforced concrete beam using embedded piezoelectric transducer: numerical simulation. Measurement **192** (2022)
19. Dumoulin, C., Karaiskos, G., Sener, J.Y., Deraemaeker, A.: Online monitoring of cracking in concrete structures using embedded piezoelectric transducers. Smart Mater. Struct. **23**(11), 115016 (2014)
20. Mohamad, F.A.J., et al.: A preliminary investigation on the correlation between the arrival time of ultrasonic signals and the concrete condition. Commun. Comput. Inf. Sci. **1912** (2024)

Predictive Maintenance in Aerospace: Leveraging Machine Learning for Optimized Filter Lifespan

Syaimak Abdul Shukor[1](✉), Jia Hui Tang[1], Iqraq Kamal[2], Mohd Faizzuan Mohammad[3], Zalinda Othman[1], Shahnorbanun Sahran[1], Nor Samsiah Sani[1], Mohd Ridzwan Yaakub[1], and Azizi Abdullah[1]

[1] Center for Artificial Intelligence Technology, Fakulti Teknologi Dan Sains Maklumat Universiti Kebangsaan Malaysia Selangor, Bangi, Malaysia
syaimak@ukm.edu.my
[2] Aerospace Malaysia Innovation Centre, Kajang, Selangor, Malaysia
[3] Spirit Aerosystems Malaysia Sdn Bhd, Subang, Selangor, Malaysia

Abstract. Predictive maintenance (PdM) models leverage advanced data analytics and machine learning techniques to predict equipment failures and optimize maintenance schedules, enhancing operational efficiency and minimizing downtime. Maintaining stringent safety and reliability standards is paramount in the aerospace industry, particularly within high-technology sectors such as aeronautics and propulsion systems. This study highlights the transition from traditional preventive maintenance to proactive and predictive maintenance strategies, with a specific focus on the maintenance of painting booth filters—a critical yet underexplored aspect of aerospace maintenance. The objective of this paper is to compare the performance of various machine learning approaches in enhancing the predictive maintenance model for this application. This research employs machine learning (ML) methodologies to advance PdM by accurately predicting filter lifespan and scheduling timely maintenance. Ten algorithms: Random Forest (RF), Gradient Boosting (GB), Ridge Regression (RR), CatBoost (CB), XGBoost (XGB), Light Gradient Boosting Machine (LGBM), Multilayer Perceptron (MLP), Convolutional Neural Network (CNN), Long Short-Term Memory (LSTM), and CNN-LSTM—were evaluated based on their forecasting performance using metrics such as Mean Square Error (MSE), Mean Absolute Error (MAE), Mean Absolute Percentage Error (MAPE), and Symmetric Mean Absolute Percentage Error (SMAPE). The results demonstrate that Ridge Regression (RR) achieved superior performance, closely followed by MLP and LGBM, despite RR and LGBM exhibiting slightly longer inference times. While CNN demonstrated the shortest inference time, it lacked forecasting accuracy. Consequently, MLP emerged as the optimal model, balancing accuracy and efficiency. This research highlights the potential of integrating ML algorithms in PdM. This study establishes a foundation for future advancements in predictive maintenance technology, underscoring its transformative potential in industrial applications.

Keywords: Predictive maintenance · Machine Learning · Deep Learning · Filter Lifespan Prediction

1 Introduction

Engineered objects are entities designed, created, or innovated to perform specific tasks within an operation or service, ensuring seamless operational synergy and smooth performance. However, due to wear and tear, extensive utilization, and ageing, asset variation and degradation are inevitable, necessitating maintenance when they fail to achieve the expected performance. Delays in addressing these maintenance needs in a production environment can lead to potentially catastrophic operational downtimes and late deliveries of goods and services, impacting both monetary and corporate image [1]. Therefore, strategic maintenance is essential to minimize expected or unexpected downtime and ensure optimal operational efficiency.

The aerospace industry, a high-technology sector encompassing research, project manufacturing, and commercialization, deals with aeronautics integrated with propulsion systems, infrastructures, equipment, spaces, and missiles. It adheres to strict safety and reliability standards during production, ensuring high airworthiness while meeting operational efficiency [2, 3]. Strategic maintenance is integral to achieving these objectives. Generally, two widely practised maintenance philosophies are reactive and proactive [4]. The former is a legacy approach, divided into Corrective Maintenance (CM) and Emergency Maintenance (EM), where assets are restored, replaced, or readjusted to perform their intended function efficiently upon failure and fault occurrence. Both are distinguishable by their urgency level and impact on safety or performance, with EM having higher urgency and impact. This approach incurs forced operational downtime, disrupting the timely delivery of goods and resulting in monetary losses. Studies by Vathoopan et al. (2018) [5] and Menegon & Isatto (2023) [6] suggest incorporating Digital Twin (DT) technology in CM to virtualize physical assets for monitoring. However, maintenance remains reactive, performed upon failure, and setting up DT requires thorough investigation, refined selection, and compliance with industry-specific requirements and architectural planning to ensure faithful representation of physical assets, the establishment of data communication, and data processing modules [7]. These are preliminary requirements to guarantee the proper operation of DT.

Proactive maintenance is classified as preventive maintenance and predictive maintenance. Both are executed before failure, with preventive maintenance occurring at prescheduled intervals to balance maintenance costs and failure risks while minimizing asset failure during operation. However, this approach becomes questionable when manufacturing operations face random disturbances in the production environment [4], characterized by inconsistent workload fluctuations and failure rate distributions. For instance, an unexpected workload hike can lead to asset failure during live operation and downtime, while a reduced workload can lead to asset underutilization and over-maintenance. Exempting prescheduled maintenance under such circumstances risks operational disruption during subsequent workloads.

Recently, the global predictive maintenance (PdM) market grew by 11% from 2021 to 2022, reaching USD 5.5 billion, with a Compound Annual Growth Rate (CAGR) of 17% anticipated until 2028. With unplanned breakdown costs exceeding USD 100,000 per hour, the priority to accurately predict asset failure has increased, and advanced analysis approaches, such as AI-driven analysis, are gaining prominence [8]. This emerging type of maintenance is adaptively determined by continuously obtaining real-time insights on assets' conditions and environments rather than assuming their conditions based on historical baselines or component suppliers' data. PdM and its associated concepts have received attention for the past 12 years, reporting a positive ROI in 95% of predictive maintenance adopters, with 27% reporting amortization in less than a year [8]. There are two main subclasses of predictive maintenance: condition-based monitoring (CBM) and remaining useful life (RUL). CBM triggers maintenance when specific conditions or thresholds are met, while RUL estimates when the asset will likely fail in future time steps. Both are widely employed alongside machine learning across various industries, including aircraft and aeronautical [9], automotive [10], renewable energy, and microelectronic manufacturing [11]. As the aerospace industry aims to achieve high airworthiness standards while optimizing operational efficiency, embracing predictive maintenance strategies would be advantageous. These strategies offer advanced insights into assets' points of failure and real-time conditions, determining their fitness for continued production activities. This minimizes cascading drawbacks from unplanned or delayed maintenance execution, such as operational downtime, which leads to postponement of workload completion and delivery of goods and services, significantly impacting monetary gains and corporate image.

While extensive predictive maintenance engineering research within the aerospace industry, such as maintenance of aircraft engines, bearings, hydraulics, pneumatics, and fuselage [9], has been conducted, the maintenance of painting booth filters, including panel, pocket, and wall filters, has received limited attention. Painting is a crucial step in the aerospace industry, serving as an essential process to safeguard aircraft construction materials from corrosion [2]. Aero part painting processes are executed within a paint booth, a controlled environment with strict requirements for maintaining optimal differential pressure, airflow, and temperature to ensure high-quality painting results. This involves the painters' techniques and expertise in preventing contamination from airborne paint particles and dust. Maintaining optimal paint booth pressure while ensuring environmental safety involves filtering airborne paint particles and dust before extracting them from the paint booth through the exhaust vent. As filters become clogged, prompt replacement is essential to guarantee consistent filtration capacity.

However, the decision to maintain the filter in paint booths heavily relies on expert engineers' domain knowledge and experience and the availability of paint booth operators to perform maintenance activities. Timely execution of filter replacement is crucial, as any negligence or delay can lead to poor painting quality due to saturated filters inhibiting smooth airflow within the paint booth. Although filter replacement is necessary, tardy or point-of-failure replacement halts ongoing painting operations, inevitably delaying the completion of scheduled workloads and component delivery. Additionally, congested filters strain the paint booth motors, deteriorating their lifespan and efficiency in maintaining the optimal environment. This further risks airborne paint particles and dust seeping into the environment, causing contamination

Machine Learning and Deep Learning methods are employed because they can handle vast amounts of data from industrial machines, including sensors, environments, and processes, due to the diversity of data sources consisting of complex data structures [12, 13]. The data contains essential features or parameters for predicting maintenance, and ML can identify which parts are most meaningful for predictive purposes. If essential elements can be identified, they can serve as precautionary measures to monitor the machine's condition and be more alert to the cause of damage. The data for feature selection is obtained from the sensors that can classify the machine's condition. While operating the machine, ML and DL algorithms can identify, monitor, and analyse system parameters. With the PdM method, these monitored parameters will warn the machine operator before system failure occurs [14]. By leveraging machine learning and deep learning for predictive maintenance, machine maintenance scheduling time can be suggested, maintenance processes can be managed efficiently, and production quality can be improved.

This paper uses machine learning to introduce the transformation from preventive maintenance to predictive maintenance for online paint booth filters. The main objective of this work is to compare ten different machine learning and deep learning algorithms utilized to perform multi-time series forecasting, predicting the filter life based on the forecasting activities of eight sensor data (7 differential pressure sensor and 1 temperature sensor) into two-time steps for the future based upon their values at several previous time steps. The machine learning algorithms employed in this work include Random Forest (RF), Gradient Boosting (GB), Round Robin (RR), CB, XGBoost (XGB), Light Gradient Boosting Machine (LGBM), Multilayer Perceptron (MLP), Convolutional Neural Network (CNN), Long Short-Term Memory (LSTM), and CNN-LSTM. Upon deployment, the model's performance is validated by comparing real-time sensor values with those forecasted by machine learning and deep learning models. The differences between actual and forecasted values are quantified using MSE, MAE, MAPE, and SMAPE.

This paper focuses on the machine learning algorithm in the predictive maintenance domain by addressing an underexplored area in aerospace operations: the maintenance of paint booth filters. It demonstrates the transformation from preventive to predictive maintenance using machine learning techniques. The key contribution lies in comprehensively evaluating and comparing ten distinct ML algorithms. Based on eight sensor parameters, these models are utilized for multi-time series forecasting on paint booth filter lifespan. The need to compare ten methods comes from the complexity and variability of sensor data in industrial operations, where each algorithm has distinct strengths in handling various data structures, feature dependencies, and time-series forecasting challenges. A systematic comparison ensures the identification of the appropriate algorithm for reliable predictions under varying operational conditions. Evaluating the models using the performance metrics will provide insights for optimizing maintenance schedules, minimizing downtime, and ensuring consistent production quality. This comparative analysis validates ML models' effectiveness and highlights their practicality in real-world aerospace manufacturing environments.

The remainder of this paper is organized as follows: Sect. 2 thoroughly discusses predictive maintenance approaches driven by various industries' conventional machine learning and deep learning models. Section 3 highlights and discusses the methodology, data analysis, the comparison results based on the performance evaluation, and finally, Sect. 4 concludes the paper with future work.

2 Related Work

First, Predictive-based maintenance approaches received great interest in various domains, including the Aerospace industry. The predictive capability of maintenance is attributed to the utilization of sensors that provide real-time statuses of the assets or any key indicators, and the machine learning and deep learning models in providing insights on in-advanced maintenance execution before the actual point of failure. The machine learning and deep learning models such as Bayesian Networks, Classification Trees, Long-Short Term Memory (LSTM), Artificial Neural Networks (ANN), Random Forests (RF), Support Vector Machine (SVM), One-Class SVM (OC-SVM), XGBoost, LSTM-Auto encoders (LSTM-AE), Convolutional AE were trained under the different approaches-supervised, unsupervised and semi-supervised learning as deemed appropriately by the author [15].

With the vision of correcting the equipment manufactured flaws prior to pushing into production and provision of explain-ability on the contributing factors to impending failures, Bayesian Network and Classification Trees with Gini criterion were used in the PdM of industrial equipment trained on production error data with rolling window cross-validation [16]. The models then predict the production error that would fail and require critical inspection errors in later in-field inspections. Random under-sampling (RUS) and over-sampling (ROS) were incorporated into the training pipeline to alleviate class imbalance, and RUS increased both models' accuracy by over 50%, compared to its absence. Conceding that model explain-ability is crucial in enhancing the understanding of its predicted results and the main causing factor for a defect in equipment among relevant stakeholders and operators. However, only rule-based and probabilistic-based explanations were explored. In contrast, other explanation types, such as SHAP values and LIME algorithm provided by other "black-box" models as referenced by the paper, such as LSTM and ANN, are underexplored in the industrial equipment manufacturing field.

Under the absence of labelled normal and anomalous data, Jyh-Yih et al. (2020) [17] conducted a 2-stage density-based spatial clustering of applications with noise (DBSCAN) classification and statistical process control analysis to analyse the features' strength in distinguishing or classifying normal and anomalous data for wind turbine fault diagnosis. Domain experts verified the classification results before training RF and CART decision tree algorithms with entropy in predicting faulty wind turbines. ANN with Dropout and Bilateral LSTM (Bi-LSTM) with Dropout and Rectified Linear Unit (ReLu) activations in between trained under a semi-supervised approach were utilized to predict power generated by wind turbines [18]. Then, statistical tests such as standard deviation, Monte Carlo dropout, and few-shot dropout were used to assess the deviation of the predicted from the actual output that later served as features of a 1D Convolution to classify data samples as normal or prefault.

Makridis et al., (2020) [19] detected abnormal operation and condition of the Crosshead Bearing of the generic vessel and main engine by first training various models such as LSTM, OC-SVM, weighted XGBoost and an equal-voting ensemble of these models to forecast the sensors' values for the next time-step. It involves flagging samples as anomalies if the deviation between the predicted and actual exceeds 3.5 times the standard deviation, the usage of the static window in marking anomalies, and a static threshold derived from the anomalies score, respectively. Additionally, Weighted Permutation Entropy (WPE), a complexity measuring technique, was utilized, and its best parameters were found with an exhaustive grid search for anomaly detection. It was found that the equal voting of ensemble models yields a steadier detection of anomalies than the performance of the individual models. In the case of classifying anomalous motor EPFAN machines, a researcher developed a SVM and RF trained on the consolidated periodical sensor data with downtime machine and maintenance activities report data and have their hyper-parameters fined-tuned through exhaustive grid search [20]. The availability of existing manufacturing-related data, as utilized by the latter study, is an advantage over its formers as it exempts the complexity of fine-tuning models in identifying anomalies before treating the PdM task as a supervised problem nor calculating the deviation of forecasted future parameters to its actual value and applying statistical measurement or models to flag it as an anomaly or otherwise. Nonetheless, the formers' approaches are feasible whenever there is a lack of comprehensively labelled data, which are often complex and expensive to acquire in real-world situations. Additionally, utilizing an exhaustive grid search to find the models' best hyper-parameters can be computationally intensive under circumstances where the dataset size and number of features are extensive.

Bampoula et al., (2021) [21] predicted three possible health statuses (high, medium, low) of a rolling mill machine in the steel manufacturing industry case study via a supervised deep learning approach – LSTM- AE. The data is sliced according to each health status before being trained with a dedicated LSTM- AE network to ensure the network learns temporal data related to the corresponding machine's status. The final prediction output depends on which LSTM-AE network resulted in the highest accuracy score. Dynamic Predictive Maintenance Scheduling (DPMS) method was proposed by [22] that uses deep AE in extracting representation features of system degradation from the sensor data before feeding it to a deep forest network, which is a decision tree ensemble in computing system failure probabilities. The failure probabilities are further utilized to compute the costs of different maintenance decisions before scheduling maintenance and inventory activities. This approach not only signifies whenever a failure occurs but also provides insights on the appropriate maintenance or inventory action to execute. Nevertheless, this prognostic and health management approach would only be practical if maintenance or inventory actions exist to perform under a specific environmental condition.

Deep learning models exhibit a much more complex architecture than conventional machine learning and demonstrate better capability in handling non-linearity. However, its performance advantage over the conventional approach might only be apparent when training with a large dataset. Furthermore, it has a higher need for an extensive dataset to avoid overfitting. Thus, utilizing deep learning models may be less optimal when data sources are scarce. Additionally, training complex deep learning models such as LSTM has a higher demand for computation resources and time to achieve accuracy that is comparable to ANN, which has a simpler neural network architecture even though LSTM is better at capturing long sequential relationships [18].

Each different approach adopted in identifying the anomalous behaviour of key indicators, and the choice of machine learning model and deep learning architecture for predictive maintenance have their respective pros and cons. They are highly dependent on the amount and availability of problem-specific data such as maintenance data, industry process data, real-time key indicator data, and computational time and resources. This paper takes full consideration of these key aspects. Having machine learning and deep learning models trained on maintenance key indicators in any industry holds the potential to minimize the reliance on costly human expertise and experience in determining the optimal timing for maintenance by providing data-driven predictions. This not only promotes in-advanced scheduling of maintenance to avoid operation breakdown and its cascading drawbacks but also minimizes unnecessary maintenance involving human labour and resources. The timely and accurate execution of maintenance is a priority for industries obligated to adhere to strict safety and reliability regulations during production or post-production as seen in the Aerospace and Aeronautical industries, ensuring a high level of airworthiness while operating within the objectives of aircraft availability and cost [3].

Predictive-based maintenance approaches have garnered significant interest across various domains, notably the aerospace industry. These approaches leverage real-time data from sensors and the predictive power of machine learning (ML) and deep learning (DL) models to address maintenance needs pre-emptively before asset failure. Advanced models such as Bayesian Networks, Classification Trees, Long-Short Term Memory (LSTM), Artificial Neural Networks (ANN), Random Forests (RF), Support Vector Machine (SVM), Bayesian Networks and Classification Trees and others have been applied in both supervised and unsupervised learning contexts. Despite the promise of predictive maintenance (PdM) strategies, challenges still need to be addressed, particularly regarding the availability and quality of labelled data, computational resources, and the interpretability of complex models. These methods predict failures and provide actionable insights for maintenance and inventory decisions, underscoring the importance of explainability and real-world applicability. Ultimately, the successful adoption of PdM in industries such as aerospace hinges on the balance between leveraging advanced ML and DL models and addressing the practical constraints of data availability, computational demands, and the need for clear, actionable insights. This approach has shifted from reactive to predictive maintenance promises to reduce operational downtimes, optimize maintenance schedules, and ensure compliance with stringent safety and reliability standards, thus enhancing operational efficiency and cost-effectiveness.

3 Methodology

The proposed architecture offers a detailed framework for predictive maintenance in filter replacement, beginning with data acquisition, the cornerstone of the PdM process. This stage encompasses collecting sensor data and historical records, which are invaluable for assessing equipment condition and performance. Subsequent data pre-processing ensures data quality and consistency by removing missing values or outliers and standardizing and normalizing the data. Exploratory Data Analysis (EDA) then unveils patterns and relationships within the dataset, guiding the selection of appropriate modelling techniques. Feature selection, a critical step, identifies the most pertinent features for predicting equipment failures, reducing the dataset's dimensionality and enhancing model efficiency. For instance, redundant features can be discarded to streamline the dataset. Before developing the PdM model, further pre-processing ensures the data is primed for analysis.

The project aims to mitigate unexpected breakdowns in painting operations by predicting the necessity for filter replacement using differential pressure sensors installed in the paint booth. Traditionally reliant on expert judgment, the process now leverages predictive maintenance to forecast differential pressure sensor values and trigger automatic filter replacement when thresholds are exceeded. Figure 1 illustrates the proposed architecture of the PdM system for this project. The proposed architecture is based on the analysis of previous studies and the current data availability and analysis. Data collection begins with the acquisition of process data from available sources and sensor

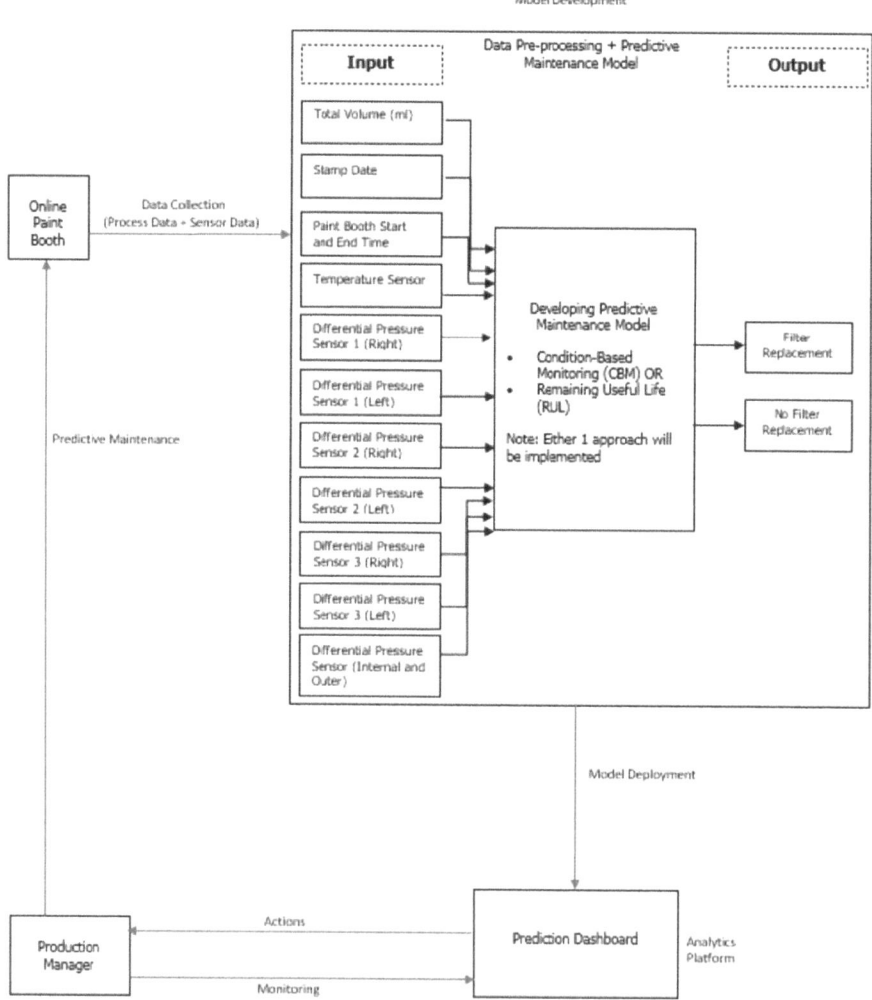

Fig. 1. Proposed Architecture for PdM System

data collected via sensors installed in the online paint booth for the duration of August 2023 to May 2024.

These data sources provide valuable insights into the operational condition of the equipment. Eleven (11) specific inputs train the model, designed to predict the need for filter replacement independently of human intervention. Predictive maintenance employs ML and DL algorithms to forecast equipment failures, facilitating timely maintenance decisions. The training phase prepares the models with known data, while the testing phase evaluates their performance on unseen data. Models commonly used in PdM include Bayesian Networks, Classification Trees, LSTM, ANN, RF, SVM, OC-SVM, XGBoost, LSTM-AE, and Convolutional AE. These models, selected for their ability to handle continuous variables, are optimized using Random Search and HyperOpt, with mean square error as the cost function.

Model evaluation, a crucial step, assesses performance using metrics such as MSE, MAE, MAPE, SMAPE, and MASE. These metrics, measure the models' accuracy and efficiency. The generalizability of the ten models is evaluated based on these metrics and prediction duration on the test set. Once developed, tested, and evaluated, the models are deployed in a production environment and integrated with industrial machinery or maintenance systems. These deployed models continuously monitor equipment health, providing maintenance recommendations and alerts to ensure timely actions. This PdM data flow leverages data-driven insights, enhancing maintenance strategies' efficiency and effectiveness.

3.1 Results from Training Process

All models were trained on an Intel(R) Xeon(R) CPU @ 2.20 GHz using the Scikit-Learn and Skforecast frameworks for machine learning models and the Tensor Flow framework for neural network or deep learning models. The optimal hyperparameters, training configurations, and training durations for the best-performing models are detailed in Table 1. The models' training times, listed in ascending order, were Ridge Regression (RR), LightGBM (LGBM), CatBoost (CB), Gradient Boosting (GB), XGBoost (XGB), Random Forest (RF), Convolutional Neural Network (CNN), Multi-Layer Perceptron (MLP), Long Short-Term Memory (LSTM), and CNN-LSTM.

Post-training, the generalizability of the models was evaluated using metrics such as mean square error (MSE), mean absolute error (MAE), mean absolute percentage error (MAPE), symmetric mean absolute percentage error (SMAPE), mean absolute scaled error (MASE), and prediction duration, as summarized in Table 2. Ridge Regression (RR) achieved the lowest MSE (1.66×10^{-3}), closely followed by Gradient Boosting (GB) and LightGBM (LGBM). Other models exhibited slightly higher MSE scores, ranging from 2.7×10^{-4} to 4.8×10^{-4}. Regarding MAE, MLP and LSTM were also top performers alongside LGBM, indicating superior forecasting accuracy. LSTM recorded the lowest MAPE, with CNN-LSTM and MLP following. Given MAPE's tendency to heavily penalize overestimates, SMAPE was employed for a more balanced evaluation, where MLP excelled, followed by CNN and CNN-LSTM. Adding a small constant value (ε) mitigated issues when actual values approached zero. MASE compared models to a naive forecast, revealing that deep learning models (MLP, LSTM, CNN-LSTM) had MASE scores below 1, indicating superior pattern recognition. GB had the shortest inference time (2.66 s), with RR and CB also performing efficiently. RR, GB, and LGBM are projected to excel in future forecasting based on MSE and MAE. Conversely, MLP, LSTM, and CNN-LSTM are anticipated to perform well on MAPE, SMAPE, and MASE.

Table 1. Model hyperparameters utilized during model training with specific training configuration and time (seconds)

Model	Model hyperparameters	Training Configuration	Training Time (Seconds)
RR	lags: 4, alpha: 0.6	–	0.106
RF	lags: 9, n_estimators: 105, criterion:mean_squared_error,	–	6.317
GB	lags: 13, n_estimators: 125, loss: squared_error	learning rate: 0.05	2.124
CB	lags: 12, random_strength: 0.7, n_estimators: 125, grow_policy: Depthwise, boosting_type: Plain	learning rate: 0.05	1.885
XGB	lags: 10, n_estimators: 90, booster: gbtree, importance_type: total_gain	–	5.203
LGBM	lags: 3, n_estimators: 115, boosting_type: gbdt, importance_type: total_gain	learning rate: 0.05	0.294
MLP	lags: 6, Dense filters: 128, Dense Activation: Tanh	Batch size: 64, Adam optimizer with Exponential Rate Decay with the initial learning rate: 0.001, decay steps: 18, decay rate: 0.96; Early stopping on validation loss with patience 10	9.204
CNN	lags: 6, 1st Conv1D filters: 16, 1st Conv1D activation: Tanh, 2nd Conv1D filters: 512, 2nd Conv1D Activation: Tanh	Batch size: 64, Adam optimizer with Exponential Rate Decay with the initial learning rate: 0.005, decay steps: 15, decay rate: 0.96; Early stopping on validation loss with patience 10	9.409

(*continued*)

Table 1. (*continued*)

Model	Model hyperparameters	Training Configuration	Training Time (Seconds)
LSTM	lags: 6, LSTM filters: 32	Batch size: 64, Adam optimizer with Exponential Rate Decay with the initial learning rate: 0.05, decay_steps: 5, decay_rate: 0.97; Early stopping on validation loss with patience 10	37.87
CNN-LSTM	lags: 6, Conv1D filters: 16, Conv1D Activation function: ReLu, LSTM filters: 64	Batch size: 64, Adam optimizer with Exponential Rate Decay with the initial learning rate: 0.005, decay_steps: 18, decay_rate: 0.96; Early stopping on validation loss with patience 10	42.350

Table 2. The MSE, MAE, MAPE, and MASE performance of 10 models and inference time on the test set

Model	Evaluation Metrics					Inference Time (Seconds)
	MSE	MAE	MAPE	SMAPE	MASE	
RR	$\mathbf{1.66 \times 10^{-3}}$	$\mathbf{2.20 \times 10^{-2}}$	3.88×10^{13}	1.237	2.00	**3.22**
RF	1.93×10^{-3}	2.35×10^{-2}	3.32×10^{13}	1.204	2.14	10.34
GB	$\mathbf{1.69 \times 10^{-3}}$	$\mathbf{2.10 \times 10^{-2}}$	3.35×10^{13}	1.237	1.91	**2.66**
CB	2.03×10^{-3}	2.27×10^{-2}	3.02×10^{13}	1.261	2.06	**4.76**
XGB	2.32×10^{-3}	2.38×10^{-2}	2.83×10^{13}	1.315	2.16	6.30
LGBM	$\mathbf{1.71 \times 10^{-3}}$	2.34×10^{-2}	4.12×10^{13}	1.232	2.13	5.48
MLP	1.94×10^{-3}	2.34×10^{-2}	$\mathbf{8.75 \times 10^{6}}$	**0.621**	**0.551**	6.35
CNN	1.97×10^{-3}	2.42×10^{-2}	8.77×10^{6}	0.633	0.565	6.51
LSTM	2.00×10^{-3}	2.34×10^{-2}	$\mathbf{8.13 \times 10^{6}}$	0.656	**0.552**	6.51
CNN-LSTM	2.14×10^{-3}	2.50×10^{-2}	$\mathbf{8.33 \times 10^{6}}$	0.640	0.564	6.51

3.2 Model Evaluation

For model evaluation in production, performance is validated by comparing real-time sensor values with those forecasted by the models for the duration of February 2024 to May 2024. Differences between actual and forecasted values are quantified using MSE, MAE, MAPE, and SMAPE. Evaluations are conducted in the same environment as training (Intel(R) Xeon(R) CPU @ 2.20 GHz). The evaluation process includes:

- Extracting daily real-time sensor data files (.txt) from local storage.
- Concatenating files, forward-filling missing values, and resampling data to 30-min intervals.
- Filtering data to exclude training data, preventing false optimism.
- Forecasting sensor values using pre-processed and resampled data.
- Calculate MSE, MAE, MAPE, and SMAPE by comparing forecasted and actual values.

Table 3 summarizes the models' forecasting performance. RR provides the most accurate forecasts, followed by MLP and LGBM, achieving lower MSE, MAE, MAPE, and SMAPE. However, MLP has an edge in inference time, 81% and 92% faster than RR and LGBM, respectively. Although CNN and CNN-LSTM excel in inference time, their accuracy is lower than that of RR and LGBM. Therefore, MLP strikes the best balance between forecasting accuracy and inference time. *Inference times* are estimates based on a different processing unit (NVIDIA GeForce RTX4060). Despite the promising results for RR, MLP, and LGBM, qualitative aspects, particularly threshold-based alerts, must be considered. An issue with the threshold value for differential pressure post-pocket filters led to false positives, which were rectified by updating the threshold.

Table 3. Models' performance on production sensor data

Model	Evaluation Metrics				Inference Time on Intel * Xeon ® CPU @ 2.20 Hz (Seconds)
	MSE	MAE	MAPE	SMAPE	
RF	4.72×10^{-3}	3.00×10^{-2}	3.83×10^{13}	1.35	279.69
GB	5.63×10^{-3}	2.95×10^{-2}	3.33×10^{13}	1.37	36.09
RR	$\mathbf{2.63 \times 10^{-3}}$	$\mathbf{2.40 \times 10^{-2}}$	$\mathbf{4.27 \times 10^{13}}$	**1.32**	29.17
CB	5.38×10^{-3}	2.90×10^{-2}	3.31×10^{13}	1.37	74.79
XGB	5.90×10^{-3}	3.19×10^{-2}	3.50×10^{13}	1.43	59.56
LGBM	$\mathbf{3.09 \times 10^{-3}}$	$\mathbf{2.64 \times 10^{-2}}$	$\mathbf{4.57 \times 10^{13}}$	**1.33**	70.87
MLP	$\mathbf{2.99 \times 10^{-3}}$	$\mathbf{2.62 \times 10^{-2}}$	$\mathbf{4.18 \times 10^{13}}$	1.38	**5.60**
CNN	3.37×10^{-3}	2.90×10^{-2}	4.33×10^{13}	1.41	**5.24**
LSTM	3.88×10^{-3}	2.98×10^{-2}	3.85×10^{13}	1.46	7.30
CNN-LSTM	4.25×10^{-3}	3.08×10^{-2}	3.75×10^{13}	1.45	**7.19**

The critical insight from the models' quantitative performance metrics—specifically MSE, MAE, MAPE, and SMAPE—is that RR exhibits superior performance, followed by MLP and LGBM due to their lower error values compared to other models. However, RR and LGBM significantly disadvantage inference time relative to models like MLP, CNN, LSTM, and CNN-LSTM. Given the importance of inference time for real-time applications, CNN shows the lowest inference time, followed by MLP and CNN-LSTM. Despite this, CNN and CNN-LSTM underperform in forecasting accuracy. Thus, the MLP model strikes an optimal balance between forecasting accuracy and efficiency. It is important to note that the current models' performance is based on a limited duration of sensor data, and their ability to handle new data fluctuation patterns remains to be evaluated. Furthermore, the inference times presented in Table 3 are approximate estimates for the given set of sensor data.

While the quantitative performance of models such as RR, MLP, and LGBM appears promising, their qualitative aspects must also be considered. The models must reliably trigger maintenance alerts based on whether the forecasted sensor values exceed established thresholds for filter replacement. Under the conditions where models were trained on sensor data, MLP emerged as the most balanced model regarding forecasting accuracy and inference time, achieving lower MSE, MAE, MAPE, and SMAPE within a shorter duration than other models.

Despite initial inaccuracies in sensor readings due to installation issues, the model holds significant potential for improvement. Retraining the model with updated and accurate data can substantially enhance predictive accuracy regarding filter behaviour. This retraining process provides the model with corrected sensor readings that accurately reflect the filters' performance and condition. As the model learns from this accurate dataset, it will become better equipped to predict genuine filter replacement needs, reducing false positives and improving maintenance scheduling. Moreover, incorporating new data will help the model adjust to actual fluctuation patterns, leading to more reliable and precise forecasting in future applications.

4 Conclusion

The primary contribution of this paper is developing a PdM method using ten machine learning and deep learning algorithms that can monitor the state of the filters in real time and predict filter faults in advance, setting it apart from other related works. This work used datasets related to the condition monitoring of painting booth filters to evaluate the effectiveness of various machine learning techniques within a practical Predictive Maintenance context. Ten distinct machine learning and deep learning algorithms were employed: RF, GB, RR, CB, XGB, LGBM, MLP, CNN, LSTM, and CNN-LSTM. A comparative analysis of these algorithms was conducted by training them on varying-sized datasets and evaluating their accuracy on test data. The analysis and findings offer significant insights into the effectiveness of these machine-learning techniques in the PdM context. Comprehensive testing and evaluation revealed that the Multilayer Perceptron (MLP) model provided the best balance of accuracy and speed for real-time predictions. Implementing the predictive model has optimized maintenance schedules, significantly reducing filter replacement frequency and associated costs. Future work should focus on fine-tuning model parameters, incorporating additional data factors, and assessing real-time performance to enhance the model's performance further. Moreover, integrating real-time monitoring and predicting the filters' Remaining Useful Life (RUL) can further optimize maintenance strategies, leading to even greater efficiency and cost savings. This research demonstrates the potential of machine learning in PdM and sets a foundation for future advancements in predictive maintenance technology.

Acknowledgement. This work was supported by AMIC, SPIRIT and UKM (FOF23-VRP22 & TT–2023–009).

References

1. Mohammed, B.-D., Kumar, U., Prabhakar, M.D.: Introduction to Maintenance Engineering: Modelling, Optimization, and Management. Wiley, Hoboken (2016)
2. Arnedo, M.S.: Fundamentals of aerospace engineering: an introductory course to aeronautical engineering. Createspace Independent Publishing Platform, Madrid (2014)
3. Scott, M.J., Verhagen, W.J., Bieber, M.T., Marzocca, P.: A systematic literature review of predictive maintenance for defence fixed-wing aircraft sustainment and operations. Sensors **22**(18), 7070 (2022)
4. Ben-Daya, M., Duffuaa, S.O., Raouf, A., Knezevic, J., Ait-Kadi, D.: Handbook of Maintenance Management and Engineering. Springer, London (2009)
5. Vathoopan, M., Johny, M., Zoitl, A., Knoll, A.: Modular fault ascription and corrective maintenance using a digital twin. IFAC (Int. Fed. Autom. Control) PapersOnLine **51**(11), 1041–1046 (2018)
6. Menegon, J., Isatto, E.L.: Digital twins as enablers of structure inspection and maintenance. Gestão Produção **30**(e4922) (2023)
7. Dong, Z., Zhelei, X., Yian, Z., Junhua, D.: Overview of predictive maintenance based on digital twin technology. Heliyon **9**, e14534 (2023)
8. Brügge, F.: Predictive maintenance & asset performance market report 2023–2028. Retrieved from IOT analytics (2023). https://iot-analytics.com/predictive-maintenance-market/
9. Stanton, I., Munir, K., Ikram, A., El-Bakry, M.: Predictive maintenance analytics and implementation for aircraft: challenges and opportunities. Syst. Eng. J. Int. Counc. Syst. Eng. **26**(2), 216–237 (2023)
10. Theissler, A., Pérez-Velázquez, J., Kettelgerdes, M., Elger, G.: Predictive maintenance enabled by machine learning: use cases and challenges in the automotive industry. Reliab. Eng. Syst. Saf. **215**(107864), 1–21 (2021)
11. Fitrayudha, A., Wijaya, S.K.: Predictive maintenance for aircraft engine using machine learning: trends and challenges. Int. J. Aviat. Sci. Eng. **3**(1), 37–44 (2021)
12. Calabrese, M., Cimmino, M., Fiume, F., et al: SOPHIA: an event-based IoT and machine learning architecture for predictive maintenance in industry 4.0. Information (Switzerland) **11** (2020)
13. Bekar, E.T., Nyqvist, P., Skoogh, A.: An intelligent approach for data pre-processing and analysis in predictive maintenance with an industrial case study. Adv. Mech. Eng. **12** (2020)
14. Qolomany, B., Mohammed, I., Al-Fuqaha, A., et al.: Trust-based cloud machine learning model selection for industrial IOT and smart city services. IEEE Internet Things J. **8**, 2943–2958 (2021)
15. Becherer, M., Zipperle, M., Karduck, A.: Intelligent choice of machine learning methods for predictive maintenance of intelligent machines. Comput. Syst. Sci. Eng. **35**(2), 81–89 (2020)

16. Burmeister, N., Frederiksen, R.D., Høg, E., Nielsen, P.: Exploration of production data for predictive maintenance of industrial equipment: a case study. IEEE Access **11**, 102025–102037 (2023)
17. Jyh-Yih, H., Yi-Fu, W., Kuan-Cheng, L., Mu-Yen, C., Jenneille, H.H.: Wind turbine fault diagnosis and predictive maintenance through statistical process control and machine learning. IEEE Access **8**, 23427–23439 (2020)
18. Bosch, C., Simon-Carbajo, R.: Perspective Chapter: Computation of Wind Turbine Power Generation, Anomaly Detection, and Predictive Maintenance. IntechOpen (2023)
19. Makridis, G., Kyriazis, D., Plitsos, S.: Predictive maintenance leveraging machine learning for time-series forecasting in the maritime industry. In: 23rd International Conference on Intelligent Transportation Systems (ITSC), pp. 1–8. IEEE, Rhodes (2020)
20. Kusumaningrum, D., Kurniati, N., Santosa, B.: Machine learning for predictive maintenance. In: Proceedings of the International Conference on Industrial Engineering and Operations Management, pp. 2348–2356. IEOM Society International, Sao Paulo (2021)
21. Bampoula, X., Siaterlis, G., Nikolakis, N., Alexopoulos, K.: A deep learning model for predictive maintenance in cyber-physical production systems using LSTM auto encoders. Sensors **21**(972) (2021)
22. Yu, H., Chen, C., Lu, N., Wang, C.D.: Deep auto-encoder and deep forest-assisted failure prognosis for dynamic predictive maintenance scheduling. Sensors **21**, 8373 (2021)

Safety Control with Conflict Resolution for Multi Robots in Flexible Manufacturing System

Jakaisa Riskhalifah Bhuwana[1], Muhammad Zakiyullah Romdlony[1](✉), Angga Rusdinar[1], Norhaliza Abdul Wahab[2], and Muhammad Azhar Ismail[1]

[1] Telkom University, Jl. Telekomunikasi Terusan Buah Batu, 40257 Bandung, Indonesia
zakiyullah@telkomuniversity.ac.id
[2] Universiti Teknologi Malaysia, 81310 Skudai, Johor Bahru, Johor, Malaysia

Abstract. An approach for integrating multi-autonomous mobile robot (AMR) systems that work in flexible manufacturing environments while avoiding deadlocks and livelocks, as well as obstacles that are not registered on the map server with the Control Lyapunov Barrier Function (CLBF). Dijkstra's algorithm is utilized for path planning and CLBF as the control method for AMRs to ensure that it runs safely to a determined station without colliding with unregistered obstacles. The setup assigns the AMRs a mission retrieving goods from the station and transporting them to another station. Additionally, the map server does not have any previously registered obstacles. A communication and coordination algorithm among AMRs is designed to prevent deadlocks, livelocks, and collisions among them. The efficacy of the proposed control and algorithm is demonstrated in a ROS simulation. The simulation is divided into two scenarios: one with two AMRs in a 16 m × 9 m workspace with unregistered obstacles, and another without any obstacles. The simulation result shows that the latter scenario has completed more missions by 26.6% than the former.

Keywords: FMS · Multi AMR · CLBF · ROS

1 Introduction

Introduction In this rapidly developing era, market needs are also changing dramatically, causing conventional manufacturing systems not to be able to find a balance point between demand and production [1, 2]. Which affects manufacturing profit margins. How to speed up the manufacturing cycle and increase the utilization rate of production resources is the key to the development of the intelligent manufacturing industry [3]. Flexible manufacturing systems offer features that can be a solution to these problems because FMS provides flexibility features that can change according to situations quickly [4]. The manufacturing paradigm has evolved to embrace flexibility in response to diverse production demands [5]. The system will become more complex than a conventional manufacturing system due to its increased flexibility [6]. Autonomous mobile robots can increase material handling flexibility and enable distributed control processing in flexible manufacturing systems [7].

AMR is a mobile robot that can move independently or autonomously indoors or outdoors [8, 9]. Autonomous mobile robot (AMR) systems are an important aspect of low-level to middle-level manufacturing, including flexible manufacturing systems, as well as the service industry, where they deliver goods such as mail, laundry, and food to hospitals [10]. Flexible production networks based on autonomous mobile robots (AMRs) have the potential to increase industrial flexibility and productivity [11]. AMRs are renowned for their ability to work continuously, safely, and efficiently, delivering various types of loads without human intervention, so multi-AMR systems have become an important part of the automation of manufacturing facilities and warehousing systems [12, 13].

AMRs may experience deadlock and livelock behavior, which is when they become trapped in local equilibria or periodic orbits inside a multi-robot system [14]. When two or more AMRs become trapped in a non-running process due to mutual waiting, the system halts [15–17]. Meanwhile, livelock occurs when AMRs continue to change positions or actions to avoid collisions or conflicts but remain stuck in movement patterns that prevent them from reaching their destination [18]. Implementing a communication system among AMRs to facilitate coordination is a crucial aspect of resolving deadlocks [19].

This research aims to develop a multi-AMR decentralized control system capable of inter-AMR communication, item delivery in an FMS environment, and avoidance of deadlocks, livelocks, and obstacles that are not registered on the map server. This project will create localization, navigation, and control systems, as well as communication between AMRs, in the ROS environment during the various stages of AMR design. ROS enables easier data analysis [20]. Since this AMR uses LiDAR sensors and encoders for localization, Adaptive Monte Carlo Localization (AMCL) will be employed. In the framework of Robot Operating System (ROS), Adaptive Monte Carlo Localization (AMCL) accurately locates mobile robots in unknown territories, with accuracy depending on the number of particles and the surrounding environment [21, 22]. For navigation in this research, the Dijkstra method is employed. By choosing the unvisited vertex with the smallest distance and updating its neighbors with shorter distances, Dijkstra's method determines the shortest path [23, 24]. And for the AMR control system, it will use the Control Lyapunov Barrier Function (CLBF), The CLBF approach substantially enhances the stability and safety of mobile robot systems by avoiding obstacles and obtaining near-perfect alignment with the origin [25].

2 Methods

This chapter discusses path planning, control for obstacle avoidance, and conflict resolution.

2.1 Path Plan and Control Avoidance

Path planning is the process by which a robot looks for the best path from one site to another while avoiding obstacles along the way [26, 27]. Path planning and obstacle avoidance are the foundations of autonomous mobile robotics, allowing robots to achieve their destination while avoiding obstacles and traveling along optimum paths based on distance, time, or energy [28]. This subsection discusses path planning using the Dijkstra method and obstacle avoidance using the Control Lyapunov Barrier Function method.

2.2 Control Lyapunov Barrier Function

Control Lyapunov Barrier Function is a collision-free controller that can be created by combining control Lyapunov functions with control barrier functions, therefore avoiding unwanted equilibria at the safe set's border [29]. Control Lyapunov Barrier Function (CLBF) is a control equation that combines and integrates CLF and CBF, allowing the utility of the two control functions to be achieved with a simpler control equation [30].

CLF will control AMR from any location to the destination point. This is because CLF has a positive definite equation, which implies it has a negative value for the root value, causing the state value to decrease with time. CBF will manage AMR to avoid an unsafe state and ensure safety. Following the parameter in reference [30], the CLBF pseudocode is constructed in algorithm 1 with AMR target coordinate at (0, 0) and the radius of the obstacle set to 0,6 m, and adding 0,2 m for margin safety and the result is 0,8 m.

CLBF
1 **define** x_1 and x_2 as AMR x-axis coordinates, and y-axis coordinates
2 **define** x_{1D} and x_{2D} coordinates of the obstacle's center point
3 **define** r as distance between AMR and obstacle's center point
4 **while** $x_1 \ne 0$ and $x_2 \ne 0$ **do**
5 $\quad L_f V = 0$
6 $\quad L_g V = [\, 2x_1 + x_2 \quad 2x_2 + x_1]$
7 $\quad L_f B = 0$
9 $\quad$ **if** r < 0.8 **do**
10 $\quad\quad L_g B = [\, 2x_1 - 2x_{1D} \quad 2x_2 - 2x_{2D}]$
11 $\quad$ **else do**
12 $\quad\quad L_g B = 0$
13 $\quad$ **end if**
14 $\quad L_f W = L_f V + \lambda * L_f B$
15 $\quad L_g W = L_g V + \lambda * L_g B$
16 $\quad b = \text{transpose}(L_g W)$
17 $\quad a = L_f W$
18 $\quad u = -\dfrac{a + \sqrt{a^2 + \gamma\, \|b\|^4}}{b^T b}\, b$
19 **end while**

Algorithm 1. CLBF pseudocode

CLBF is used for controlling the AMR. Dijkstra's output is simply the path plan, which contains coordinates that will be followed by CLBF. When an obstacle on the AMR's path is not recognized on the map server, the CLBF control system will automatically avoid it while still guiding the AMR to its destination. Once the LiDAR no longer detects an unregistered obstacle, the AMR will resume to its original path.

AMR is equipped with LiDAR, which can identify obstacles that are not registered on the map server. LiDAR can determine the distance and angle of an unregistered obstacle to the AMR at any angle. The angle and distance of the obstacle relative to the AMR will be calculated using trigonometry to determine its position relative to the AMR. Next, the obstacle position data that is still related to the AMR is combined with the current

AMR location to generate global obstacle position data. The global obstacle position's x and y axes will be used as input for CLBF.

2.2.1 Dijkstra Path Plan

Path planning using Dijkstra on AMR will plot the map provided to AMR by generating a very tiny grid that represents the working environment surrounding AMR. This grid will eventually be utilized as a node by Dijkstra and assigned a weight depending on the accessibility of the environment, which will be determined using the distance necessary to travel through the grid and obstacles on the map. This method will select the path that AMR will travel by calculating the cost of the grid that will be traversed, such that the path with the lowest cost is picked, which is also the shortest path in terms of distance, making the most effective path [31]. The Dijkstra method in route planning finds an optimum path in an area by utilizing the grid-map structure to construct an accurate approximation of the best path [32].

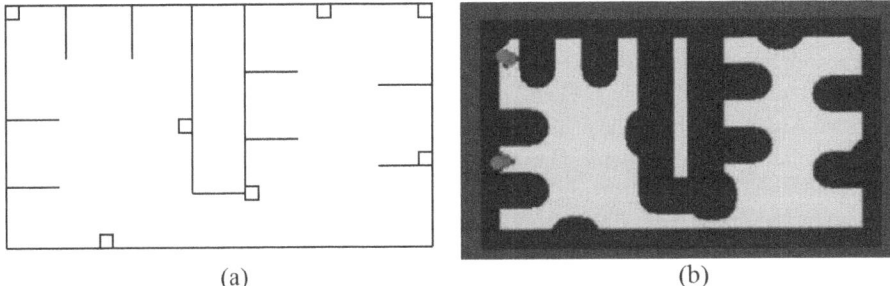

Fig. 1. Registerd Map, Map Server (a), Costmap (b).

The costmap picture in Fig. 1a will serve as a map server. As shown in Fig. 1a, the partitions and walls are not that thick. But in Fig. 1b, it has been given a safety margin. This is critical because if the safe boundary with the wall or partition is too close, Dijkstra takes the closest path without estimating the area of the AMR itself. Because in this research the Dijkstra's method only considers a point, the AMR will crash because the specified path is too close to a wall or partition. To avoid collisions, this safe limit must be bigger than the AMR radius.

2.3 Conflict Resolution

In this research, deadlocks and livelocks can be resolved by creating a conflict resolution for AMR, so that AMR can cooperate with one other. This conflict resolution algorithm is meant if there is a conflict between AMRs. When AMRs compete for an area, or in the instance of this research, they pass through a resource or corridor and stations that are only accessible by one AMR. The first location protocol is based on priority [33, 34] and first gets into the resource [35]. The pseudocode for the conflict resolution algorithm is constructed in Algorithm 2.

```
Conflict Resolution
1   define s as distance between AMRs
2   define c as distance between path plan
3   define j as resource being occupied
4   define m₁ and m... as priority of each robot
5   if s < 3 do
6   |   if c < 1 do
7   |   |   if j = 1 do
8   |   |   |   command the robot to wait
9   |   |   else if m₁ < mₓ do
10  |   |   |   command the robot to yield
11  |   |   end if
12  |   end if
13  end if
```

Algorithm 2. Conflict resolution algorithm

First, AMR will determine whether other AMRs enter the private zone. The private zone is defined as a three-meter radius around the AMR. If one AMR detects that another AMR has entered the private zone, each AMR will check to see if their trajectories will intersect. Whether a collision occurs, AMR will figure out whether the resource, or in this example, the corridor and station, is occupied or whether another AMR is nearby. If the resource is full, AMR will enter the "wait" command, which will stop AMR in its tracks and wait until there is a "clear" order from AMR who was using the resource.

Next, AMR will check the priority for each AMR. If the AMR has a lower priority, it will enter the "yield" command. This command will direct the AMR to a location where there is minimal conflict, allowing the AMR with a higher priority to enter without being blocked by the AMR currently in front of the resource. Because the resource entrance is narrow, this is important to prevent deadlocks. If the AMR with high priority has been finished and is at a safe distance, then the AMR will transmit a "clear" signal, and the AMR with low priority will continue its mission normally.

Other AMRs will subscribe to other AMR coordinates or /amcl_pose topics to obtain the coordinate location for every AMR. The coordinates of the other AMRs will be examined so that they can determine which resources are being utilized by each AMR, as well as so that an AMR can detect when another AMR enters its private zone. Apart from subscribing to coordinate topics, AMR will also publish and subscribe to priority or /prior topics so that AMR can know the priorities of other AMRs so that they can determine whether the AMR has low priority or high priority. Fellow AMRs will then subscribe to the route plan or /Globalplanner/plan topics. This topic covers information about the path plans for each robot. Information about other AMR travel plans is required so that each AMR knows the destination and route that the other AMR is taking. The information contained in the path plan is the coordinates that the AMR will go through, that information must be processed so that other AMRs can determine which resource the other AMR will visit, and which will be used.

3 Design and Implementation Setup

This study will create an FMS system that uses two AMRs to hand some missions, which include the pickup of goods in the station and delivering them to its destination point. Then, this AMR is equipped with LiDAR for localization and navigation to detect obstacles that are not registered on the map server. The map server will be sent to the AMR, allowing it to navigate without having to examine the region. Priority negotiations require an inter-AMR communication mechanism to avoid deadlocks or livelocks. In addition, the Control Lyapunov Barrier Function (CLBF) approach is used to avoid barriers that have not yet been recorded on the map server while moving towards the objective. The purpose of this research is to create a multi-AMR with a decentralized control system that can communicate with one another and deliver items in an FMS environment while avoiding dead locks and livelocks, as well as obstacles that are not registered on the map server (Fig. 2).

3.1 Design System

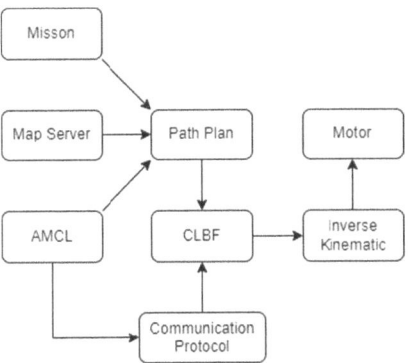

Fig. 2. Diagram block

Here's an overview of how the AMR will move and interact with the environment. First, the mission will be the input for the Dijkstra parameters. Then, using map server and localization, Dijkstra will decide the overall path, such as which corridors and terminals to take. Next, Dijkstra's output is in the form of coordinates, which are used as input for CLBF. CLBF will then follow the path plan that has been given. However, if the path plan provided by Dijkstra runs across an obstacle that is not on the map server. The LiDAR sensor will identify any obstacle that is not on the map server, and CLBF will avoid it and continue along the planned course. The CLBF control system outputs the speed of each axis. So inverse kinematics is required, this stage converts data from the speed of each axis to the speed of each motor on the AMR. The location provided by AMCL will then be processed so that AMR can coordinate effectively and avoid livelocks and deadlocks. The procedure will be repeated till the user terminates it.

3.2 ROS Design

ROS facilitates the development of ideas on robotics subjects by allowing researchers to run simulations [36]. The ROS system for this research will be separated into numerous

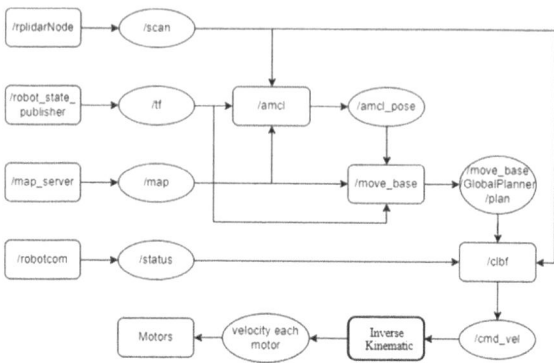

Fig. 3. ROS diagram block

sections. The first portion will go over localization and LiDAR. The second section covers path planning, move_base, and CLBF. Figure 3 shows an overview diagram.

In Fig. 3 there are several processes. The square-shaped ones are nodes, and the oval-shaped ones are topics. The first process is at the /rplidarnode node. This node contains coding that will take and process the rplidar into scan data and is called topic /scan, topic /scan contains information about each distance and angle from the lidar to objects around it. Then there is the /robot_state_publisher node, this node contains coding about the transfer function of the AMR itself, and issues a /tf topic whose contents are the transfer function. After that, there is the /map_server node, this node contains coding information about the map that will be used, and a published topic with the name /map which contains information about the map that will be used. Then there is the /robotcom node, this node contains the conflict resolution algorithm coding between AMR, and the publish topic /status, which contains the contents of the conflict resolution algorithm.

Then there is the /amcl node, this node contains coding for AMR localization, this node requires information from the topics /map, /scan, and /tf the output from this node is information about the AMR coordinates on the map that has been given, which is called the topic /amcl_pose. The next node, /move_base, is a node that contains the code to produce a path plan using the Dijkstra method. This node requires information from the topics /amcl_pose, /tf, and /map, which will publish the topic /move_base/GlobalPlanner/plan which contains the coordinates of the resulting path plans. Then the /clbf node is the node that will process information from the topics /move_base/GlobalPlanner/plan, /status, /scan, and /amcl_pose, into speed on the x-axis and on the y-axis AMR, with the topic /cmd_vel. /cmd_vel will enter inverse kinematics that are outside ROS, which will produce speed information for each motor.

3.3 Simulation and Environment Setup

The first experiment is designed to demonstrate the safety aspect of the Control Lyapunov Barrier Function (CLBF) as a control method for Autonomous Mobile Robots (AMRs). The experiment is conducted in Fig. 4 where the robot navigates from its initial coordinate $(-4, -7)$ or in station 1, to its target coordinate $(0, 0)$ or station 5 in the presence of circle unsafe state centered at (−1, −4), detailed information about the specific coordinates is illustrated in Fig. 5. The author attempts to simulate CLBF with results and parameters as follows, the equations used in this experiment are adapted from [30].

$$V(x) = x_1^2 + x_1 x_2 + x_2^2$$
$$L_g V = [2x_1 + x_2 x_1 + 2x_2] \tag{1}$$

where x_1 and x_2 represent the coordinates of the robot along the x-axis and y-axis, respectively. The unsafe region is defined as a circular area with a radius of r meters around the point $(-x_{1\mathcal{D}}, -x_{2\mathcal{D}})$, and $\mathcal{D}$ is a subset of $\mathcal{X}$.

$$\mathcal{D} = \{x \in \mathcal{X} | (x_1 - x_{1\mathcal{D}})^2 + (x_2 - x_{2\mathcal{D}})^2 < r^2\} \tag{2}$$

where r is the radius of the obstacle, $x_{1\mathcal{D}}$ and $x_{2\mathcal{D}}$ is coordinates of the obstacle's center point. For the application procedure see Algorithm 1 and the equation model in Eq. (36) in [30]. For more clarity there is a similar procedure in [30] in sub Sect. 6.2.

The second experiment is to examine the conditions before and after applying the conflict resolution algorithm. The third experiment is Flexible Manufacturing System (FMS) warehouse design intended for testing AMR performance is represented. A serval partitions is used to separate different stations and provide distinct differences between stations areas. Each terminal has several stations, and in this setup, each station can be used as loading and unloading area depend on the mission, and two terminal is separated by corridor. To simulate real-world challenges unregistered obstacles in the form of randomly placed boxes are provided within the terminal areas. These obstacles are not preregistered in the map server requiring the AMRs to dynamically detect and avoid them while executing their missions. The map of the environment is shown in Fig. 4.

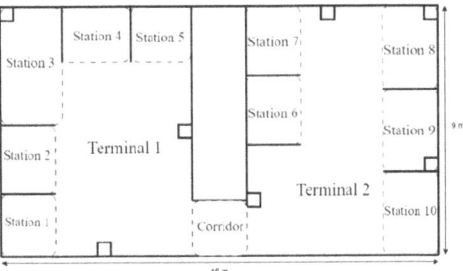

Fig. 4. Map of the Environment

4 Result

Following Fig. 5 is the first experiment simulation results demonstrating CLBF can guarantee the safety of the AMR and keep it on its way to its destination.

The simulation in Fig. 5 shows that AMR departs from station 1 to station 5. But it encounters an unregistered obstacle, represented as a black circle in the middle of

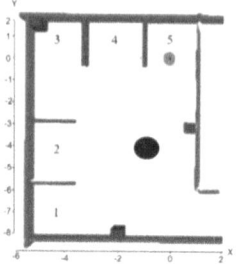

Fig. 5. CLBF simulation

the workspace. In addition, the AMR is represented as a grey circle with a dot at its center. The edges of the workspace are equipped with registered obstacles, which are visualized by square shapes near the walls. The trajectory of the AMR can be clearly seen from the green line, which shows that it successfully avoids the unregistered obstacle and reaches station 5 without any problem.

The second experiment is to examine the conditions before and after applying the conflict resolution algorithm. The following is an example if the conflict resolution algorithm is not implemented.

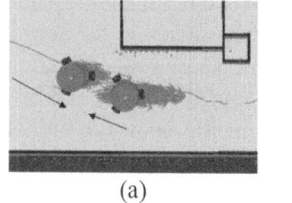   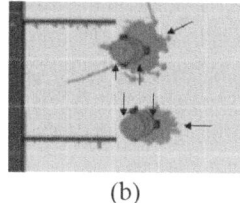   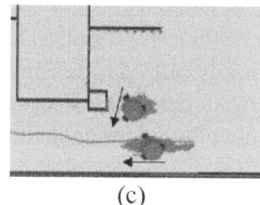
(a)   (b)   (c)

Fig. 6. AMR without conflict resolution algorithm, Deadlock in Corridor (a), Livelock in Station(b), Livelock in Corridor(c).

It can be seen in Fig. 6a. Depicting that when one AMR wants to leave the corridor, and another AMR wants to enter the corridor, because there is no conflict resolution algorithm, there is no AMR to give in, so a deadlock occurs. In Fig. 6b. it can be seen that there are two AMRs fighting over one station, because this experiment has a safety feature so that the two AMRs don't crash, they will avoid each other, but because no one yields, a live lock occurs. In Fig. 6c. The same thing at Fig. 6b. is happening, but with corridors.

The following are the few simulation results if a conflict resolution algorithm is applied to the system.

It can be seen from Fig. 7a. That there are two AMRs in different directions, AMR wants to leave the corridor while the other AMR wants to enter the corridor. In the Fig. 7b. Because the two AMRs have entered the private zone of another AMR, a conflict resolution algorithm occurs, where the AMR that is in the corridor sends a "wait" message to the AMR that wants to enter the corridor. So that AMR who wanted to enter the corridor stopped. Figure 7c. Shows that when the AMR which was previously in the corridor has

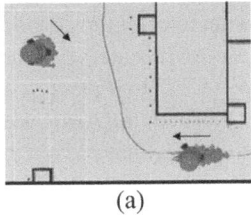 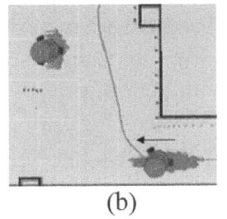 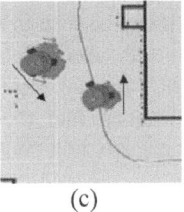

Fig. 7. AMR with conflict resolution algorithm, Conflict Detection (a), AMR with low priority wait (b), Mission Continued (c).

left the corridor, the AMR sends a message to the AMR who is waiting to continue its mission. Thus, an algorithm like this can avoid deadlocks.

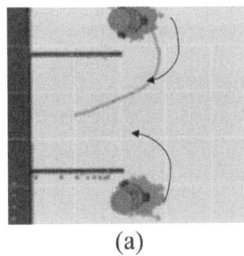 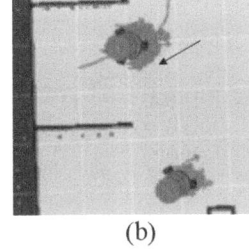 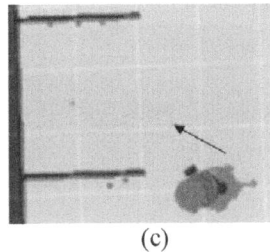

Fig. 8. AMR with conflict resolution algorithm, Conflict Detection (a), AMR with low priority yield(b), Mission Continued(c).

It can be seen from Fig. 8a. That there are two AMRs who want to compete for the station. The Fig. 8b. Shows that an AMR with a high priority will order an AMR with a low priority to "yield", then the AMR with a low priority will "yield" to a place with minimal conflict. Then Fig. 8c. Shows that when the AMR with high priority has finished using its station, and the station is empty, the AMR with high priority gives a message to the AMR with low priority to continue its mission. Thus, this conflict resolution algorithm can avoid live lock.

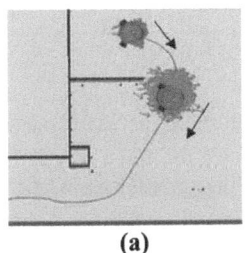 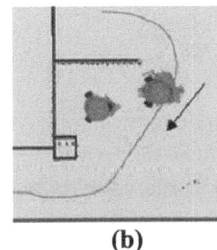 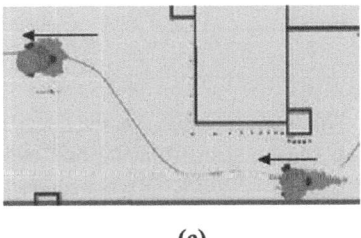

Fig. 9. AMR with conflict resolution algorithm, Conflict Detection (a), AMR with low priority yield (b), Mission Continued (c).

It can be seen in the Fig. 9a. That there are two AMRs who want to enter the corridor from the same directions. In the Fig. 9b. Can be seen, when an AMR enters another AMR's private zone, the AMR with high priority is ordered to "yield" to a place with minimal conflict, then when the AMR with high priority is in the corridor, the AMR with low priority will "wait". In the Fig. 9c. When the AMR with high priority has passed the corridor, the AMR with high priority will order the AMR with low priority to continue its mission. Thus, this conflict resolution algorithm like this can avoid deadlocks.

In the third experiment the simulation used in this set up uses two AMRs, which are given missions in the form of picking up goods and delivering them randomly, which will be given a time of 3, 5, and 8 min. With the second AMR having the highest priority. This will be carried out on a smooth and flat area with a width and length of 16 m × 9 m, equipped with a corridor with a width of 2 m and also having 10 stations as an additional feature. The following in Fig. 10 is a picture of the workspace.

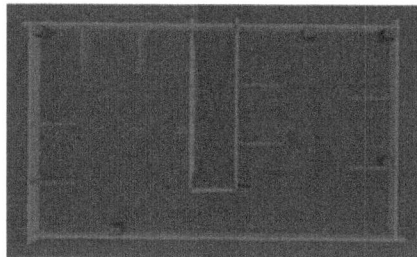

Fig. 10. Workspace 16 m × 9 m

Table 1. Simulation Result with two AMRs in 16 m × 9 m workspace.

	3 minutes		5 minutes		8 minutes	
	AMR1	AMR2	AMR1	AMR2	AMR1	AMR2
First Experiment	4 Missions	5 Missions	8 Missions	8 Missions	12 Missions	13 Missions
	9 Missions		16 Missions		25 Missions	
Second Experiment	3 Missions	4 Missions	6 Missions	7 Missions	9 Missions	12 Missions
	7 Missions		13 Missions		21 Mission	
Third Experiment	5 Missions	6 Missions	8 Missions	9 Missions	13 Missions	14 Missions
	11 Missions		17 Missions		27 Missions	
Average Missions	4 Missions	5 Missions	7 Missions	8 Missions	11,3 Missions	13 Missions
	9 Missions		15,3 Missions		24,3 Missions	

And seen in Table 2 is the success rate data from one AMR in the same workspace.

It can be seen from the first experiment that AMR2 completes the most missions compared to AMR1. This is because AMR2 has a higher priority than AMR1, so AMR1 gives in more and prioritizes AMR2 first. It can be seen from Table 1 and Table 2 that adding twice as many robots does not mean increasing the double number of missions completed. There are several factors that influence the number of missions completed because the missions given in this experiment are given randomly, so each trial has a different and random path length, thus affecting the AMR travel time. The next factor is

Table 2. Simulation Result with one AMR in 16 m × 9 m workspace

	3 min	5 min	8 min
First Experiment	6 missions	10 missions	17 missions
Second Experiment	5 missions	8 missions	13 missions
Third Experiment	5 missions	9 missions	14 missions
Average Missions	5,3 missions	9 missions	14,6 missions

interference between AMRs. One AMR working in one space does not need to interfere with another AMR. When more than one AMR is working in the workspace, a communication system is needed to avoid deadlocks, livelocks, and collisions with other AMRs. The communication system in this set up will control AMR queuing, and yield.

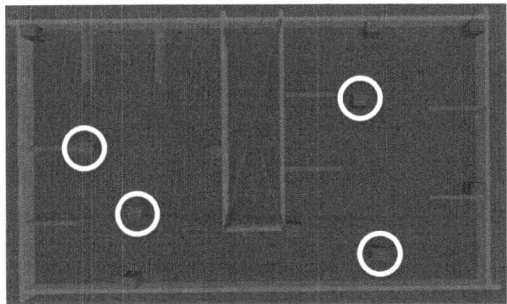

Fig. 11. Map image of 16 m × 9 m terrain with unregistered obstacles

In the following experiment, we'll add obstacles that are not registered on the map server to the 16 m × 9 m workspace. Four additional obstacles have been added, which are in the terminal area of the workspace and not in the resource area. This obstacle can be put anywhere if it does not cover the resource area or its safety margin, allowing the Dijkstra to locate a safe entrance while the CLBF avoids the obstacle.

In Fig. 11, the white circle is an obstacle that is not registered on the map server. When AMR runs and encounters the obstacle while the coordinates provided by the Dijkstra algorithm are within the obstacle, CLBF will prevent AMR from crashing into the obstacle. And with the help of LiDAR, AMR will register the obstacle in the map server, so that AMR will call the Dijkstra route again, which has considered the obstacle, and a new route will appear where AMR will avoid the obstacle safely.

It can be seen from Table 3 that the comparison of the presence and absence of obstacles that are not registered in the workspace greatly influences the success rate of the mission carried out by AMR. The following is a comparison of two AMRs carried out in a 16 m × 9 m area without any obstacles that are not registered on the map server with workspaces that have obstacles that are not registered on the map server.

The difference in mission success can be observed in a 16 m × 9 m workspace with an unregistered obstacle compared to one without an unregistered obstacle. Percentage

Table 3. Simulation result with two AMRs in 16 m × 9 m workspace with unregistered obstacle

	3 min		5 min		8 min	
	AMR1	AMR2	AMR1	AMR2	AMR1	AMR2
First Experiment	2 Missions	4 Missions	4 Missions	6 Missions	6 Missions	9 Missions
	6 Missions		10 Missions		15 Missions	
Second Experiment	5 Missions	4 Missions	9 Missions	6 Missions	16 Missions	12 Missions
	9 Missions		15 Missions		28 Mission	
Third Experiment	2 Missions	3 Missions	4 Missions	6 Missions	6 Missions	10 Missions
	5 Missions		10 Missions		16 Missions	
Average Missions	3 Missions	3,6 Missions	5,6 Missions	6 Missions	9,3 Missions	10,3 Missions
	6,6 Missions		11,6 Missions		19,6 Missions	

difference between the two experiments, in the third minute it is 26,6%, in the fifth minute it is 24,1%, and in the eighth minute it is 19,3%.

$$\frac{q-t}{t} * 100\% \tag{3}$$

where q Average Missions without unregistered obstacle, and t is Average Missions with unregistered obstacle Eq. 3 is the percentage formula, the percentage decrease in mission success rates achieved by AMR is reducing with time. This is because when AMR is first run, it encounters multiple new obstacles and registers these obstacles.

5 Conclusions

This study successfully created a multi-AMR with a decentralized control system, enabling it to interact with each other and transport items in an FMS setting, all while preventing deadlocks, livelocks, and obstacles not registered on the map server. In this research, CLBF with AMR is applied to control and avoid obstacles, followed by Dijkstra to find an optimal route plan. The conflict resolution algorithm used in this research is also based on priority and who uses it first. This research is tested using a scenario of delivering goods from one station to another, workspaces in this simulation are 16 m × 9 m dimensions. In this research, the mission will be given randomly in the simulations Rviz and Gazebo. With the average result, 9 missions in the third minute, 15,3 missions in the fifth minute, and 24, 3 missions in the eighth minute. And then was additionally tested by introducing multiple unregistered obstacles on the map server with a workspace of 16 m × 9 m and using Rviz and Gazebo simulations. According to the average test results provided by multiple obstacles that are not registered on the map server, the third minute is 6,6 missions, the fifth minute is 11,6 missions, and the eighth minute is 19,6 missions. The percentage difference between the two experiments is in the third minute, it is 26,6 %, in the fifth minute it is 24, 1%, and in the eighth minute it is 19,3%, These percentages indicate that simulation without multiple unregistered obstacles outperforms simulation with multiple unregistered by the respective values.

Acknowledgement. This research was supported by Ministry of Education, Culture, Research, and Technology of Republic Indonesia under project 106/E5/PG.02.00.PL/2024; 043/SP2H/RT-MONO/LL4/2024; 086/LIT07/PPM-LIT/2024.

References

1. Brecher, C.: Lecture notes in production engineering advances in production technology (2015). http://www.springer.com/series/10642
2. Yin, Y., Stecke, K.E., Li, D.: The evolution of production systems from Industry 2.0 through Industry 4.0. Int. J. Prod. Res. **56**(1–2), 848–861 (2018). https://doi.org/10.1080/00207543.2017.1403664
3. Liu, Q., Wang, N., Li, J., Ma, T., Li, F., Gao, Z.: Research on flexible job shop scheduling optimization based on segmented AGV. CMES – Comput. Model. Eng. Sci. **134**(3), 2073–2091 (2023). https://doi.org/10.32604/cmes.2022.021433
4. Udhayakumar, P., Kumanan, S.: Task scheduling of AGV in FMS using non-traditional optimization techniques. Int. J. Simul. Model. **9**(1), 28–39 (2010). https://doi.org/10.2507/IJSIMM09(1)3.139
5. Azangoo, M., Taherkordi, A., Blech, J.O., Vyatkin, V.: Digital twin-assisted controlling of AGVs in flexible manufacturing environments. In: IEEE International Symposium on Industrial Electronics. Institute of Electrical and Electronics Engineers Inc. (2021). https://doi.org/10.1109/ISIE45552.2021.9576361
6. Kusiak, A.: Flexible manufacturing systems a structural approach. Int. J. Prod. Res. **23**(6), 1057–1073 (1985). https://doi.org/10.1080/00207548508904765
7. Milberg. I.J., Lutz, D.-I.P.: Integration of autonomous mobile robots into the industrial production environment by (1987)
8. Lin, C., Hsieh, P., Wang, S.H., Chiang, H.H.: Automatic data acquisition of indoor environment using the autonomous mobile robot Taiwan. In: 2021 IEEE International Conference on Consumer Electronics-Taiwan (ICCE-TW), pp. 1–2 (2021)
9. Morris, C., Chauhan, V.: Design of a Low-Cost Autonomous Mobile Robot for Outdoor Applications. In: ASME International Mechanical Engineering Congress and Exposition, vol. 86670 (2022)
10. Ganesharajah, T., Hall, N.G., Sriskandarajah, C.: Design and operational issues in AGV-served manufacturing systems (1998)
11. Fragapane, G., Ivanov, D., Peron, M., Sgarbossa, F., Strandhagen, J.O.: Increasing flexibility and productivity in Industry 4.0 production networks with autonomous mobile robots and smart intralogistics. Ann. Oper. Res. **308**(1–2), 125–143 (2022). https://doi.org/10.1007/s10479-020-03526-7
12. Draganjac, I., Petrović, T., Miklić, D., Kovačić, Z., Oršulić, J.: Highly-scalable traffic management of autonomous industrial transportation systems. Robot Comput. Integr. Manuf. **63** (2020). https://doi.org/10.1016/j.rcim.2019.101915
13. Čech, M., et al.: Autonomous mobile robot technology for supplying assembly lines in the automotive industry. Acta Logistica **7**(2), 103–109 (2020). https://doi.org/10.22306/al.v7i2.164

14. Weng, B., Chen, H., Zhang, W.: On the convergence of multi-robot constrained navigation: a parametric control lyapunov function approach. In: 2022 International Conference on Robotics and Automation (ICRA), pp. 4972–4978 (2022)
15. Du, N., Hu, H.: A robust prevention method for automated manufacturing systems with unreliable resources using petri nets. IEEE Access **6**, 78598–78608 (2018). https://doi.org/10.1109/ACCESS.2018.2885116
16. Du, N., Hu, H.: Robust deadlock detection and control of automated manufacturing systems with multiple unreliable resources using petri nets. IEEE Trans. Autom. Sci. Eng. **18**(4), 1790–1802 (2021). https://doi.org/10.1109/TASE.2020.3019684
17. Čapkovič, F.: Dealing with deadlocks in industrial multi agent systems. Future Internet **15**(3) (2023). https://doi.org/10.3390/fi15030107
18. Mogul, J.C., Ramakrishnan, K.K.: Eliminating receive livelock in an interrupt-driven kernel. ACM Trans. Comput. Syst. **15**(3), 217–252 (1997)
19. Stetsyura, G.G.: Fast decentralized algorithms for resolving conflicts and deadlocks in resource allocation in data processing and control systems. Autom. Remote. Control. **71**(4), 708–717 (2010). https://doi.org/10.1134/S0005117910040119
20. Araújo, A., Portugal, D., Couceiro, M.S., Rocha, R.P.: Integrating arduino-based educational mobile robots in ROS. J. Intell. Robot. Syst. Theory Appl. **77**(2), 281–298 (2015). https://doi.org/10.1007/s10846-013-0007-4
21. Wasisto, I., Istiqomah, N., Trisnawan, I.K.N., Jati, A.N.: Implementation of mobile sensor navigation system based on adaptive monte carlo localization. In: 2019 International Conference on Computer, Control, Informatics and its Applications (IC3INA), pp. 187–192. IEEE 2019
22. Chung, M.-A., Lin, C.-W.: An improved localization of mobile robotic system based on AMCL algorithm. IEEE Sens. J. **22**(1), 900–908 (2022). https://doi.org/10.1109/JSEN.2021.3126605
23. Chandak, A., Bodhale, R., Burad, R.: Optimal shortest path using HAS, A star and Dijkstra algorithm. Imperial J. Interdisc. Res. **2** (2016)
24. Srivastava, R., Singh, P.: A novel optimized travel planner. J. Inform. Electr. Electron. Eng. (JIEEE) **3**(1), 1–17 (2022). https://doi.org/10.54060/JIEEE/003.01.007
25. Widiarta, G.H., Romdlony, M.Z., Rosa, M.R., Trilaksono, B.R.: Control lyapunov — barrier function implementation for mobile robot model with hardware in the loop. In: 2021 International Conference on Mechatronics, Robotics and Systems Engineering (MoRSE), pp. 1–6, Bali (2021)
26. MacFarlane, A.G.I, Karcanias, K., Lumelsky, V.J., Stepanov, A.A., Stepanov, A.A.: The analysis of sampleddata control systems with a with periodic feedback. IRE Trans.4urornor. Contr. YOI. J. Conrr. **32**, 33–74 (1980)
27. Joshi, V.A., Banavar, R.N., Hippalgaonkar, R.: Design and analysis of a spherical mobile robot. Mech. Mach. Theory **45**(2), 130–136 (2010). https://doi.org/10.1016/j.mechmachtheory.2009.04.003
28. Agarwal, D., Bharti, P.: A-review-on-comparative-analysis-of-path-planning-and-collision-avoidance-algorithms. Int. J. Mech. Aerosp. Ind. Mechatronic Manuf. Eng. **12**, 608–624 (2018)
29. Reis, M.F., Aguiar, A.P., Tabuada, P.: Control barrier function-based quadratic programs introduce undesirable asymptotically stable equilibria. IEEE Control. Syst. Lett. **5**(2), 731–736 (2021). https://doi.org/10.1109/LCSYS.2020.3004797
30. Romdlony, M.Z., Jayawardhana, B.: Stabilization with guaranteed safety using control lyapunov-barrier function. In: Automatica, pp. 39–47. Elsevier Ltd. (2016). https://doi.org/10.1016/j.automatica.2015.12.011

31. Yu,Y., Wang, M.: The path planning algorithm research based on cost field for autonomous vehicles. In: Proceedings of the 2012 4th International Conference on Intelligent Human-Machine Systems and Cybernetics, pp. 38–41. IHMSC 2012 (2012). https://doi.org/10.1109/IHMSC.2012.105
32. Ammar, A., Bennaceur, H., Châari, I., Koubâa, A., Alajlan, M.: Relaxed dijkstra and a* with linear complexity for robot path planning problems in large-scale grid environments. Soft. Comput. **20**(10), 4149–4171 (2016). https://doi.org/10.1007/s00500-015-1750-1
33. Gonzilez De Mendivil, J.R., Farifia, F., Alastruey, C.F., Garitagoitia, R.: A safe distributed deadlock resolution algorithm (1998)
34. Lu, W., Yang, Y., Wang, L., Xing, W., Che, X.: A novel priority-based deadlock detection and resolution algorithm in mobile agent systems. In: Proceedings - DMS 2016: 22nd International Conference on Distributed Multimedia Systems. Knowledge Systems Institute Graduate School, pp. 61–68 (2016). https://doi.org/10.18293/DMS2016-014
35. Lomet, D.B.: Subsystems of processes with deadlock avoidance. IEEE Trans. Softw. Eng. **3**, 297–304 (1980)
36. Estefo, P., Simmonds, J., Robbes, R., Fabry, J.: The robot operating system: package reuse and community dynamics. J. Syst. Softw.Softw. **151**, 226–242 (2019)

Investigating Interdisciplinary Research Impact: A Framework for Integrating Linguistic and Citation Information

Masanao Ochi[1](✉), Masanori Shiro[2], Jun'ichiro Mori[3], and Ichiro Sakata[3]

[1] Faculty of Science and Technology, Oita University,
700 Dannoharu, Oita City, Japan
`masanao.oochi@gmail.com`
[2] Human Informatics and Interaction Research Institute,
National Institute of Advanced Industrial Science and Technology,
Umezono 1-1-1 Central2, Tsukuba, Ibaraki, Japan
[3] Department of Technology Management for Innovation, Graduate School of Engineering, The University of Tokyo, Hongo 7-3-1, Bunkyo, Tokyo, Japan

Abstract. Interdisciplinary research has become a crucial part of modern scientific progress, but predicting the future impact of studies across multiple research fields remains a significant challenge. Several studies have shown that integrating linguistic information and citation networks can effectively predict the citation counts of research papers. This paper proposes a model that fuses linguistic information and citation networks, which applies even to large-scale data from the academic literature, to classify high-impact documents. Our model applies to large-scale academic datasets, enabling the classification of citation counts in broader academic fields such as interdisciplinary areas. In our experiments, we used a large dataset of scholarly papers in multiple research fields to evaluate the classification accuracy of the citation counts using our model. The experimental results show that our classification accuracy improves by 0.022 points compared to models using only citation networks. This result indicates that our model is valid across multiple research fields. Furthermore, we clustered the embeddings we obtained and demonstrated that highly cited papers are biased at the boundaries of the clusters. This bias indicates that highly cited papers are more interdisciplinary than less cited ones.

Keywords: citation analysis · scientific impact · graph neural network · BERT · Transformer

1 Introduction

In recent years, the importance of interdisciplinary research has increased as scientific progress increasingly relies on integrating insights from diverse fields.

Despite the growing recognition of its significance, predicting the long-term impact of interdisciplinary research remains challenging due to the complex nature of cross-disciplinary interactions and the limitations of traditional bibliometric indicators. Predicting future research impact, especially in interdisciplinary contexts, is crucial to allocate resources and foster innovation efficiently. However, existing methods often rely on specialized indicators designed for specific domains, which can overlook the broader context of interdisciplinary knowledge exchange.

Previous studies on predicting the impact of the scholarly literature have used domain-specific features [1–3,5,9,13–15] or citation-based link prediction in custom networks [12,17,19]. Although these approaches have provided insights, they fail to capture the full complexity of how research influences various disciplines over time. Advances in deep learning, particularly with models such as the Transformer [16], have opened new possibilities for integrating different data types and developing more general-purpose models. Initially developed for natural language processing, the Transformer model has been successfully applied to areas such as image processing and network science, demonstrating its versatility in modeling diverse data types.

In this paper, we propose a model to predict the impact of interdisciplinary research. Since citation networks are known to be important in predicting citation counts, we employ a Transformer-based framework that combines language models and citation networks [11]. Specifically, because applying Graph Neural Networks to large-scale networks is a significant challenge, we propose a model that incorporates NAGphormer [6], which can handle large-scale Graph Neural Networks. Experimental results using the Web of Science suggest that the NAGphormer model can classify the top cited papers published in 2016 with an accuracy of 0.637, running on 68,879,825 papers in 6.4 h using a single NVIDIA A100 GPU, with the loss decreasing each epoch.

Next, we combined ten small datasets [11] into one dataset and verified whether the model could classify the most cited papers in interdisciplinary fields. The proposed fusion model achieved a classification accuracy of 0.801, 0.022 points higher than the base NAGphormer model. This result suggests that incorporating linguistic information can improve the classification accuracy of highly cited papers in interdisciplinary areas. Furthermore, we found that papers that became highly cited were located farther from the centroids of the clusters. Based on these results, our proposed method is feasible for large-scale data from the academic literature. It effectively identifies groups of articles with a high impact in interdisciplinary fields.

The contributions of this paper are as follows.

- We demonstrated a feasible predictive model for large-scale datasets of academic literature of more than 68 million papers.
- We showed that papers with the highest citation counts are biased toward interdisciplinary areas in the embedding results.
- We suggest that applying the proposed model can detect high-impact research in interdisciplinary areas.

The remainder of this paper is structured as follows. Section 2 reviews related work on impact prediction and citation networks. Section 3 details the architecture of our proposed model. Section 4 presents the experimental setup and results. Section 5 discusses our findings, and Sect. 6 concludes with the contributions of this research and future directions.

2 Related Works

This section categorizes and summarizes recent research reports on Transformer models related to our study. We also clarify the position of our research, which focuses on detecting high-impact studies in interdisciplinary fields.

The Transformer model [16], an Encoder-Decoder model that uses Attention, allows for large-scale training due to its low computational complexity and ability to perform parallel computations. Applications of the Transformer model to academic literature data are also progressing. First, there is the SciBERT model [4], based on the BERT model trained on text data from the academic literature. SciBERT focuses on obtaining general embeddings of massive academic datasets at the word level. In contrast, the SPECTER model [7] attempts to obtain embeddings at the paper level rather than the word level. SPECTER acquires paper-level embeddings using pairs of papers with citation relationships as positive examples.

Some studies predicted future citation counts for articles [3,5,13,15]. Among them, Stegehuis et al. and Cao et al. predicted citation counts in the distant future by considering the citations within 1 to 3 years after publication. In contrast, Sasaki et al. directly indicated the number of citations three years after publication [13]. This study also evaluates a task that predicts citation counts five years after publication, similar to the work of Sasaki et al. Previous efforts to predict metrics have created various features and used them as input to models.

Our previous research showed that citation networks are more important than linguistic information to predict the impact of papers [10]. We proposed a model that integrates Transformers for citation networks and linguistic information, and we verified its effectiveness on small data sets [11]. However, we have not sufficiently tested its applicability to larger datasets because applying GNNs to large networks is a significant challenge. The NAGphormer model has emerged [6]. The NAGphormer model makes saving information about node interactions possible by aggregating information according to the number of hops from the target node. In this study, by incorporating this model into our previously verified fusion framework, we propose a model that can run on a larger scale than before.

3 Method

We have used a framework that can learn end2end by merging linguistic and citation information among the various data possessed by the academic literature using Transformer [11]. In this paper, we introduce NAGphormer [6], which applies to large-scale graphs, into this framework and propose a new model applicable to impact prediction in large-scale interdisciplinary fields. In this section, we first describe the NAGphormer that we introduce in our model, followed by a model in which the Transformer fuses linguistic information and citation information.

3.1 NAGphormer

Existing Graph transformers treat nodes as independent tokens and build a single sequence of all node tokens to train the transformer model. The NAGphormer, on the other hand, aggregates neighborhood features from different hop counts by Hop2Token into different representations for each node, thus generating a sequence of token vectors as a single input. Therefore, NAGphormer can be trained in a mini-batch fashion and handle large graphs. Furthermore, compared to graph convolutional networks, one of the advanced graph neural networks (GNNs), NAGphormer has been mathematically shown to learn more informative node representations from multi-hop neighborhoods. A model of NAGphormer modified for our task is shown in Fig. 1.

3.2 Integration of Linguistic and Citation Information

In this paper, we construct a model that learns end-to-end by fusing language information and citation information from large-scale academic literature data to detect research that will have a high impact in interdisciplinary fields in the future. Therefore, we propose this method as shown in Fig. 2. We employ a framework that fuses Transformer models for network and language processing [11], connecting SciBERT [4] as the Transformer for language information and NAGphormer as the Transformer for citation information.

We first select a target paper in Fig. 2. The target paper is a node randomly sampled from the citation network. Our proposed method learns a classification task that predicts whether the target paper will likely become a top-cited paper.

We input two pieces of information—the title and abstract of each paper—into a pre-trained SciBERT, as shown in Fig. 2. We tokenize each and input them as a sequence of words, just like in BERT.

Centering on the target paper, we use Hop2Token to tokenize based on the distance from the target paper. We assign a vector to each token corresponding to its distance from the target paper and input these tokens into NAGphormer. These tokens enable efficient learning of the classification task. Finally, we perform a classification task predicting whether the target paper will likely become top-cited.

4 Experiment and Results

This section outlines the experiments conducted to evaluate the proposed model. We describe the datasets used in the experiments and present the results, comparing the proposed method with baseline models. Our focus is on two key

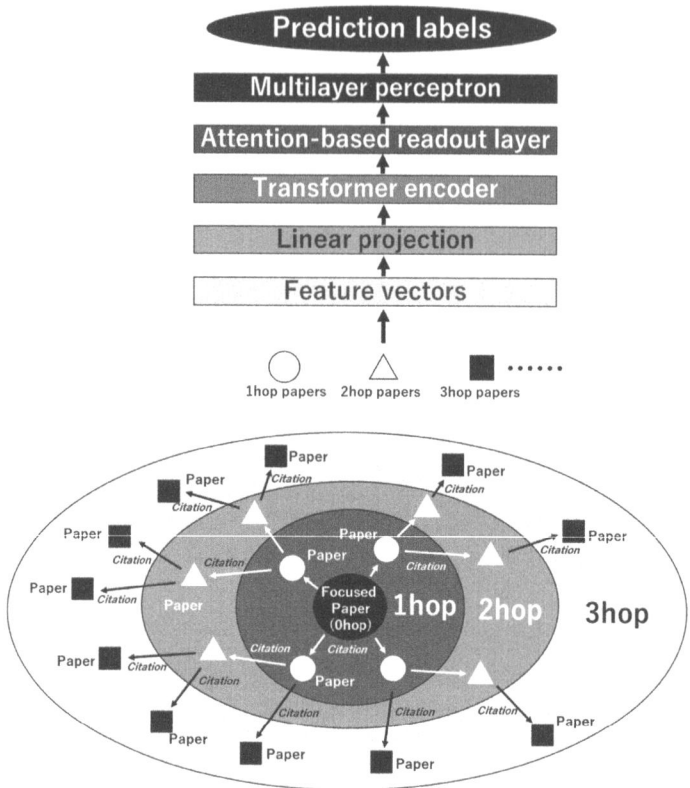

Fig. 1. NAGphormer (Modified from the original figure to fit our task [6])

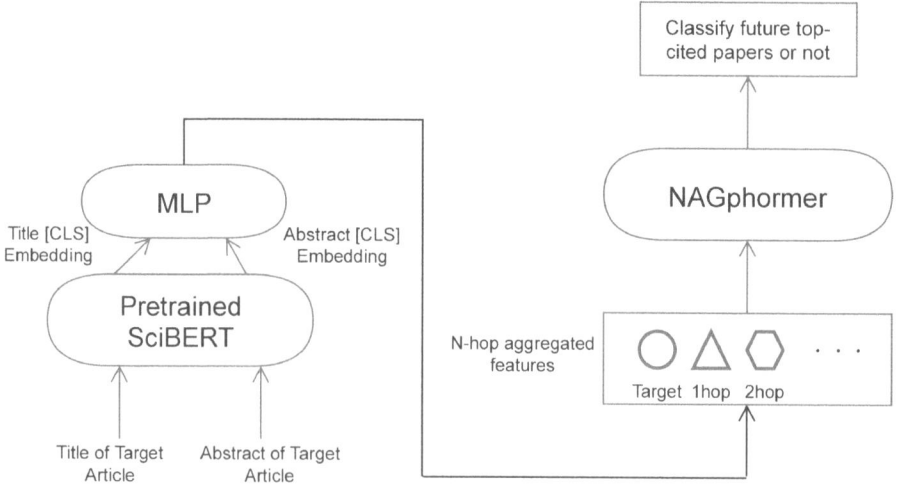

Fig. 2. Proposed model(NAGphormer+Pretrained SciBERT model)

aspects: the model's scalability and its ability to identify high-impact papers in interdisciplinary fields.

4.1 Scientific Literature Dataset

We used two datasets derived from the Web of Science[1], one of the largest academic literature databases, to evaluate our proposed model:

1. **Large-scale dataset**: This dataset contains 68,879,825 papers published in 2016. We use this dataset to test the scalability and efficiency of the model in handling massive academic data.
2. **Small-scale integrated dataset**: This combines ten smaller datasets covering various research areas, drawn from our previous study [11]. We use this dataset to verify the model's ability to classify highly cited papers in interdisciplinary fields.

Each paper in the dataset includes citation network information (i.e., the relationships between papers) and linguistic information from the titles and abstracts. The large dataset is particularly useful for testing the model's performance in a more generalized academic setting. In contrast, the small integrated dataset provides a controlled environment to examine the model's specific capacity for interdisciplinary research.

4.2 Verification of Viability on Large Data Sets

To test the scalability of the NAGphormer model on large datasets, we applied it to classify the top 20% most-cited papers published in 2016 based on citation counts five years after publication. This experiment assessed the model's performance when handling a massive academic literature dataset.

We experimented using a single NVIDIA A100 GPU, which took approximately 6.4 h to run. Figure 3 shows that the loss decreased steadily over each epoch, indicating successful model convergence. The final classification accuracy was 0.637, demonstrating that the NAGphormer model can handle large-scale datasets efficiently while maintaining reasonable classification performance.

This result confirms that the proposed method can effectively process massive citation networks and predict citation impact.

---

[1] Web of Science https://www.webofknowledge.com.

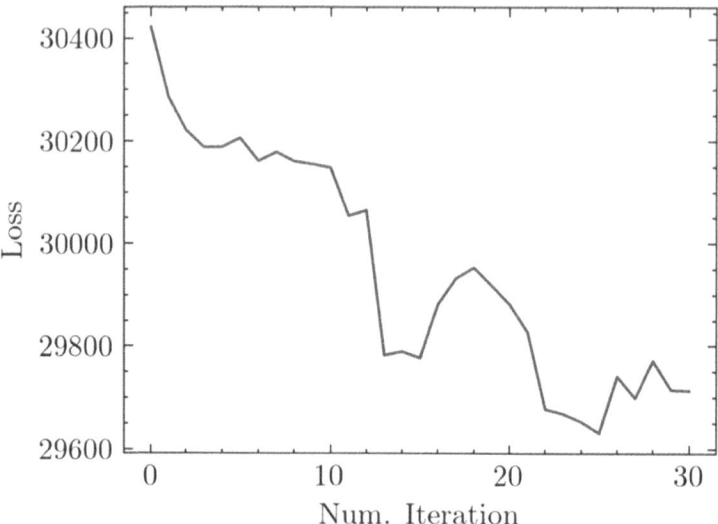

Fig. 3. Loss transition for each epoch.

4.3 Classification Results

We then evaluated the model's ability to classify highly cited papers in interdisciplinary fields using the small-scale integrated dataset. This experiment focused on whether the fusion of linguistic and citation information could improve classification accuracy in detecting high-impact research across multiple disciplines.

We compared the performance of three models:

1. **NAGphormer**: A model based solely on citation network information.
2. **SciBERT**: A model based solely on linguistic information from the papers' titles and abstracts.
3. **NAGphormer + SciBERT**: The proposed model combines citation and linguistic information.

As shown in Table 1, the proposed fusion model achieved the highest classification accuracy, scoring 0.801. This result is 0.022 points higher than the base NAGphormer model (0.779) and 0.011 points higher than SciBERT (0.790). These results demonstrate that integrating linguistic data with citation networks can significantly improve the prediction of high-impact research in interdisciplinary areas.

The improved performance is likely due to the complementary nature of citation and linguistic information. While citation networks capture relationships between papers, linguistic data provide insights into the content and novelty of the research, which can be particularly important in interdisciplinary fields where traditional citation patterns may not fully reflect a paper's impact.

Table 1. Classification Results.

	Accuracy
NAGphormer	0.779
SciBERT	0.790
NAG+Sci	**0.801**

5 Discussion

This section presents the findings of our experiments, beginning with the model's performance and then an analysis of the clustering metrics, which are depicted in Fig. 4. The figure illustrates the measurements used to evaluate the relationship between interdisciplinarity and citation impact, precisely the Distance from the Nearest Cluster Centroid (Table 2, Fig. 5) and the Averaged Distance from the Two Closest Clusters (Table 3, Fig. 6). These metrics help quantify how top-cited interdisciplinary papers are positioned within and across research clusters.

5.1 Model Performance and Contributions

The proposed model, which combines linguistic information via SciBERT and citation network data through NAGphormer, achieved a classification accuracy of 0.801, as shown in Table 1. Our proposed model outperformed models using only citation networks (0.779) or linguistic data (0.790), emphasizing the importance of integrating both data sources.

The success of the fusion model highlights the complementary nature of citation networks and linguistic content. Citation networks reveal the structural positioning of papers, indicating how knowledge flows between research works, while linguistic data capture the novelty and relevance of the paper's content. Integrating both enables the model to more effectively identify high-impact papers, particularly in interdisciplinary fields where traditional citation-based approaches may miss broader contextual insights. This approach efficiently recognizes interdisciplinary papers that might not conform to conventional citation patterns but are influential across multiple research domains.

5.2 Distance from Nearest Cluster Centroid

As shown in Fig. 4, we measured the Distance from the Nearest Cluster Centroid to examine the correlation between a paper's interdisciplinarity and its citation impact. This metric assesses how far a paper is from the cluster centroid it primarily belongs. The results, detailed in Table 2 and Fig. 5, show that top-cited papers are, on average, located farther from their nearest cluster centroid (0.1423) compared to non-top-cited papers (0.1351).

This greater distance suggests that highly cited papers are more interdisciplinary and positioned at the periphery of established research clusters. These

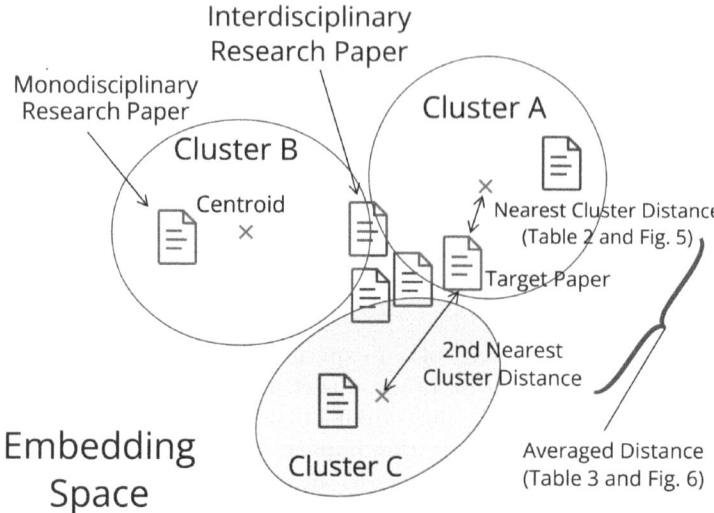

Fig. 4. Embedding concept of interdisciplinary and monodisciplinary research.

results are consistent with the analysis of Alfredo et al. [18]. Papers at these boundaries are more likely to integrate insights from multiple fields, positioning them as connectors of diverse research areas. This result indicates their relevance to broader academic communities, enhancing their potential for higher citation impact. Thus, measuring the distance from the nearest cluster centroid provides a quantifiable means of assessing the interdisciplinary nature of impactful research.

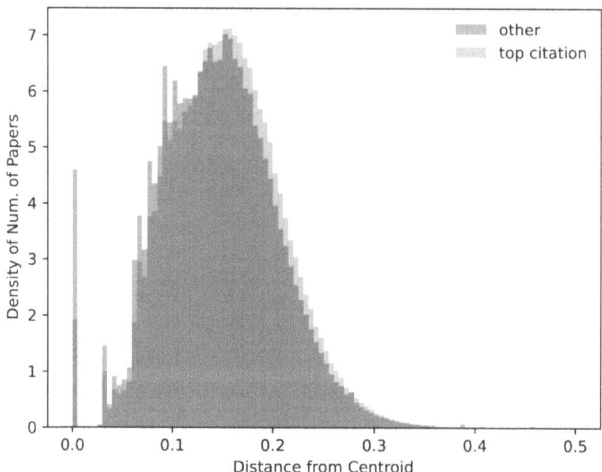

Fig. 5. Comparison distance from each cluster centroid between top citation papers and others.

Table 2. Distance from nearest cluster centroid.

	Num. Papers	Distance
top-citation	693,301	**0.1423** ± 0.0560
not-top	3509,774	0.1351 ± 0.0553

5.3 Averaged Distance from the Two Closest Clusters

The Averaged Distance from the Two Closest Clusters, illustrated in Fig. 4, provides additional insight into how interdisciplinary research spans multiple clusters. This metric measures the average distance between a paper and its two closest clusters, indicating how papers bridge different domains. As shown in Table 3 and Fig. 6, top-cited papers exhibit a more considerable averaged distance (0.2382) compared to non-top-cited papers (0.2284).

This increased distance from multiple clusters further reinforces the idea that highly cited papers operate at the intersection of different disciplines. By being positioned between two or more clusters, these papers draw on broader knowledge areas, contributing to their interdisciplinary nature. The ability to bridge distinct research domains makes these papers more likely to be recognized and cited by scholars from various fields. This insight is crucial for understanding how interdisciplinary research drives scientific innovation, as it demonstrates the value of connecting previously siloed areas of inquiry.

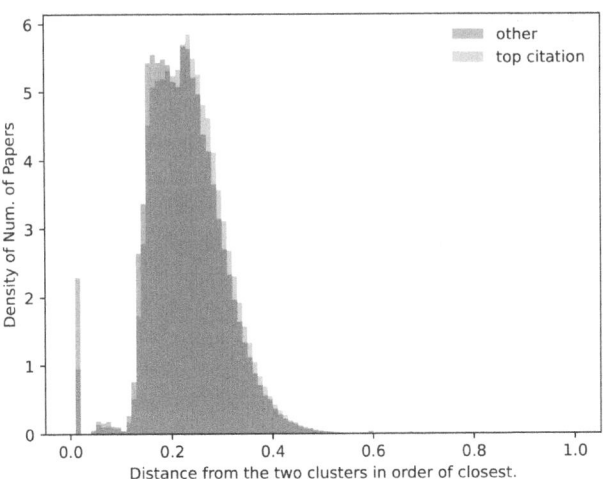

Fig. 6. Comparison distance from the two clusters in order of closest between top citation papers and others.

Table 3. Distance from the two clusters in order of closest.

	Distance
top-citation	**0.2382** ± 0.0717
not-top	0.2284 ± 0.0775

6 Conclusion

In this paper, we proposed a model that fuses linguistic information and citation networks to classify high-impact papers, applicable even to large-scale academic literature data. We validated the proposed model as follows. First, we verified its feasibility on a large dataset. NAGphormer is one of the graph neural network models for large graphs, and it is our base model. It was necessary to confirm whether this model could function with a massive scholarly dataset. Using an extensive academic literature dataset of 68,879,825 records extracted from the Web of Science, we observed that the loss decreased with each epoch, training required 6.4 h on an NVIDIA A100, and the classification accuracy was 0.637. This result indicates that the proposed model can operate effectively on large-scale academic literature datasets.

Next, we combined ten small datasets into one dataset and verified whether the model could classify papers in the top citation group within interdisciplinary fields. The classification accuracy of the proposed fusion model was 0.801, 0.022 points higher than the base NAGphormer model. This result suggests that incorporating linguistic information can improve the classification accuracy of highly cited papers in interdisciplinary areas. Additionally, we found that papers that belonged to the top citation group were located farther from the centroids of the clusters. Based on these results, the proposed method is feasible for large-scale academic literature data and effectively identifies groups of papers with high impact in interdisciplinary fields.

Several limitations exist to the proposed method and the evaluation results obtained this time. First, we did not directly apply the proposed method to large-scale academic literature datasets. We aim to verify the effectiveness of the proposed method on even larger datasets. Since the proposed model is end-to-end, it can rapidly increase the number of tasks. We intend to test the effectiveness of the proposed method in multiple tasks, not just citation count classification. Furthermore, because academic literature data contains information on figures and tables, we would like to evaluate the integration of methods such as ViT[8].

Acknowledgements. This article is based on results obtained from a project, JPNP20006, commissioned by the New Energy and Industrial Technology Development Organization (NEDO) and supported by JSPS KAKENHI Grant Number JP21K17860, JP21K12068 and 24K05080. The funders had no role in the study design, data collection and analysis, publication decision, or manuscript preparation.

References

1. Acuna, D.E., Allesina, S., Kording, K.P.: Predicting scientific success. Nature **489**(7415), 201–202 (2012). https://doi.org/10.1038/489201a
2. Ayaz, S., Masood, N., Islam, M.A.: Predicting scientific impact based on h-index. Scientometrics **114**(3), 993–1010 (2018)
3. Bai, X., Zhang, F., Lee, I.: Predicting the citations of scholarly paper. J. Informet. **13**(1), 407–418 (2019)
4. Beltagy, I., Lo, K., Cohan, A.: Scibert: pretrained language model for scientific text. In: EMNLP (2019)
5. Cao, X., Chen, Y., Liu, K.R.: A data analytic approach to quantifying scientific impact. J. Informet. **10**(2), 471–484 (2016)
6. Chen, J., Gao, K., Li, G., He, K.: Nagphormer: a tokenized graph transformer for node classification in large graphs. In: Proceedings of the International Conference on Learning Representations (2023)
7. Cohan, A., Feldman, S., Beltagy, I., Downey, D., Weld, D.S.: Specter: document-level representation learning using citation-informed transformers. In: ACL (2020)
8. Dosovitskiy, A., et al.: An image is worth 16×16 words: transformers for image recognition at scale. In: 9th International Conference on Learning Representations, ICLR 2021, Virtual Event, Austria, 3–7 May 2021. OpenReview.net (2021). https://openreview.net/forum?id=YicbFdNTTy
9. Miró, Ò., et al.: Analysis of h-index and other bibliometric markers of productivity and repercussion of a selected sample of worldwide emergency medicine researchers. Emerg. Med. J. **34**(3), 175–181 (2017)
10. Ochi, M., Shiro, M., Mori, J., Sakata, I.: Which is more helpful in finding scientific papers to be top-cited in the future: content or citations? case analysis in the field of solar cells 2009. In: Mayo, F.J.D., Marchiori, M., Filipe, J. (eds.) Proceedings of the 17th International Conference on Web Information Systems and Technologies, WEBIST 2021, 26–28 October 2021, pp. 360–364. SCITEPRESS (2021). https://doi.org/10.5220/0010689100003058
11. Ochi, M., Shiro, M., Mori, J., Sakata, I.: Integrating linguistic and citation information with transformer for predicting top-cited papers. In: Marchiori, M., Mayo, F.J.D., Filipe, J. (eds.) Web Information Systems and Technologies - 18th International Conference, WEBIST 2022, Valletta, Malta, 25–27 October 2022, Revised Selected Papers. Lecture Notes in Business Information Processing, vol. 494, pp. 121–141. Springer, Heidelberg (2022). https://doi.org/10.1007/978-3-031-43088-6_7
12. Park, I., Yoon, B.: Technological opportunity discovery for technological convergence based on the prediction of technology knowledge flow in a citation network. J. Informet. **12**(4), 1199–1222 (2018)
13. Sasaki, H., Hara, T., Sakata, I.: Identifying emerging research related to solar cells field using a machine learning approach. J. Sustain. Dev. Energy Water Environ. Syst. **4**, 418–429 (2016). https://doi.org/10.13044/j.sdewes.2016.04.0032
14. Schreiber, M.: How relevant is the predictive power of the h-index? A case study of the time-dependent hirsch index. J. Informet. **7**(2), 325–329 (2013)
15. Stegehuis, C., Litvak, N., Waltman, L.: Predicting the long-term citation impact of recent publications. J. Informetr. **9** (2015). https://doi.org/10.1016/j.joi.2015.06.005

16. Vaswani, A., et al.: Attention is all you need. In: Guyon, I., et al. (eds.) Advances in Neural Information Processing Systems, vol. 30. Curran Associates, Inc. (2017). https://proceedings.neurips.cc/paper/2017/file/3f5ee243547dee91fbd053c1c4a845aa-Paper.pdf
17. Yan, E., Guns, R.: Predicting and recommending collaborations: an author-, institution-, and country-level analysis. J. Informet. **8**(2), 295–309 (2014)
18. Yegros-Yegros, A., Rafols, I., D'Este, P.: Does interdisciplinary research lead to higher citation impact? The different effect of proximal and distal interdisciplinarity. PLOS ONE **10**(8), 1–21 (2015). https://doi.org/10.1371/journal.pone.0135095
19. Yi, Z., Ximeng, W., Guangquan, Z., Jie, L.: Predicting the dynamics of scientific activities: a diffusion-based network analytic methodology. Proc. Assoc. Inf. Sci. Technol. **55**(1), 598–607 (2018)

Identification and Control of Industrial Hydraulic Actuators with a Hammerstein-Wiener Model Approach

Nur Husnina Mohamad Ali[1], Hazriq Izzuan Jaafar[1,2(✉)], Rozaimi Ghazali[1,2], Muhamad Fadli Ghani[3], Chong Chee Soon[4], and Zulfatman Has[5]

[1] Fakulti Teknologi dan Kejuruteraan Elektrik, Universiti Teknikal Malaysia Melaka, Melaka, Malaysia
hazriq@utem.edu.my
[2] Center for Robotics and Industrial Automation, Universiti Teknikal Malaysia Melaka, Melaka, Malaysia
[3] Department of Control and Mechatronics Engineering, Faculty of Electrical Engineering, Universiti Teknologi Malaysia, Johor, Malaysia
[4] Department of Engineering and Built Environment, Tunku Abdul Rahman University of Management and Technology (TAR UMT), Penang Branch Campus, Pulau Pinang, Malaysia
[5] Electrical Engineering Department, University of Muhammadiyah Malang, Malang, Indonesia

Abstract. In motion control applications, the precise regulation of Industrial Hydraulic Actuators (IHA) is essential for properly ascertaining the position of the actuator rod. This research used a Hammerstein-Wiener (HW) model to characterize the system, integrating a dead zone to account for nonlinearities and using an open-loop methodology. A grey-box modelling technique is used for estimating the discrete-time model of the system and identifying its parameters. The System Identification toolbox in MATLAB functions as the principal instrument for model estimation, using input-output data acquired via experimental procedures. The validation of the input-output data demonstrated a promising best-fit percentage of 88.21%, indicating a robust connection between the HW model and the real system. To assess the trajectory tracking efficacy of the system, Proportional-Integral-Derivative (PID) and Fractional Order PID (FOPID) controllers developed and evaluated using both simulation and experimental data based on a multi-sinusoidal trajectory. The findings indicate that the FOPID controller surpasses the traditional PID controller, producing reduced Mean Square Error (MSE) and Root Mean Square Error (RMSE) values. In simulations, the FOPID control technique enhanced MSE by 32.8% and RMSE by 18.0% relative to the PID controller, although the experimental findings indicated improvements of 31.9% and 17.5%, respectively. The results highlight the superior efficacy of the FOPID controller in improving trajectory tracking performance owing to its greater complexity and flexibility. Future research may concentrate on developing more sophisticated controllers, including intelligent or nonlinear variants, to enhance the tracking precision of the IHA system.

Keywords: Hammerstein-Wiener Model · Dead-Zone · PID · FOPID

1 Introduction

Industrial Hydraulic Actuators (IHA) are crucial in manufacturing and industrial sectors such as aircraft, rolling mills, and mobile lifting systems [1–3]. They provide exact control of high torque and fast reaction, and their growing use in these domains is attributed to the desire for improved force and position regulation. IHA systems combine the benefits of electrical and hydraulic components.

IHA are favoured in advanced technological applications due to their distinctive advantages, including the ability to generate high forces at rapid speeds, robust rigidity, and fast response times [4–6]. Despite these benefits, two significant challenges are associated with IHA systems: the high nonlinearity of their dynamics and the complexity of their mathematical modelling. These nonlinearities stem from various factors including hysteresis, actuator friction, fluid compressibility, and the inherent nonlinear characteristics of hydraulic valves and actuator [5]. Such nonlinearities may result in unexpected system behaviour, complicating the control and predictability of the system's response.

Given these complexities, developing an accurate system model is crucial before designing any control strategy for position tracking. Understanding the system's dynamics through system identification methods allows for the creation of more precise models, which are essential for the design and implementation of effective control strategies.

1.1 System Identification (Linear and Nonlinear Models)

System identification is the creation of mathematical models for dynamic systems based on measured data. It is crucial in control and modelling applications, where systems can exhibit linear or nonlinear behavior. The approach varies depending on the system's nature, with linear system identification aiming to fit a linear model to the input-output data. Linear systems are characterized by the principle of superposition, where the output is directly proportional to the input. This proportionality simplifies the modelling process and makes it easier to develop predictive models for system behavior. Common types of linear models include transfer functions, state-space representations, and autoregressive models with exogenous inputs (ARX) [5, 6]. However, one of the significant limitations for linear models is their inability to accurately represent systems that exhibit nonlinearities. In many practical applications, systems may involve complex dynamics such as hysteresis, friction, or saturation, which linear models fail to capture, resulting in inaccuracies during the system identification process.

On the other hand, nonlinear system identification is necessary when the system demonstrates behaviors that cannot be approximated by linear models. Such behaviors often include phenomena like saturation, dead zone [7, 8], hysteresis, and other nonlinear characteristics that require more sophisticated modelling techniques. Nonlinear system identification involves fitting models that are capable of capturing these dynamics. Some widely used nonlinear models include the Hammerstein model, Wiener model and Hammerstein-Wiener (HW) model [9–12]. These models are known as block-oriented models that combine both linear and nonlinear elements to effectively represent the complex behavior of the system [8]. The HW model, for instance, incorporates both nonlinear and linear components, making it particularly suitable for systems that display both types of behavior. This model is often applied in the context of discrete-time (DT) systems [7–12], as it allows for detailed representation of nonlinear phenomena while maintaining computational tractability. Given the highly nonlinear nature of the IHA system, the use of a nonlinear modelling approach is crucial for achieving an accurate representation of the system's dynamics.

Modelling the IHA system with a nonlinear model is particularly important for facilitating the subsequent design of control strategies. By accurately capturing the system's inherent nonlinearities during the identification process, researchers are better equipped to develop controllers that can be effectively implemented in real-world hardware. This approach not only improves the precision and robustness of the control system but also ensures that the identified model can handle the complex dynamics of the IHA, thereby enhancing overall system performance.

1.2 Linear Control Strategy

The linear control method is one of the most prevalent and essential techniques in control systems, mostly owing to its simplicity and ease of implementation. The premise is that the system's behavior may be characterized by a series of linear differential equations, facilitating a systematic and mathematically rigorous framework for controller development. Despite the increasing complexity of modern systems, linear control strategies continue to be extensively used in industry today, particularly for systems that can be well-approximated by linear models.

The strength of linear control methods lies in their reliability and effectiveness across a broad range of practical applications. Linear controllers are often easier to design, tune, and offer a level of predictability that is essential for many industrial processes. However, in cases where the system exhibits significant nonlinear behavior, linear control strategies may need to be supplemented or replaced by more advanced nonlinear control techniques to achieve the desired level of performance.

In the context of IHA systems, traditional linear control methods such as the Ziegler-Nichols tuning method [13, 14] are frequently used for position control. This method provides a straightforward approach for tuning PID controllers, which are widely regarded as an effective control technique for trajectory tracking in IHA systems [14–16]. The PID control technique is particularly popular due to its simplicity, ease of implementation, and ability to provide satisfactory performance in many linear control applications.

Beyond conventional PID controllers, more advanced methods such as the FOPID controller have gained attention as an extension of the conventional PID framework. Unlike standard PID controllers, where the differentiation and integration terms are limited to integer orders [16, 17], FOPID generalizes these operations to fractional (non-integer) orders. This increased flexibility enables more accurate tuning of the controller's dynamics, resulting in improved control performance in systems necessitating finer adjustments. The fractional order framework enables better control over memory effects and system dynamics, making FOPID a more effective option in complex and nonlinear systems like IHA [18–21], where conventional PID controllers may fall short [16, 19].

This paper's scientific contributions are as follows: (i) The creation of a nonlinear HW model for precise IHA modelling, considering the system's asymmetric dead zone characteristics; and (ii) The design and execution of both PID and FOPID controllers on the proposed HW model. These controllers are used in simulation and experimental studies to assess and examine the tracking performance of the IHA system.

2 Modelling and Experimental Setup

2.1 Mathematical Modelling of Industrial Hydraulic Actuators

A mathematical model of the IHA system, as depicted in Fig. 1, has been formulated using the fundamental principles of system dynamics. The model incorporates key parameters such as mass, damping, and stiffness. The derivation of the model is presented as follows:

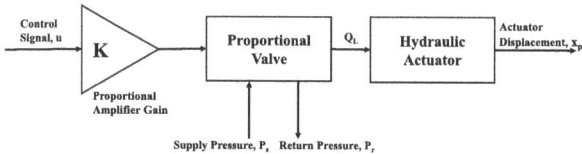

Fig. 1. Block diagram of IHA [20]

$$Q_L = A_p \dot{x}_p + \frac{V_t}{4\beta_e} + f(P_L) \quad (1)$$

$$\dot{P}_L = \frac{M_t \ddot{x}_p}{A_p} \quad (2)$$

Equation (3) is a transfer function that combines Eqs. (1) and (2), forming a comprehensive mathematical model of the IHA system.

$$\frac{M_t \ddot{x}_p}{A_p} = \frac{4\beta_e}{V_t}(Q_L - A_p \dot{x}_p - C_{tp} P_L) \quad (3)$$

The differential equation in Eq. (3) is transformed using a Laplace transform, resulting in the transfer function of the IHA system expressed in Eq. (4). In this context, $X_p(s)$ represents hydraulic cylinder displacement and $U(s)$ denotes the input voltage.

$$\frac{X_p(s)}{U(s)} = \frac{b_1}{s(s^2 + a_1 s + a_0)} \quad (4)$$

The discrete-time model is obtained by transform the continuous-time model represented in Eq. (4) using a zero-order hold. This transformation is expressed as follows:

$$G(z) = \frac{x_p(k)}{u(k)} = \frac{b_1 z^2 + b_2 z + b_3}{z^3 + a_1 z^2 + a_2 z + a_3} \quad (5)$$

The discrete-time model is integrated into the HW model, incorporating a delay function.

2.2 Hammerstein-Wiener Model with Asymmetric Dead Zone

Figure 2 illustrates a nonlinear HW model designed as a linear dynamic model positioned between two static nonlinear functions. In the absence of output nonlinearity, the model reduces to a Hammerstein model, whereas a linear dynamic system followed by a nonlinear static function constitutes a Wiener model. The Hammerstein, Wiener, and HW models have garnered significant recognition because to their relative simplicity and effectiveness in describing many nonlinear dynamic systems [10].

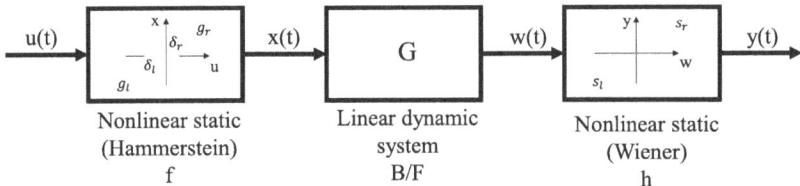

Fig. 2. Structure of HW model

Equation (6) provides a mathematical representation of the input nonlinearity in the static Hammerstein configuration, characterized by an asymmetric dead zone. In this context, $u(t)$ represents input while $x(t)$ denotes output, which also serves as an input to the linear subsystem. The parameter g_l and g_r indicate the slops of linear segments, while δ_l and δ_r define the thresholds of the dead zone.

$$x(t) = \begin{cases} g_l & \text{if} \quad u(t) \leq \delta_l \\ 0 & \text{if} \quad \delta_l < u(t) < \delta_r \\ g_r & \text{if} \quad x \geq \delta_r \end{cases} \quad (6)$$

The linear subsystem G is expressed as:

$$w(t) = \frac{B(z^{-1})}{F(z^{-1})} x(t) \quad (7)$$

where the polynomial of $B(z^{-1})$ and $F(z^{-1})$ is defined as

$$B(z^{-1}) = b_1 z^{-1} + \cdots + b_i z^{-nb} \quad (8)$$

$$F(z^{-1}) = 1 + f_1 z^{-1} + \cdots + f_j z^{-nf} \quad (9)$$

where both n_b and n_f are predetermined values. Based on Eq. (5), both the numerator and denominator of the linear model will consist of three elements since the system is three orders in discrete-time model, accompanied by a delay function of one. Therefore, i and j do not need to be estimated because this method uses grey box identification as shown as in equation below:

$$G(z) = \frac{w(k)}{x(k)} = \frac{b_1 z^{-1} + b_2 z^{-2} + b_3 z^{-3}}{1 + f_1 z^{-1} + f_2 z^{-2} + f_3 z^{-3}} \quad (10)$$

The nonlinear static (Wiener) component signifies the result of the nonlinearity model. Within the framework of the IHA system, the piston may extend and retract within a range of -100 mm to 100 mm. Thus, this block is seen as a saturation function, representing the operational constraints of piston movement. This methodology ensures that the simulation model faithfully mirrors the behavior of the actual system.

$$y(t) = h \begin{cases} s_r = 100 \\ s_l = -100 \end{cases} \tag{11}$$

2.3 Experimental Setup

Figure 3 presents the control schematic diagram for the system, marking the initial step in the identification process of this project. The data obtained from the experimental setup is subsequently analyzed and compared with the simulation data in the time domain. This analysis allows for the observation and comparison of response patterns, thereby facilitating the validation of the results.

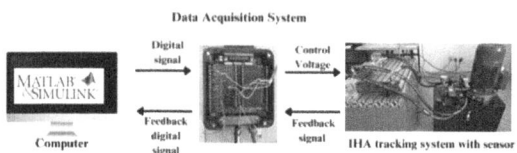

Fig. 3. Control schematic diagram of IHA workbench

Figure 4 depicts the hardware and software combination used for data gathering. The IHA system workbench in the laboratory is a real-time control system using MATLAB Simulink and the Data Acquisition system interface to regulate the velocity of a double acting cylinder with a hydraulic power source. The hydraulic system functions at a nominal pressure of 5 MPa and delivers a flow capacity of 16.6 L/min. Additionally, real-time data of the cylinder's displacement are acquired by a wire displacement sensor, facilitating accurate monitoring and regulation of the system's operation.

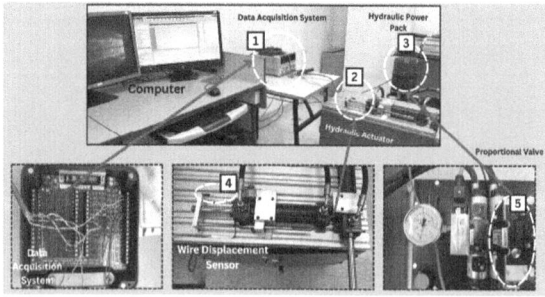

Fig. 4. IHA experimental Setup

2.4 Controller Design (PID and FOPID)

In contrast to the conventional PID controller which is characterized by three parameters: proportional k_p, integral k_i, and derivative k_i while the FOPID incorporates two additional parameters: integrating order, γ and the derivative order, μ. These parameters are integrated into the gains for the integral and derivative components of the controller. Generally, the transfer function of a standard PID can be expressed as follows:

$$G(s)_{PID} = k_p e(t) + k_i \int_0^t e(t)dt + k_d \frac{de(t)}{dt} \quad (12)$$

where k_p represents the proportional gain, k_i denotes the integral gain and k_d indicates derivative gain. The transfer function of FOPID controller is derived by integrating the supplementary orders, resulting in the following expression:

$$G(s)_{FOPID} = k_p e(t) + k_i D^{-\gamma} e(t) + k_d D^{\mu} e(t) \quad (13)$$

where the order of γ and μ does not necessarily correspond to an integer number. It was previously demonstrated that by introducing integral and derivation orders, which form the FOPID controller, also known as the PID controller, the performance of traditional controllers can be improved. However, from a computational perspective, the incorporation of additional parameters in the FOPID controller complicates the process of parameter acquisition, leading to increased computational time.

3 Results and Discussion

3.1 System Identification of IHA

For the identification process, the input and output data were imported into the System Identification Toolbox, and trend filtering was applied to improve data quality, as illustrated in Fig. 5. The filtered data were then split into two segments: the first part was used for model parameter estimation, while the second part was reserved for validation to evaluate the model's performance. The experiment spanned 40 s, with a sampling time of 0.01 s, ensuring comprehensive data collection throughout the identification process. The multi-sine input that used for the input data collection with amplitude of 0.5, 1 and 1.2, and frequencies of 0.05, 0.2 and 1 (Hz), respectively. By applying multiple sine waves at varying frequencies, the IHA was effectively excited over a broad frequency range, facilitating an accurate characterization of its dynamic behavior.

The offline estimation process, utilizing a grey-box approach was conducted for the nonlinear model of a third-order system derived in the previous section. Table 1 presents the results obtained from three distinct data sets, including their respective parameters as determined during the non-recursive identification process. As described in Eq. (14), this represents the linear subsystem within the HW model. Each parameter has been estimated during the identification process. Specifically, the dead-zone range between -0.3 to 0.5 (based on the system) and the saturation range between -100 mm to 100 mm (based on the double acting cylinder displacement) were utilized to represent the real

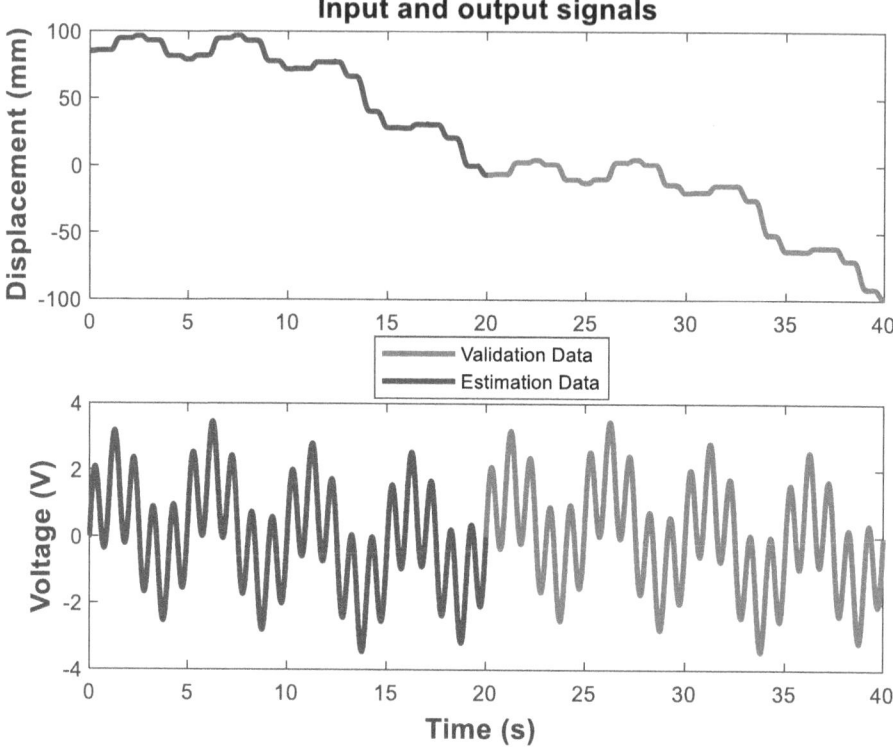

Fig. 5. Input and output data for system identification

Table 1. Estimated parameter of the data set

Parameter	b_1	b_2	b_3	f_1	f_2	f_3
Data Set	0.0365	0.1265	−0.1630	−1.0113	−0.9774	0.9887

system for simulation purposes. This simulated model serves as the basis for controller design prior to its implementation on the actual system.

$$G(z) = \frac{w(k)}{x(k)} = \frac{0.0365z^{-1} + 0.1265z^{-2} - 0.1630z^{-3}}{1 - 1.0113z^{-1} - 0.9774z^{-2} + 0.9887z^{-3}} \quad (14)$$

The validation of the IHA model in the system identification process was demonstrated through the best-fit output result between the simulated and validation models, achieving 88.21%. These results reinforce the accuracy and reliability of the estimated model, confirming its capability to effectively capture the system's underlying dynamics and produce reliable predictions based on the experimental data (Fig. 6).

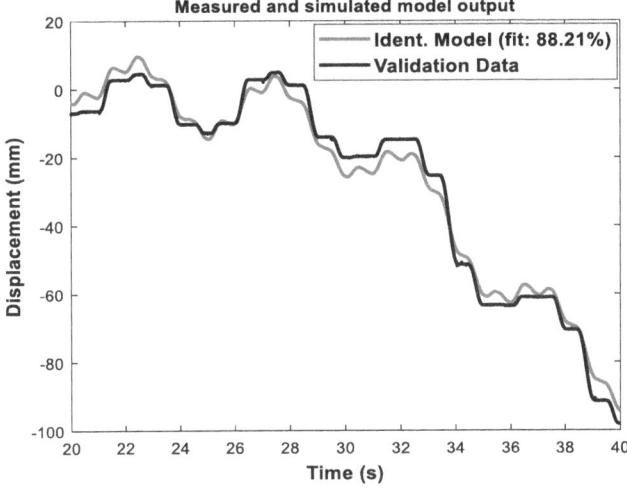

Fig. 6. Measured and simulated output for fit validation process

3.2 Trajectory Tracking of the Designated Controller

The study tested two kinds of controllers for precise trajectory tracking with a multi-sinusoidal reference signal. Figures 7 and 8 illustrate the tracking performance of the PID and FOPID controllers in simulation and experimental work, respectively. A comprehensive analysis of Root Mean Square Error (RMSE) and Mean Square Error (MSE) assessments were performed to evaluate the performance of the controllers presented in tabulated form.

3.2.1 Trajectory Tracking of PID and FOPID Controller in Simulation Work

Figure 7 depicts the trajectory tracking efficacy of the PID and FOPID controllers in simulation. A detailed analysis of the output findings indicates that the FOPID controller more precisely adheres to the intended multi-sinusoidal trajectory. Table 2 further substantiates this, indicating that the MSE and RMSE values for the FOPID controller are much lower than those of the PID controller. The total performance improvement for MSE and RMSE is 32.8% and 18.0%, respectively.

To evaluate the capability and performance of the designated controllers, an analysis was conducted using histograms of the recorded tracking error data as shown in Fig. 8. The bin width for the histograms in both simulation and experimental setups was set to 0.03 mm. The results based on error analysis, indicate that the simulation errors are significantly smaller than those observed in the experiments, primarily due to the presence of noise in the experimental data. The analysis further indicates that the PID controller has a higher error, with values approaching 0.4 mm compared to the FOPID controller where the error distribution is within 0.2 mm. Additionally, the PID controller results in 41 bins, whereas the FOPID controller generates 38 bins reflecting a narrower error distribution. These findings clearly demonstrate that the FOPID controller outperforms the PID controller offering enhanced tracking performance and reduced error.

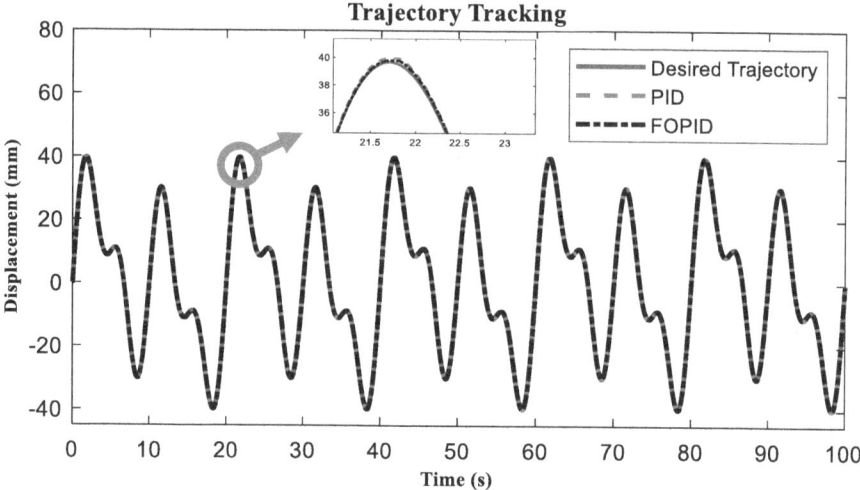

Fig. 7. Trajectory tracking response of PID and FOPID controller during simulation works

Table 2. Tracking performance of controllers

Control Approach	MSE	RMSE
PID	0.0457	0.2138
FOPID	0.0307	0.1753

3.2.2 Trajectory Tracking of PID and FOPID Controller in Experimental Work

Figure 9 depicts the trajectory tracking efficacy of the PID and FOPID controllers throughout the experiment. An in-depth analysis of the output findings indicates that the FOPID controller more precisely adheres to the intended multi-sinusoidal trajectory. Table 3 further substantiates this, indicating that the MSE and RMSE values for the FOPID controller are much lower than those of the PID controller. The total performance improvement for MSE and RMSE is 31.9% and 17.5%, respectively.

As previously discussed in Fig. 8, the trajectory tracking error observed in the experimental setup is higher compared to the simulation results due to the presence of noise. Figure 10 further illustrates that the PID controller produces a larger tracking error of approximately 0.9 mm, whereas the FOPID controller demonstrates a reduced error of 0.7 mm. Additionally, the histogram analysis reveals that the PID controller results in 89 bins, compared to 73 bins for the FOPID controller, indicating a broader error distribution. These findings align with the simulation results, reinforcing the conclusion that the FOPID controller enhances the efficiency and performance of the conventional PID in trajectory tracking applications.

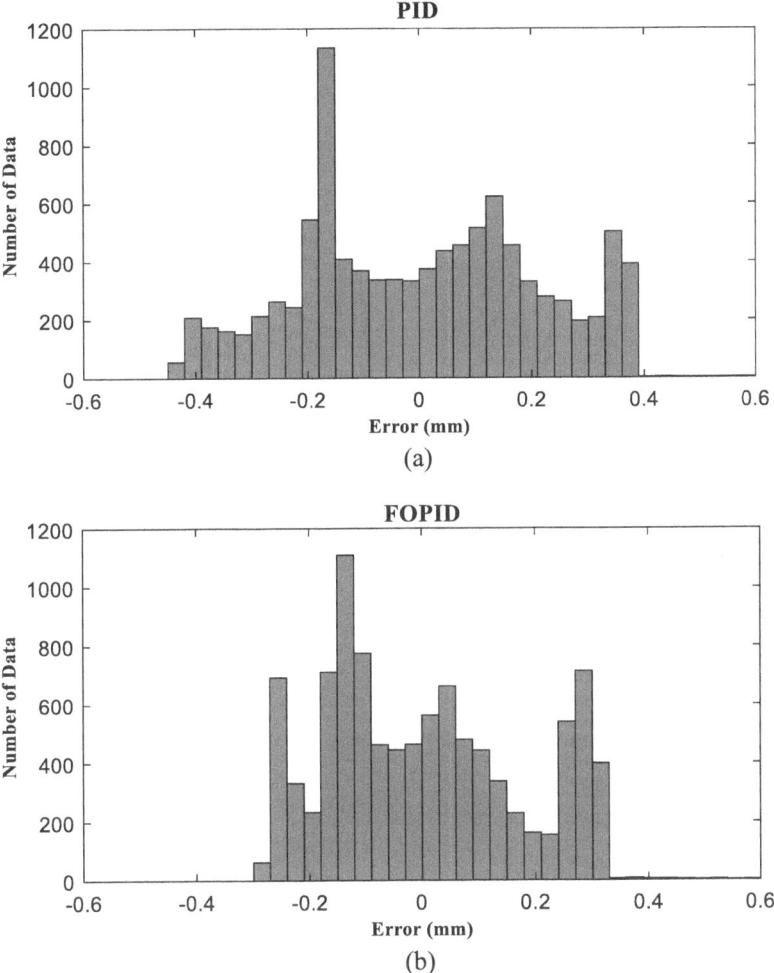

Fig. 8. Histogram Tracking Error in Simulation (a) PID (b) FOPID

The simulation and experimental findings indicate that the HW model precisely reflects the dynamic behavior of the IHA system, producing output results that closely align with the system's real performance. Furthermore, the results validate that the FOPID controller surpasses the traditional PID controller in effectiveness, due to the incorporation of non-integer parameters that improve control accuracy.

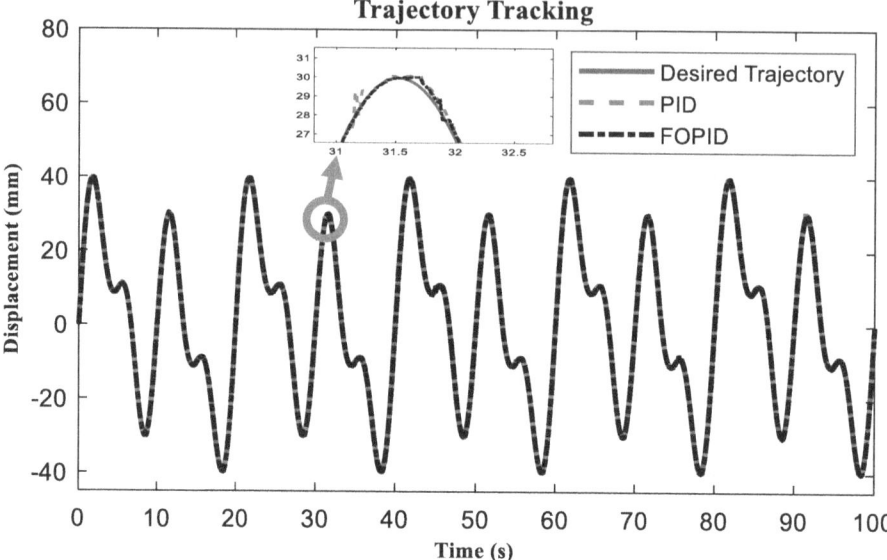

Fig. 9. Trajectory tracking response of PID and FOPID controller during experimental works

Table 3. Tracking performance of controllers

Control Approach	MSE	RMSE
PID	0.0570	0.2388
FOPID	0.0388	0.1971

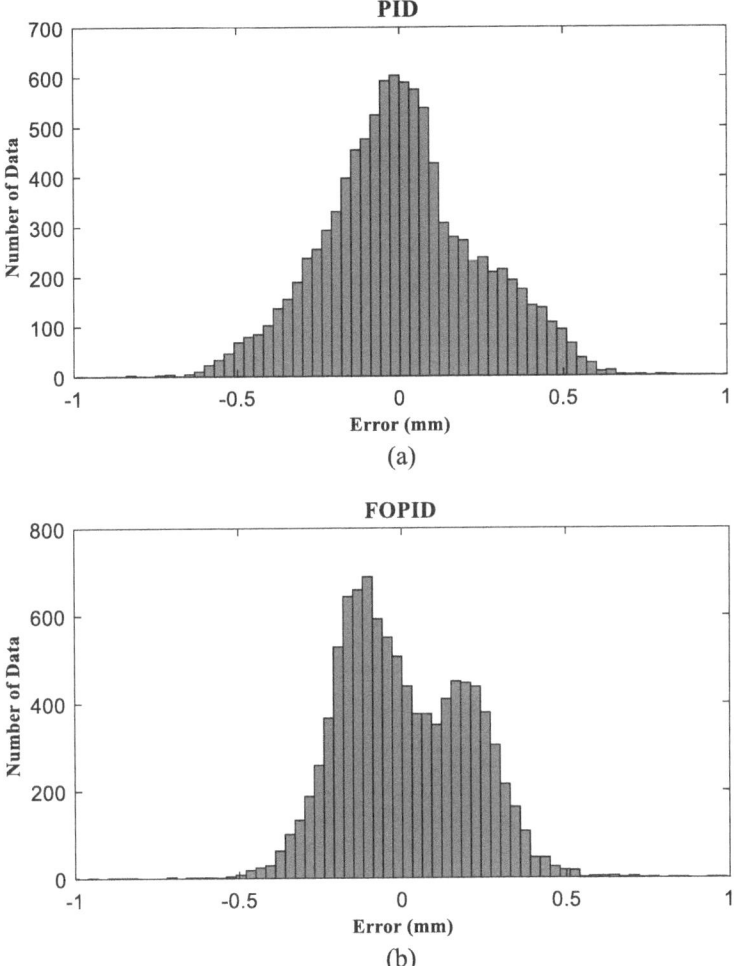

Fig. 10. Histogram Tracking Error in Experiment (a) PID (b) FOPID

4 Conclusion

In conclusion, the IHA system has been effectively modeled, and its parameters approximated using a HW model that incorporates the system's asymmetric dead-zone nonlinearities. The model was verified by graphical and fit analysis, attaining a best-fit percentage of 88.21%, therefore satisfying the requisite requirements without exhibiting any statistical discrepancies. This validation verifies the efficacy of the suggested nonlinear model, rendering it appropriate for following control design methodologies. The advantage of this HW model lies in its ability to accurately capture the dynamic behavior of the IHA system, simplifying the control design process. Furthermore, both FOPID and PID controllers were used in modelling and experimental studies to evaluate the tracking efficacy of the IHA system. The findings indicate that the FOPID controller surpasses the conventional PID controller, producing reduced MSE and RMSE values. The FOPID controller demonstrated simulated enhancements of 32.8% in MSE and 18.0% in RMSE, while experimental findings revealed improvements of 31.9% and 17.5%, respectively, in comparison to the PID method. Notwithstanding these encouraging findings, there is opportunity for improvement, especially considering the system's highly nonlinear characteristics, which persistently pose obstacles and provide poor control outputs. Future research should concentrate on formulating more sophisticated control methods, such as intelligent or nonlinear controllers, and use optimization techniques to refine the gain values of PID and FOPID controllers, therefore improving the tracking performance and overall efficacy of the IHA system.

Acknowledgement. The authors wish to thank Universiti Teknikal Malaysia Melaka and Ministry of Higher Education (MoHE) for their support. This research is funded under FRGS grant (FRGS/1/2021/TK0/UTEM/02/12).

References

1. Zhang, T., Fan, Y., Qiu, H.: Research on electro-hydraulic position servo synchronous control system based on adaptive robust control. J. Phys. **2760**, 1–7 (2023)
2. Nguyen, M.H., Ahn, K.K.: A novel trajectory adjustment mechanism – based prescribed performance tracking control for electro-hydraulic systems subject to Disturbances and Modeling uncertainties. Appl. Sci. **12**(12), 1–24 (2022)
3. Xu, B., Shen, J., Liu, S., Zhang, J.: Research development of electro-hydraulic control valves oriented to Industry 4.0: a review. Chin. J. Mech. Eng. **30**(29), 1–20 (2020)
4. Dörr, M., Leitenberger, F., Wolter, K., Matthiesen, S., Gwosch, T.: Model-based control design of an EHA position control based on multicriteria optimization. Machines **10**(12), 1–18 (2022)
5. Yousef, M.A., Rabie, M.G., Rateb, R.A.: System identification and controller design for a typical electro-hydraulic servo motor. In: International Telecommunications Conference, pp. 1–4, IEEE, Alexandria (2021)
6. Nguyen, M.H., Dao, H.V., Ahn, K.K.: Active disturbance rejection control for position tracking of electro-hydraulic servo systems under modeling uncertainty and external load. Actuators **10**(20), 1–22 (2021)
7. He, H., Huang, Y.: Integrated modeling and adaptive parameter estimation for hammerstein systems with asymmetric dead-zone. IEEE Trans. Ind. Electron. **70**(5), 4942–4951 (2023)
8. Li, L., Ren, X., Lv, Y.: Parameter estimation for control of hammerstein systems with dead-zone nonlinearity, innovative techniques and applications of modelling, identification and control. Lect. Notes Electr. Eng. **467**, 109–118 (2018)
9. Jinxing, S., Hongxin, C., Ke, F., Hong, Z., Huanliang, L.: Parameter identification and control algorithm of electrohydraulic servo system for robotic excavator based on improved hammerstein model. Math. Prob. Eng., 1–9 (2020)
10. Jespersen, S., Yang, Z., Hansen, D.S., Kashani, M., Huang, B.: Hammerstein-wiener model identification for oil-in-water separation dynamics in a de-oiling hydrocyclone system. Energies **16**(20), 1–32 (2023)
11. Li, S., et al.: Identification and H∞ robust control of wireless power transfer system by hammerstein model. IEEE Trans. Power Electron. **39**(7), 8883–8893 (2024)
12. He, H., Na, J., Wu, J., Huang, Y., Xing, Y.: Fixed-Time adaptive parameter estimation for hammerstein systems subject to dead-zone. IEEE Trans. Ind. Electron. **71**(4), 3862–3872 (2024)

13. Fan, Y., Shao, J., Sun, G., Shao, X.: Proportional–integral–derivative controller design using an advanced lévy-flight salp swarm algorithm for hydraulic systems. Energies **13**(2), 1–20 (2020)
14. Kumawat, A.K., Kumawat, R., Rawat, M., Rout, R.: Real time position control of electrohydraulic system using PID controller. Materials Today: Proceedings **47**, 2966–2969 (2021)
15. Feng, H., Ma, W., Yin, C., Cao, D.: Trajectory control of electro-hydraulic position servo system using improved PSO-PID controller. Autom. Constr. **127**, 103722 (2021)
16. Ye, Y., Yin, C.B., Gong, Y., Zhou, J.J.: Position control of nonlinear hydraulic system using an improved PSO based PID controller. Mech. Syst. Signal Process. **83**, 241–259 (2017)
17. Negara, A.T., Ardiyanto, I., Whayunggoro, O.: Tuning of fractional-order PID controller for electro-hydraulic servo valve system. In: International Conference on Information and Communications Technology, pp. 659–662. IEEE, Yogyakarta (2019)
18. Lu, Y., Liang, Y.: Fractional order modeling for hydraulic cylinder system controlled by electro-hydraulic proportional valve. In: China Automation Congress, pp. 5259–5264. IEEE, Chongqing (2023)
19. Wos, P., Dindorf, R.: Nonlinear modeling and parameter identification for electro-hydraulic servo system. In: 20th International Carpathian Control Conference, pp. 1–5. IEEE, Krakow-Wieliczka (2019)
20. Ali, M.H.N., Ghazali, R., Jaafar, I.H., Ghani, M.F., Soon, C.C., Has, Z.: Comparison study between open-loop and closed-loop identification for industrial hydraulics actuator system. Int. J. Mech. Eng. Rob. Res. **13**(5), 516–521 (2024)
21. Wang, N., Weng, J., Li, Z., Tang, X., Hou, D.: Fractional-order PID control strategy on hydraulic-loading system of typical electromechanical platform. Sensors **18**(9), 1–17 (2018)

Machine Learning and NLP-Based Approach for Constraint Acquisition Problems

Abdessalam Bahafid[1,2](✉), Zakarya Erraji[1], Naira Abdou Mohamed[1,2], Anass Allak[1], Kamel Gaanoun[1], and Imade Benelallam[1,2]

[1] SI2M Labs, INSEA, Rabat, Morocco
{a.bahafid,zerraji,nabdoumohamed,aallak,kgaanoun, i.benelallam}@insea.ac.ma
[2] ToumAI Analytics, Rabat, Morocco

Abstract. Constraint Programming (CP) serves as a widely used platform in both industry and research, simplifying the modeling and resolution of complex constraint satisfaction and optimization problems. However, the automated model reformulation platforms that aims to assist the users in modeling and solving constraint problems remains restricted. This level of restriction is primarily caused by the mathematical modeling and programming skills required to handle various types and forms of Nondeterministic Polynomial Complete (NP-complete) decision problems, including challenges related to planning, scheduling, and resource optimization. In this context, we present our research, which aims to integrate Natural Language Processing (NLP) with constraint acquisition via a conversational agent. The objective is to enhance the intuitiveness of the constraint programming paradigm, enabling a more diverse range of users to interact with it effectively. The study focuses on the development of an intelligent constraint acquisition system that allows novice users to communicate their problem in natural language. The system interprets the provided information, formulates the given problem as a constraint network, solves it, and presents alternate solutions to the user when available.

Keywords: Constraint Acquisition Problems · Natural Language Processing · Deep Learning

1 Introduction

For thousands of years, people have spoken to one another using natural language. And as knowledge grows in complexity, communication becomes more challenging; since the early phases of technological evolution in the first half of the twentieth century (1950s), the concept of computers comprehending ordinary languages and engaging in conversations with humans has been a recurring theme in science fiction. Nowadays, approximately 80% of generated and collected data is unstructured [1], with the majority consisting of natural language text, which preserves what has been said and has long been an important medium for preserving human knowledge, posing a significant challenge for computer interpretation.

Constraint Programming (CP) paradigm is known to be a powerful tool for solving complex constraint satisfaction and optimization problems. However, it is well known that developing a good model demands a significant level of expertise and the application of specialized techniques, as highlighted in [2]. This inherent complexity poses a barrier for non-expert users who seek solutions to their problems without undue complications; In other hand, the Deep Learning (DL) algorithms based on dense vector representations have recently made impressive improvements in Natural Language Processing (NLP) related tasks such as part of speech tagging, parsing, and machine translation. So far, it has gave excellent performance and results.

In this work, we are interested in investigating whether Deep Learning (DL) and Natural Language Processing (NLP) methods, can be used to acquire and extract information that can be beneficial to prune the formalization of the Constraint Satisfaction Problem (CSP) related problems expressed in natural language.

We present a substantial contribution to the field by addressing two crucial aspects. Firstly, we concentrate on the annotation of the data, employing a meticulous approach to improve its quality and relevance. This process clears the way for more accurate and insightful analyses. Secondly, our work introduces the first model that adeptly manages the intricate link between CSP and NLP. This innovative model not only bridges the gap between these two domains but also opens up new avenues for leveraging their synergies, thereby advancing the overall state-of-the-art in computational research.

In light of that, the remaining sections of this paper will be organized as follows. First, we present a summary of relevant literary works (Sect. 2). Then, we provide an essential background in Sect. 3, which will be used as the ground foundation for the following sections. Section 4 outlines the key contributions made by this study. Subsequently, Sect. 5 outlines the experimental phase with the results and we discuss the obtained result. Finally, we conclude our study by summarizing the findings and the contributions 6.

2 Related Works

In recent years, Deep Neural Networks (DNNs) have attracted significant attention thanks to the growth data and the capabilities of cutting-edge computing machines. Despite the widespread utilization and independent advancements in Natural Language Processing (NLP), its interaction with Constraint Satisfaction Problems (CSP) has been infrequently explored in the literature. Conducting comprehensive literature reviews in this context poses a noteworthy challenge due to the limited studies simultaneously addressing these substantial concepts.

The intricacies of natural language processing compound the complexity of this task, as machines must not only decipher the meaning of individual words in a sentence but also grasp the intended significance of the entire message. This challenge parallels the difficulties humans encounter, particularly in instances involving elements like sarcasm. Achieving a complete understanding of the overall message requires the consideration of various components of natural language, including phonetics, syntax, and semantics.

Efforts in problem acquisition seek to simplify the modeling process without delving into intricate mathematics. One of the interesting approaches in this domain is MiniZinc [3], which introduces a layer of abstraction by offering a robust set of built-in constraints, tackling the absence of a standardized modeling language for CP. They introduce one named "MiniZinc" which is an expressing modeling language adaptable to various solvers; A similar approach was proposed by [4], they develop a formal language that provides a high degree of abstraction using variables representing combinatorial objects and relationships.

One of the closed works to the suggested work is in this article [2]. The presented findings are effective in identifying the positions of variables or constraints within a text. However, this approach has limitations when precision is crucial for formulating the model, particularly in achieving a more accurate determination of entities such as variables, domains, and constraints.

3 Background

Artificial Intelligence is a difficult term to define because there are numerous definitions. One of the most well-known defines AI [5] as the intelligence demonstrated by machines, as opposed to the natural intelligence demonstrated by humans and other animals. It is intended to perform tasks such as Speech Recognition [6], NLP [7], planning [8], problem solving and many more.

NLP [5] is an area of research and application that explores how computers can communicate with humans in their very own language and scales other language-related tasks. It has many different applications in various domains, all refer to humans' unstructured natural language, as an example, it is applied to machine translation, speech recognition, Named Entity Recognition (NER), information extraction and text classification. Today's machines have the ability to analyze more language-based information than humans without exhaustion and in a continuous, unbiased way.

3.1 Constraint Satisfaction Problems

Formally, a CSP [9] is a triple $\langle X, Dom, C \rangle$ where:

- $X = \langle x_1, x_2, \ldots, x_n \rangle$ is a n-tuple of variables,
- $Dom = \langle Dom(x_1), Dom(x_2), \ldots, Dom(x_n) \rangle$ is a corresponding n-tuple of domains such that $x_i \in Dom(x_i)$,
- $C = \{C_1, C_2, \ldots, C_m\}$ is a set of constraints where each constraint C_i expresses a condition on a subset of X.

A CSP solution is a complete assignment of a value from $Dom(x_i)$ to each variable x_i that satisfies all of C's constraints.

Example of a CSP Problem. One of the classic problem is *the eight queens puzzle*. It consists of positioning eight chess queens on an 8 by 8 chessboard in a manner that prevents any of them from capturing another through the standard moves of a chess queen. The objective is to arrange the queens in such a way that they avoid conflicts. Consequently, it is imperative to organize the queens to ensure that none of them share the same row, column, or diagonal, thus determining a viable solution.

3.2 Constraint Programming

CP was originally introduced in the late 1960s, and it has made great progress in both scientific research and industry since then. Because it provides a powerful framework for addressing combinatorial problems, it has contributed in many techniques that are used and explored in several fields of computing [10], such as resource allocation, scheduling, and planning; The core idea behind CP is that instead of stating the actual process to solve a particular program, the user specifies a problem specification that contains the decision variables, alternative values for the variables, and the set of constraints on the decision variables to find workable solutions.

It is indeed categorized as a declarative programming paradigm since it simply requires the formulation of the combinatorial problems as a constraint network rather than providing a sequence of steps for the program to follow in order to acquire the solution. A constraint network solution is an assignment of variables to domain values that satisfies all of the network's constraints.

3.3 Natural Language Processing

NLP employs a variety of techniques to help computers understand human text and voice. Among these tasks are the following [11]:

- Part-of-Speech (POS) tagging or grammatical tagging is the process of assigning the part of speech (*i.e.* identification of grammatical cases) of a particular word or a piece of text based on its definition, use and context. It is mainly used in lexical analysis.
- Dependency Analysis, which is the process of extracting dependencies between entities (*i.e.* tokens) in a sentence, forming directional relationship between them. It is mainly used in statistical analysis.
- Named Entity Recognition (NER) is the process of identifying and classifying a token or a group of tokens into pre-defined categories such as names, location, and so on.
- Co-reference Resolution is the process of identifying all expressions that refer to the same entity in a text. For example, in the sentence *"Angela lives in Morocco. She is quite happy in that country."*. "Angela" and "She" are the same person, and "Morocco" and "that country" are the same location.
- Word sense disambiguation is the process of determining the meaning of a word with multiple meanings using semantic analysis to determine the word that makes the most sense in the given context.

4 Methodology

Our realization process is spliced into these following steps: Approach, Implementation Technologies, Data Acquisition & Labeling, Data Preparation, Modeling and Evaluation.

- **Approach**: we define the structure of the target platform and explain its sections.
- **Implementation Technologies**: we present the various technologies employed, as well as a full description of how they contribute to this work.
- **Data Gathering & Labeling** and **Data Preparation**: they cover the process of obtaining, labeling, and preparing the data for further analysis.
- **Modeling**: the detailed structure of each of the required platform's components is presented in this step.
- **Evaluation**: Overall evaluation of the approach.

The following Fig. 1 presents the eagle's view of the general architecture of the system we're building to response to the query of this work; it consist of three modules. Briefly explained, the application follows the following approach: first, the **Semantic Analyzer** receives the *user's utterance* as text and generates a semi-structured file with text extraction analysis performed. Second, the **Constraint Acquisition Algorithm** takes the semi-structured file as input and generates a Constraint network as a structure file that the **Constraint Programming Solver** can simply take and solve. Finally, the result is returned to the user.

In other words, the learning algorithm constructs the target constraint network, identifying the set of variables, domains, and constraints, based on the user's text or utterance description of his problem. During the execution process of the constraint acquisition algorithms, several questions might be asked of the user in model creation, which may be basic or sophisticated inquiries. Following the acquisition process, the platform solves the formal created CSP problem and returns the solutions to the user.

The system architecture adopted is based on the micro-services one which consists of dividing the application into multi-simple applications, each with its own scope of responsibility to facilitate maintenance and improvements while also allowing for the integration of specific sub-applications into certain ways depending on the needs.

Data (Problem) Gathering. As previously stated, the previously prepared datasets was not found. So, during the gathering process, we had to examine several resources, including CSPLib and the website NLP4CP, which are used as primary resources. However, we also gathered some additional problems from numerous websites, forums.

We were able to gather around **93** problem descriptions during the data collection phase. CSPs and Constraint Optimization Problems (COPs) are both included in the problem specification datasets; the only difference between them

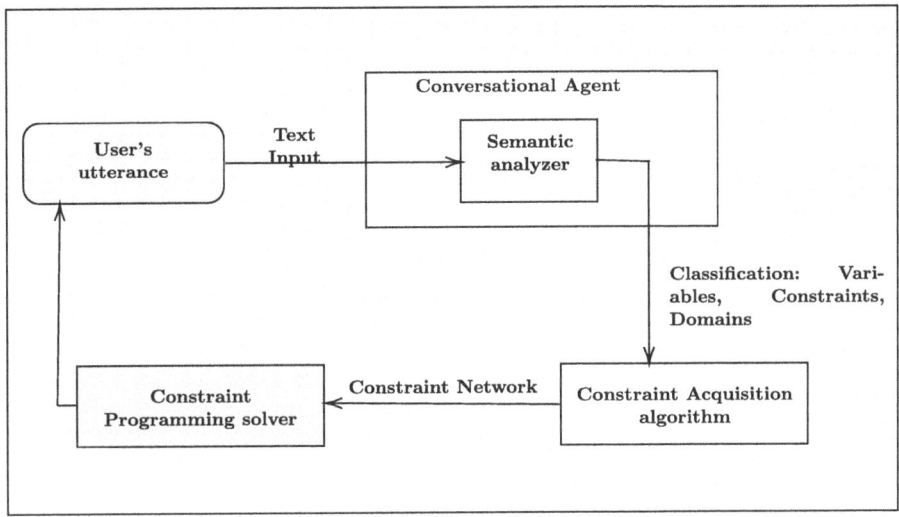

Fig. 1. Targeted acquisition system architecture

is the objective function. However, we primarily focus on overcoming the difficulties presented by CSP problems in the resolution process.

Parts of NL descriptions containing supplementary information or elements other than text needed special attention. Such situations are listed below, along with explanations of how we dealt with them manually:

- *Mathematical formulas* are translated into L^ATEX and treated as a regular text [2].
- Some of the problem specifications are duplicated with slightly different descriptions in order to simulate a simple data augmentation to diversify the data. However, we continue to calculate all different descriptions of the same problem as if they were one problem.
- Some problem specifications may have been examplificated (*i.e.* only the meaningful description text is retained).

Labeling. Data labeling is the process of identifying raw data and adding meaningful and informative labels to it, in order to provide the necessary context that machine learning models can learn from it during the supervised tasks.

The process of labeling the corpus turned out to be a challenging task to complete, as the detection of the constraints was not always easy to identify and the margin of error during the process is relatively high. As the annotations typically require a high of expertise in the field of constraint problems, as well as expertise in the domain of the problem.

The process of labeling the data of this research is carried out with the help of an annotation tool known as **Doccano**. The labeling process in our case aims to identify and categorize the four elements (*i.e.* P_NAME, VARIABLE, DOMAIN, CONSTRAINT), where are:

- **P_NAME**: This optional parameter represents the name of the current problem, serving as a helpful identifier for the user.
- **VARIABLE**: This denotes the set of variables involved in the problem.
- **DOMAIN**: This signifies the set of domains corresponding to the variables in the problem.
- **CONSTRAINT**: This represents the set of constraints that need to be satisfied within the problem.

These four categories will be used in label for the first module of our architecture (i.e. **Semantic Analyzer**) Sect. 4.2.

4.1 Data Preparation

Data preparation is a critical aspect of machine learning, essential for the success of the analysis, and this holds true for natural language processing (NLP) as well. This aspect is usually applied before any algorithm training, as it has a significant impact on the success of text analysis [12]. The main reason for this is that human spoken and written text is usually unstructured and arbitrary.

Pre-processing. The first mandatory step in the implementation is the transformation of text into tokens [13]. In the case of English language, the Tokenizer splits the text based on the presence of space symbols, then checks for language-specific exception rules, such as splitting the abbreviated word "who's" into "who" and "'s". Following that, the Tokenizer examines word prefixes, suffixes, and infixes to see if they can be divided. Thereby, special characters such as punctuation, hyphen and quotes also served as splitting criteria.

Hereafter in Fig. 2, an example of the SpaCy tokenization process applied to the English phrase "Let's go to N.Y.!". In this example, a "natural" tokenization would tend to divide the sentence based on spaces, which would result in an incorrect decomposition where terms like ""Let's"" or ""N.Y.!"" would appear. Therefore, SpaCy performs special processing for these terms between words and punctuation marks, and also separates affixes from the roots of the words.

Train and Test. This section will briefly describe the implementation's train-and-test split. According to Table 1, 20% of the data is used to test the models, while the remaining 80% is used to train them. ***Test and train instances are generated randomly.***

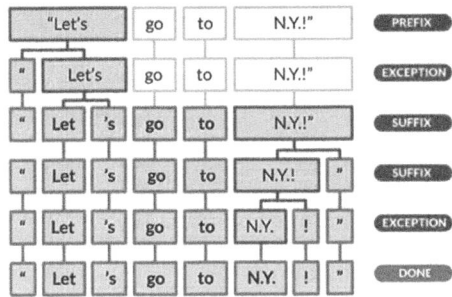

Fig. 2. Illustration of SpaCy's tokenization process

Table 1. Distribution of the training and testing datasets

Train-and-Test	Number of Samples
Number of Train Instances:	74
Number of Test Instances:	19
Total:	93
Vocabulary of Train Set:	19,774
Vocabulary of Test Set:	3,379
Total Vocabulary:	23,153

Because of the random sample selection in the Train-and-Test phase, the above-mentioned vocabulary samples are not evenly distributed and may change during actual training (*i.e.* the splicing is based on the number of instances, not the vocabulary). The number of samples, on the other hand, remains constant because it is entirely determined by the split rate we select (in this case, the **80/20** split).

4.2 Modeling

The three subsections **Semantic Analyzer**, **ConstraintAcquisition Algorithm** and **Constraint Programming Solver**, that follow will be structured as a story, with each relying on the one before it to provide the most accurate explanation possible. During these sections, we will attempt to present the input type, output type and sub-module structure used.

Module: Semantic Analyzer. The first module is **Semantic Analyzer**, which is the components that directly interacts with user input, so it is trained on the datasets that it should expect from the user. This module is an NER task based on the component `Entity Recognizer`. The goal of this task is to categorize and identify the four labels *(P_NAME, VARIABLE, CONSTRAINT, DOMAIN)*, which are then used to generate the information extraction file; This module can be fully executed from a Command Line Interface (CLI).

The module contains a pipeline that is applied to the input text. First, it tokenizes the text and performs the lexical analysis to attribute each token with tags such as Parsing and Part-of-Speech, and then it applies a custom NER pipeline trained on the previously stated datasets in order to construct and categorize the label of the problem. Finally, it generates a Semi-structured file containing the retrieved data.

Goal. Perform prepossessing of the input text and generate a semi-structured file contains information extracted during the process. This information extracted for the moment are the problem name, the variables, their domains and the constraints.

Input. The abstract description of the problem as either a text file or text input in CLI.

Output. Semi-structured files with information extraction (*e.g.* Named Entity).

Train Configurations. The model used in the process is the second version of `Transition Based Parser`, it can be trained only by providing the annotated datasets using a Command Line Interface (CLI). The model was trained during 40 epochs on the previously prepared and annotated datasets, with the application of learning rate of 10^{-3}. The training process was optimized with the Adam optimizer (weight decay: 10^{-2}). The configuration used to train this model is listed in the following Table 2:

Table 2. NER Model Configuration Parameters

Parameter/Hyper-parameter	Value
Architecture	Transition Based Parser v2
Optimizer	Adam
Learning Rate	10^{-3}
Weight Decay	10^{-2}
Regularization Technique	Dropout ($rate = 0.3$)

Module: Constraint Acquisition Algorithm. The second module is **Constraint Acquisition Algorithm**, which is the component in charge of generating the constraint network; it takes the output of the first module (*i.e.* the semi-structured file) as input and transforms it into a structured form (*in this case*: YAML is used). This module is based on `Dependency Matcher`, which is a rule-based method that uses predefined criteria to extract the information from given text.

The module takes the input and first applies the `Dependency Matcher` with the predefined criteria to find the tokens that have the exact needed information to produce the formal model; those tokens are then used in the based-rule transform approach to generate the formal model. Finally, the model is saved into a `YAML` file format.

Goal. The objective is to analyze the extracted information from the input file and generate a formalized abstract description of the problem.

Input. A semi-structured file containing the information extracted from the user input.

Output. The formal representation of the problem in YAML format.

Module: Constraint Programming Solver. The third module is **Constraint Programming Solver**, which is a Java application that takes the formal form of the input problem and converts it to the standard supported `Choco Solver` form and then solves it and display the results to the end user.

The module loads the `YAML` formal file model as an input and applies to it the parsing process to convert the model to `Choco Solver` model structure, and then takes it and applies to it internally the solving processes and returns the solution of that problem along with a detailed statistics of the process.

The resolution procedure of the problems is entirely handled by the `Choco Solver`, but it may be easily altered to rely on a other solver package or even develop a dedicated solver.

Goal. Solve the problem and display the results to the user.

Input. The formal problem presentation in YAML form.

Output. Display the results to the user.

In our experimental setup, we employed the `Choco Solver` as the solver for the Constraint Satisfaction Problems (CSPs); however, it can be effortlessly replaced with an alternative solver as required.

5 Experiment and Discussion

5.1 Evaluation

The outcomes of our approach are detailed in the subsequent section, with specific attention directed towards the first module. It is noteworthy that the remaining modules are rule-based in this implementation. Following the training of our model using 80% of the datasets, the remaining 20% is utilized to assess the model's performance on previously unseen data. The evaluation metrics for the Named Entity Recognition (NER) task are presented in Table 3 and

Table 4. The first table provides a comprehensive overview of the global evaluation results, while the second table delves into the evaluation results for each label in the datasets.

Table 3. Global Evaluation Results for NER Task

Precision	Recall	F-score
89.27	**88.59**	**88.93**

Table 4. Evaluation Results for NER Task for Each Label

(Label)	Precision	Recall	F-score
VARIABLE	88.75	93.42	91.03
CONSTRAINT	89.19	86.84	88.00
P_NAME	86.67	81.25	83.87
DOMAIN	96.00	96.00	96.00

The evaluation results are collected using the **Evaluate** module. The process of evaluation requires saving the trained model into a folder and transfer the test dataset into a compatible format; in order to facilitate the evaluation approach, we automated the entire process.

5.2 Discussion of Results

As of the time of writing this section, the framework built through this work is capable of extracting the CSPs from a natural language text description, formalizing it as a formal semi-structured description, and then solving it. It also outputs an intermediate readable format to aid in understanding and rectifying or extending the framework.

Regarding the performance of the obtained results, the Table 3 and Table 4 illustrate that the performance of predicting the labels is good because all evaluation metrics are above 80%. Overall, the outcomes obtained are fairly good. Despite the fact that the datasets is tiny and that possibility of manual misclassification occurrence during the dataset's labeling procedure. The reason for these results is the application of transfer learning in a package know as SpaCy by using its pretrained statistical model (*i.e.* en_core_web_sm) and fine-tuning it by applying the retraining process with the new datasets other than those used for the initial training.

For the second and the third modules, which are based on rule-based matching, they produces excellent results on unseen samples with lexical and syntactical similarities of seen samples. Although the potential of rule-based approaches, they remain typically constrained to predetermined rules and it is challenging to scales to enable alternative forms of describing the CSP problems [14].

6 Conclusion

In this study, we have illustrated the utilization of Natural Language Processing models grounded in Deep Learning to handle Constraint Problems articulated in natural language. Leveraging a micro-services architecture and incorporating hand-designed features, our framework has demonstrated a commendable level of accuracy with evaluation metrics surpassing 80%.

By constructing a modular and extensible framework aimed at simplifying the interaction between novice users and constraint problems, requiring minimal mathematical and programming expertise, our work has achieved its intended objectives. However, there remains ample room for improvement. in order to settle some technical gaps to achieve better test results and make the platform easier for novice users to use.

References

1. What Is Unstructured Data?. MongoDB (n.d.). https://www.mongodb.com/unstructured-data
2. Kiziltan, Z., Lippi, M., Torroni, P.: Constraint detection in natural language problem descriptions. In: 25th International Joint Conference on Artificial Intelligence, IJCAI 2016, pp. 744–750. International Joint Conferences on Artificial Intelligence (2016)
3. Nethercote, N., Stuckey, P.J., Becket, R., Brand, S., Duck, G.J., Tack, G.: MiniZinc: towards a standard CP modelling language. In: Bessière, C. (ed.) CP 2007. LNCS, vol. 4741, pp. 529–543. Springer, Heidelberg (2007). https://doi.org/10.1007/978-3-540-74970-7_38
4. Frisch, A.M., Harvey, W., Jefferson, C., Martínez-Hernández, B., Miguel, I.: Essence: a constraint language for specifying combinatorial problems. Constraints **13**(3), 268–306 (2008). https://doi.org/10.1007/s10601-008-9047-y
5. Saleh, Z.: Artificial Intelligence Definition, Ethics and Standards (2019). www.researchgate.net/publication/332548325_Artificial_Intelligence_Definition_Ethics_and_Standards
6. Deng, L., Hinton, G., Kingsbury, B.: New types of deep neural network learning for speech recognition and related applications: an overview. In: 2013 IEEE International Conference on Acoustics, Speech and Signal Processing, pp. 8599–8603 (2013). https://doi.org/10.1109/ICASSP.2013.6639344
7. Young, T., Hazarika, D., Poria, S., Cambria, E.: Recent trends in deep learning based natural language processing. IEEE Comput. Intell. Mag. **13**(3), 55–75 (2018). https://doi.org/10.1109/MCI.2018.2840738
8. Relich, M.: Predictive and prescriptive analytics in identifying opportunities for improving sustainable manufacturing. Sustainability **15**(9), 7667 (2023). https://doi.org/10.3390/su15097667
9. Dao, T.-B.-H.: Constraint Programming for Data Mining and for Natural Language Processing. (Ph.D. thesis). University of Orléans, France (2018). https://hal.archives-ouvertes.fr/tel-01832811
10. Bessiere, C., Koriche, F., Lazaar, N., O'Sullivan, B.: Constraint acquisition. Artif. Intell. **244**, 315–342 (2017). https://doi.org/10.1016/j.artint.2015.08.001

11. IBM. Natural Language Processing (2020). https://www.ibm.com/cloud/learn/natural-language-processing. Accessed 2023
12. Allahyari, M., et al.: a brief survey of text mining: classification, clustering and extraction techniques (2017). arXiv. https://arxiv.org/abs/1707.02919
13. Grefenstette, G.:. Tokenization. In: van Halteren, H. (ed.) Syntactic Wordclass Tagging, pp. 117–133. Springer, Netherlands (1999). https://doi.org/10.1007/978-94-015-9273-4_9
14. Trienes, J., Trieschnigg, D., Seifert, C., Hiemstra, D.: Comparing rule-based, feature-based and deep neural methods for de-identification of dutch medical records (2020). arXiv preprint arXiv:2001.05714

Improving Twitter Sentiment Analysis with a Hybrid BERT and Graph Neural Network Ensemble

Amar Taggu(✉) and Nabam Teyi

North Eastern Regional Institute of Science and Technology,
Nirjuli, Arunachal Pradesh, India
{atg,nbt}@nerist.ac.in

Abstract. This paper presents an enhanced approach to sentiment analysis on Twitter data, leveraging recent advancements in machine learning and deep learning. In addition to traditional classifiers, such as Logistic Regression, Naive Bayes, SVM, and Random Forest, we incorporate Transformer-based models like BERT to achieve improved performance. Furthermore, the work explores hybrid techniques combining machine learning models with lexicon-based methods and hyperparameter optimization. Ultimately, we incorporated an ensemble learning approach which provided the best performance by leveraging the strengths of multiple models. We tested the models using various metrics like F1-score and ROC-AUC. Our results show that ensemble models along with hybrid techniques perform much better than traditional methods at classifying tweets as positive, negative, or neutral.

Keywords: Sentiment Analysis · Machine Learning · Deep Learning

1 Introduction

Twitter is one of the most widely used microblogging platforms. Hence, it plays an extremely significant role in shaping public opinion and sentiment. Owing to millions of Twitter users across the globe, that includes celebrities, influencers, people in power etc., and billions of people influenced or affected by tweets, it becomes requisite that the actual message or appropriate psychology of the tweeter be understood by the followers. Any misinterpretation of the actual meaning of the tweet even by a single follower may result in calamitous sequence of misjudgment by following followers, ultimately resulting in an incorrect consequence of a single tweet. In general, the tweeter may have a positive state of mind, negative state of mind or a neutral state of mind while tweeting. By state of mind, we mean the intent of the tweet. Hence, the tweet readers must be intelligent enough to decipher the real state of the tweet and be able to categorize a tweet into one of the three categories of a positive tweet, a negative tweet or a neutral tweet. This analysis to classify whether the message is of positive, negative, or neutral sentiment is called as sentimental analysis.

Lately, the sentimental analysis of any message is being done automatically without human intervention by employing various Machine Learning (ML) models on the message (tweet) such as logistic regression, Naive Bayes, and support vector machines. However, recent advances have introduced deep learning models such as BERT, which greatly improve accuracy and performance. The current work aims to improve existing sentiment analysis methods by integrating these new methods. We propose an enhanced approach to analyse the unique challenges present in the tweets such as informal language, typos and use of emojis. Deep learning models are combined with traditional learning methods to achieve emotion classification.

2 Related Work

Sentiment analysis has been widely studied using various machine learning models. Batool et al. [2] proposed a keyword-based sentiment analysis approach, while Chikersal et al. [3] combined rule-based classifiers with machine learning models for improved accuracy. Hassan et al. [1] developed a bootstrapping ensemble framework that was able to build sentiment time series to reflect events eliciting strong positive and negative sentiments from users. Levallois [8] developed Umigon as a web application providing a service of sentiment detection in tweets. It indicated additional semantic features present in the tweets, such as time indications or markers of subjectivity. Mohammad et al. [11] created two state-of-the-art SVM classifiers, one to detect the sentiment of messages such as tweets and SMS (message-level task) and one to detect the sentiment of a term within a message (term-level task).

Transformer models, like BERT [4], have recently shown outstanding performance in sentiment classification tasks. Many recent studies focused on using newer methods like ensemble learning and other hybrid techniques. Zhang et al. [14] combined Convolutional Neural Networks (CNN) with Long Short-Term Memory (LSTM) networks that greatly improved the accuracy in sentiment classification. On similar lines, Liu et al. [7] used Transformer-based architectures for multilingual sentiment analysis, and this demonstrated how such architectures can work with different datasets.

In Sharma et al. [9], the authors combined lexicon-based methods with ML classifiers for enhancing sentiment classification for ambiguous tweets. In another study, Gupta et al. [5] applied an ensemble approach that achieved high accuracy in sentiment analysis.

Recent usage of attention mechanism in sentiment analysis can be seen in various works like Vaswani et al. [12] in which a Transformer architecture was introduced, which underpins many state-of-the-art models, including BERT. Sun et al. [10] applied BERT to sentiment analysis to achieve significantly better results compared to traditional recurrent neural networks.

The research of multilingual sentiment analysis is increasing more and more these days. Wu et al. [13] analysed the sentiments by running the multilingual BERT model, demonstrating the effectiveness of using a single model for diverse linguistic data.

Sentiment analysis has also made some significant changes such as employing graph-based frameworks in a bid to explore relationships among entities on tweets. Li et al. [6] employed GNNs as modeling tools to model connections between users, hashtags, and sentiments, thus providing a richer context for analyzing sentiment trends. This approach could effectively identify polarized communities on social media.

3 Dataset

Our study utilizes a collection of tweets categorized into positive, negative, and neutral sentiments. The dataset comprises 21,631 tweets for training purposes and 5,398 tweets designated for testing. We undertook comprehensive preprocessing steps, which involved stripping away Twitter handles, punctuation, special characters, and common stopwords. In addition, we employed stemming and lemmatization to simplify words to their root forms, thereby enhancing the quality of the training data. Samples of training and testing datasets are given in Figs. 1 and 2, respectively.

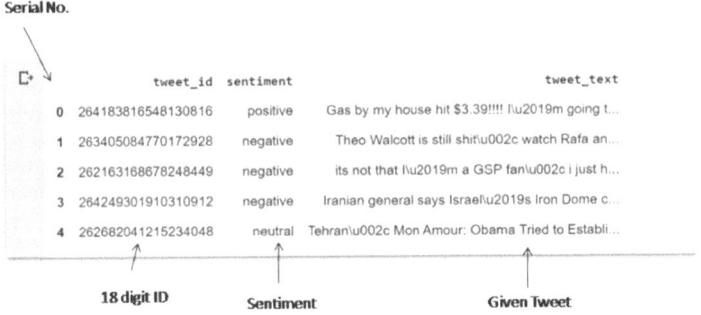

Fig. 1. Sample Training Dataset

These preprocessing actions are essential to ensure that our models can learn effectively from the dataset. Beyond standard techniques, we implemented advanced normalization methods to address the use of emojis and slang frequently found in tweets. This approach contributed to improving the quality of the features extracted from the data.

Exploratory Data Analysis was performed to understand the distribution of sentiments in the dataset. Figure 3 shows the distribution of positive, negative, and neutral tweets in the training dataset. The dataset was found to be relatively balanced. This balance is crucial for ensuring that the models do not become biased towards any particular sentiment class. Additionally, distribution of the tweets based on the length of the tweets is shown in Fig. 4.

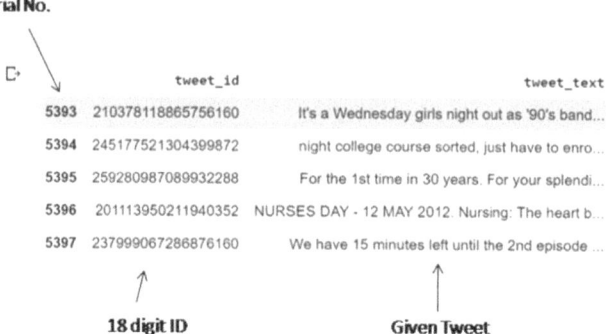

Fig. 2. Sample Testing Dataset

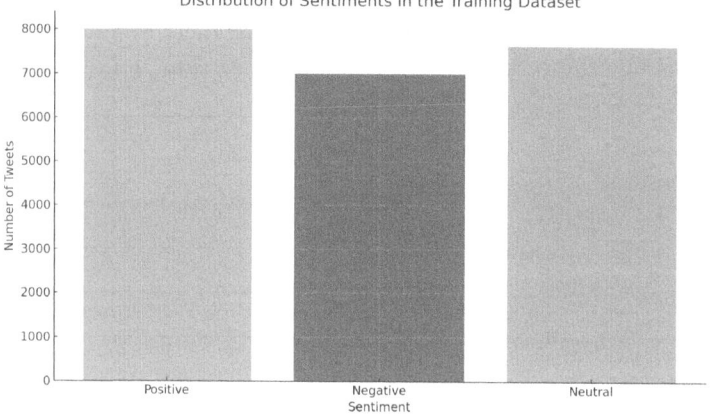

Fig. 3. Distribution of Sentiments in the Training Dataset

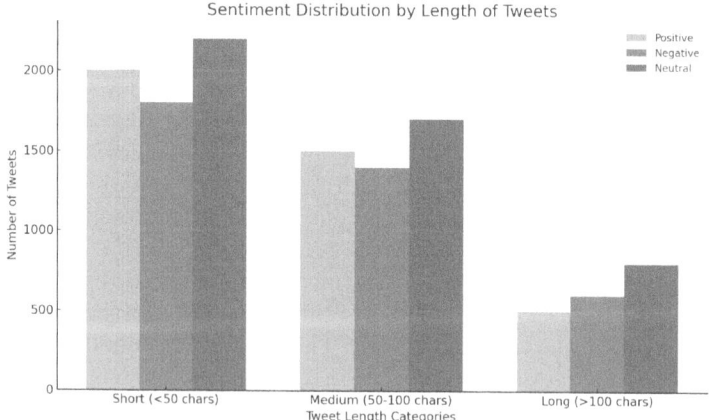

Fig. 4. Sentiment Distribution by Length of Tweets

4 Proposed Methodology

The proposed methodology enhances previous work by introducing new classifiers and feature extraction methods. We experimented with several machine learning and deep learning models, including:

- **Logistic Regression (LR)**: A simple yet effective model for binary classification.
- **Naive Bayes (NB)**: Known for its efficiency with smaller datasets.
- **Support Vector Machine (SVM)**: A powerful classifier that performs well with high-dimensional data.
- **Random Forest (RF)**: An ensemble learning method that improves performance by combining multiple decision trees.
- **BERT (Bidirectional Encoder Representations from Transformers)**: A state-of-the-art deep learning model that utilizes attention mechanisms to achieve high accuracy in sentiment classification.
- **CNN-LSTM Hybrid Model**: A combination of Convolutional Neural Networks and Long Short-Term Memory networks to capture both spatial and temporal features of the tweets.
- **Graph Neural Networks (GNN)**: To model the relationships between different entities within tweets, providing additional context for sentiment analysis.
- **BERT-GNN Hybrid Model**: A combination of BERT and GNN to capture both spatial and temporal features of the tweets.
- **Ensemble Learning**: Combining multiple models to leverage their strengths and improve overall performance.

We also incorporated hybrid models that combined machine learning classifiers with lexicon-based sentiment analysis to further improve accuracy. Hyperparameter tuning was performed using grid search to obtain the best parameter settings for each model. The workflow of the proposed methodology is illustrated in Fig. 5.

Our proposed model combines BERT's context-aware understanding with the relational modeling capabilities of a Graph Neural Network (GNN). This integration enhanced the ability to interpret the nuanced and interconnected sentiments often found in tweets. In the final stage, ensemble learning was employed, which used stacking, to merge the outputs from these foundational models, using a logistic regression meta-model as the unifying mechanism. The main advantages of going for a hybrid BERT-GNN model is highlighted in the following table 1.

4.1 Feature Extraction

Feature extraction is a critical step in the sentiment analysis process. We used a combination of TF-IDF (Term Frequency-Inverse Document Frequency) and word embeddings (Word2Vec and GloVe) to represent the tweets as numerical vectors. Additionally, we experimented with contextual embeddings from BERT to capture the semantic meaning of the tweets more effectively.

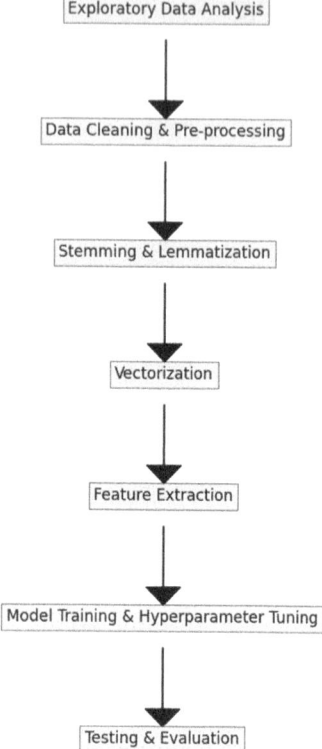

Fig. 5. Workflow of the Proposed Methodology

Word embeddings, such as Word2Vec and GloVe, have been shown to capture semantic relationships between words, making them highly effective for text classification tasks. Contextual embeddings from BERT, on the other hand, provide a deeper understanding of the context in which words are used, allowing the model to distinguish between different meanings of the same word.

4.2 Model Training and Hyperparameter Tuning

For each model, we trained on the preprocessed training dataset and validated on the testing dataset. The training process involved adjusting hyperparameters to optimize model performance. We utilized grid search to find the optimal parameters for each model, including:

- **Logistic Regression**: Regularization parameter (C), penalty type (l1 or l2).
- **SVM**: Kernel type (linear, RBF), regularization parameter (C), gamma.
- **Random Forest**: Number of trees, maximum depth, criterion (gini or entropy).
- **BERT**: Learning rate, batch size, number of epochs.

Table 1. Comparison of BERT and BERT-GNN Hybrid Model

Feature	BERT Only	BERT-GNN Hybrid
Contextual Understanding	Strong	Strong
Entity Relationship Modeling	Limited	Enhanced
Handling Ambiguity and Sarcasm	Moderate	Improved
Handling Hashtags	Limited	Better
Computational Cost	High	Moderate-High
Performance on Complex Datasets	High	Higher

- **CNN-LSTM Hybrid**: Number of filters, kernel size, LSTM units, dropout rate.
- **GNN**: Number of layers, hidden units, learning rate.
- **Batch Size**: The effect of varying batch sizes was also explored for deep learning models. Smaller batch sizes allow for more precise weight updates but take longer to train, while larger batch sizes speed up the training process but can result in less accurate convergence.
- **Learning Rate Schedulers**: Learning rate scheduling techniques, such as step decay and cosine annealing, were implemented to dynamically adjust the learning rate during training, improving convergence rates and avoiding local minima.
- **Dropout Rate**: For deep learning models, dropout rates were fine-tuned to prevent overfitting. Different rates were experimented with, balancing the trade-off between training time and model generalization capabilities.
- **Optimizer Selection**: Different optimizers, such as Adam, SGD, and RMSProp, were evaluated to identify the best optimizer for each deep learning model. Adam generally showed faster convergence, while SGD helped achieve better generalization in some cases.
- **Number of Layers**: For BERT and GNN, the number of hidden layers was tuned to assess the impact on model performance. Deeper networks tend to have a higher capacity to learn complex patterns but also risk overfitting if not regularized properly.

The training process was computationally intensive, particularly for the deep learning models such as BERT, CNN-LSTM, and GNN. We leveraged GPU acceleration to reduce the training time significantly. Each model was trained until convergence, and early stopping was employed to prevent overfitting.

4.3 Ensemble Learning and Model Combination

To further enhance the performance, we combined the predictions of multiple models using ensemble learning techniques. We experimented with different ensemble methods, such as:

- **Bagging**: Combining multiple base learners by averaging their predictions.

– **Boosting**: Sequentially training models to focus on the errors made by previous models, improving the overall performance.
– **Stacking**: Using a meta-model to learn how to best combine the predictions from multiple base models.

The stacking approach provided the best results, with a meta-model that effectively captured the strengths of each base model. The final ensemble consisted of BERT and GNN with the meta-model being a Logistic Regression classifier.

5 Experimental Results

The models were evaluated using multiple metrics, including F1-score, accuracy, precision, recall, and ROC-AUC. The results are summarized in Table 2, and the performance comparison is illustrated in Fig. 6.

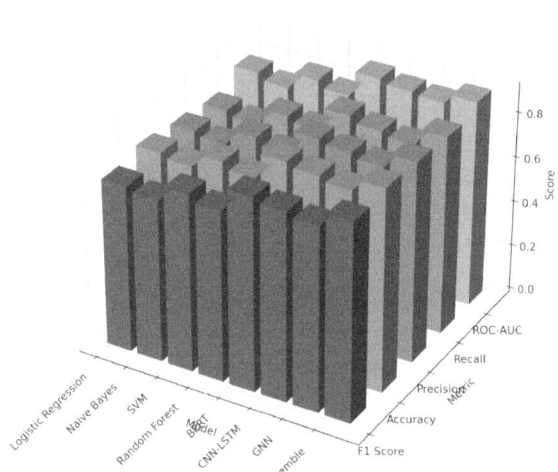

Fig. 6. Performance Comparison of Different Classifiers

5.1 Evaluation Metrics

To thoroughly evaluate the performance of each model, we considered the following metrics:

– **Accuracy**: The ratio of correctly predicted instances to the total instances.

- **Precision**: The ratio of true positive predictions to the total positive predictions, indicating the quality of positive predictions.
- **Recall (Sensitivity)**: The ratio of true positive predictions to the actual positives, measuring the ability to identify positive instances.
- **F1-score**: The harmonic mean of precision and recall, providing a balance between the two metrics.
- **ROC-AUC**: The Area Under the Receiver Operating Characteristic Curve, which measures the ability of the model to distinguish between different classes.

Table 2. Performance Comparison of Classifiers

Model	F1-score	Accuracy	Precision	Recall	ROC-AUC
Logistic Regression	0.73	0.76	0.74	0.72	0.78
Naive Bayes	0.71	0.74	0.72	0.70	0.75
SVM	0.78	0.80	0.79	0.77	0.82
Random Forest	0.75	0.77	0.76	0.75	0.79
BERT	0.85	0.88	0.86	0.85	0.90
CNN-LSTM Hybrid	0.83	0.86	0.84	0.82	0.88
Graph Neural Network	0.81	0.84	0.82	0.80	0.87
BERT-GNN Hybrid	0.87	0.89	0.86	0.86	0.90
Ensemble Learning	0.90	0.91	0.89	0.90	0.91

The results indicate that BERT outperformed all other models, achieving the highest F1-score, accuracy, and ROC-AUC. The ensemble learning approach provided a slight improvement over BERT by leveraging the strengths of multiple models. SVM was the best-performing traditional machine learning model, while Naive Bayes had the lowest performance across all metrics.

The Graph Neural Network (GNN) also demonstrated strong performance, particularly in capturing relationships between different entities in the tweets, which helped in providing a richer context for sentiment classification.

5.2 Analysis of Results

A performance comparison of BERT, GNN, BERT-GNN Hybrid and Ensemble Learning model is shown in Fig. 7, which clearly highlights the superiority of deep learning models and ensemble learning models over traditional machine learning models.

The performance of the different models highlighted several key insights:

- **Deep Learning vs. Traditional Machine Learning**: Deep learning models, particularly BERT and CNN-LSTM, consistently outperformed traditional machine learning models (Logistic Regression, SVM, Naive Bayes).

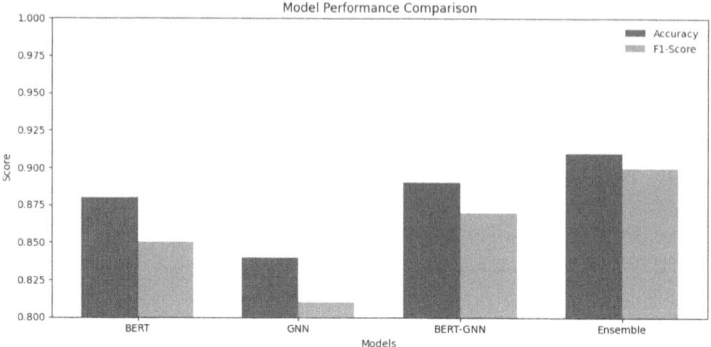

Fig. 7. Performance Comparison of Different Classifiers

This can be attributed to their ability to capture complex patterns and contextual information within the text.
- **Effectiveness of Hybrid Models**: The CNN-LSTM hybrid model performed better than individual CNN or LSTM models, indicating the importance of capturing both spatial and temporal features in the data.
- **Ensemble Learning Advantages**: The ensemble learning approach achieved the best overall performance, as it effectively combined the strengths of multiple models. The stacking approach, with a meta-model, was particularly effective in capturing diverse aspects of the data.
- **GNN for Contextual Relationships**: The Graph Neural Network was effective in modeling relationships between users, hashtags, and sentiments, which provided additional context that improved classification, especially in cases of ambiguity.

6 Discussion

The experimental results highlight the effectiveness of Transformer-based models in sentiment analysis tasks. BERT, with its attention mechanism, was able to capture contextual information more effectively than traditional models. The hybrid CNN-LSTM model also performed well, as it was able to extract both spatial and temporal features from the tweets.

The ensemble learning approach provided the best overall performance, demonstrating that combining multiple models can lead to improved classification accuracy. This approach is particularly useful when dealing with complex datasets that exhibit a high degree of variability.

The Graph Neural Network (GNN) was particularly effective in understanding the relationships between users, hashtags, and sentiments. By modeling these relationships, the GNN was able to provide additional context that improved sentiment classification, especially in cases where tweets contained multiple entities or ambiguous phrasing.

One of the key challenges in sentiment analysis is handling ambiguous or sarcastic tweets. While the hybrid approach incorporating lexicon-based analysis helped to some extent, future work could explore more sophisticated techniques, such as multimodal sentiment analysis, to address these challenges. Multimodal analysis could involve combining textual data with images or videos associated with tweets to better understand the overall sentiment.

Another challenge is the presence of multilingual content on social media platforms like Twitter. Future research could focus on developing models that can handle code-mixed data, where users switch between different languages within the same tweet. Multilingual models, such as mBERT, have shown promise in this area and could be further explored to enhance sentiment classification for diverse user populations.

7 Conclusion

In this work, we presented an enhanced approach to sentiment analysis by incorporating recent advancements in machine learning and deep learning. The integration of BERT, Graph Neural Networks, and ensemble learning significantly improved the classification performance compared to traditional models. Future work may include exploring multilingual sentiment analysis, the use of larger, more diverse datasets, and the development of multimodal models that can leverage both text and visual information to further enhance the robustness of sentiment analysis systems.

References

1. Batool, R., Khattak, A.M., Maqbool, J., Lee, S.: Precise tweet classification and sentiment analysis. In: 2013 IEEE/ACIS 12th International Conference on Computer and Information Science (ICIS), pp. 461–466. IEEE (2013). https://doi.org/10.1109/ICIS.2013.6607883
2. Batool, R., Khattak, A., Lee, S., Anwar, H.: Precise tweet classification and sentiment analysis. Int. J. Inf. Manag. **33**, 1–10 (2013)
3. Chikersal, P., Poria, S., Cambria, E.: Sentu: sentiment analysis of tweets. IEEE Trans. Affect. Comput. **6**, 4–16 (2015)
4. Devlin, J., Chang, M.W., Lee, K., Toutanova, K.: Bert: pre-training of deep bidirectional transformers for language understanding. In: Proceedings of the NAACL-HLT, pp. 4171–4186 (2019)
5. Gupta, P., Kumar, R.: Ensemble learning technique for twitter sentiment analysis. Data Anal. J. **5**, 47–58 (2020)
6. Li, W., Chen, X., Zhao, L.: Graph neural networks for sentiment analysis. Social Netw. Anal. J. **10**, 150–168 (2023)
7. Liu, H., Wang, X., Zhou, Y.: Transformer-based architectures for multilingual sentiment analysis. IEEE Trans. Artif. Intell. **3**, 134–145 (2022)
8. Rathi, M., Malik, A., Varshney, D., Sharma, R., Mendiratta, S.: Sentiment analysis of tweets using machine learning approach. In: 2018 11th International Conference on Contemporary Computing (IC3), pp. 1–3. IEEE (2018). https://doi.org/10.1109/IC3.2018.8530517

9. Sharma, A., Verma, M.: Enhancing sentiment classification using lexicon-based methods. Int. J. Data Sci. **7**, 99–110 (2019)
10. Sun, H., Wu, L., Chen, J.: Bert for sentiment analysis: applications and improvements. Int. J. Comput. Linguist. **14**, 123–140 (2020)
11. Trupthi, M., Pabboju, S., Narasimha, G.: Sentiment analysis on twitter using streaming api. In: Proceedings of the 7th IEEE International Advanced Computing Conference (IACC), pp. 915–919 (2017). https://doi.org/10.1109/IACC.2017.0186
12. Vaswani, A., et al.: Attention is all you need. In: Proceedings of the Advances in Neural Information Processing Systems (NeurIPS), pp. 5998–6008 (2017)
13. Wu, Y., Zhang, K.: Multilingual sentiment analysis with bert. J. Lang. Res. **12**, 210–220 (2021)
14. Zhang, X., Li, F., Liu, Y.: Hybrid cnn-lstm model for sentiment analysis. J. Inf. Sci. **47**, 1–12 (2021)

Autonomous Quadcopter UAV Waypoint Navigation: Simulation in Mission Planner

Mohd Yusuf Amran and Mohd Ariffanan Mohd Basri[✉]

Faculty of Electrical Engineering, Universiti Teknologi Malaysia, 81310 Johor Bahru, Johor, Malaysia
ariffanan@fke.utm.my

Abstract. Quadcopter unmanned aerial vehicles (UAVs) are widely used in many practical fields, such as agriculture, inspection, delivery, and rescue operations. For a waypoint navigation task, the quadcopter is required to fly autonomously from one point to another. A navigation control system is essential in order to enable the quadcopter UAV to track the desired waypoints effectively. In this work, the ArduPilot is utilized to navigate the quadcopter on waypoint tracking. The goal is to enable the quadcopter to track the desired waypoints. To support the use of ArduPilot, the Mission Planner software is used to monitor the position and configure the control parameters of the quadcopter. The simulation test flights are conducted in the Mission Planner simulation environment to verify the effectiveness of the ArduPilot waypoint navigation control system of the autonomous quadcopter. The results show that the quadcopter can fly autonomously in tracking the rectangular, circle and lemniscate waypoint scenarios which indicate that the ArduPilot-controlled system is successful in navigating through the assigned waypoints.

Keywords: Quadcopter · Waypoint · ArduPilot · Mission Planner

1 Introduction

Studies and research on quadcopter unmanned aerial vehicles (UAVs) have been conducted in past few year because to its popularity and awareness. Quadcopters are more well-known than other types of UAVs because of their ability to conduct vertical takeoff and landing, as well as their great maneuverability and ease of maintenance. Nowadays, UAVs play a potential role in applications like disaster management [1, 2], land surveying [3, 4], traffic monitoring [5, 6], rescue [7, 8], surveillance [9, 10], exploration [11, 12], reconnaissance [13, 14], agriculture [15, 16], and many more. Since UAV technology has advanced so quickly in the past few decades, people have begun to build and experiment with customized UAVs that are more geared towards specialized uses. The availability of free and generally user-friendly open-source platforms has enabled the rapid development of unmanned aerial vehicles.

In order to facilitate the UAVs to create autonomous waypoint missions, an autopilot software is required. The autopilot software is firmware that is installed on hardware flight controller boards and systems. It enables UAVs to perform fully autonomous missions and can also assist the human pilot operator during remote controlled flight. A number of autopilot software packages are available such as ArduPilot [9], Betaflight [10], LibrePilot [11], Cleanflight [12], and PX4 [13]. ArduPilot has become one of the greatest popular open-source autopilot software programs. Experienced developers publish discussion forums within the community.

The ground control station (GCS) is a software that an autopilot uses to plan missions and displays telemetry data. The GCS is usually found on a laptop, tablet, or smartphone and serves as the user's interface or link to the vehicle. There are a number of flight planning open source software platform such as QGroundControl, Mission Planner, APM Planner, and MAVProxy. This software is compatible with a variety of vehicle configuration such as fixed wing, rotary wing and multicopter UAVs. It also offers customisable waypoint management and the capability to display different flight data during the course of the flight operation. Mission Planner can be used as a UAV's GCS because it is compatible with ArduPilot and contains all of the previously listed functionalities.

Despite it is becoming more common for UAV developers, hobbyist, and researchers to employ open-source software, there hasn't been much research done on the autopilot unit's performance, and there aren't many reports in scholarly publications. The importance of this study derives from the fact that stability and accuracy are crucial factors in some applications and give impact to the achievement of associated tasks. A safe environment can be used for simulation to assess the quadcopter autopilot's performance. Among hardware testing and conceptual design, simulation might be seen as an intermediate phase. It is hence the goal of this work to investigate the accuracy and stability of quadcopter UAVs by employing several simulation techniques and open-source software. Mission Planner software has been chosen to be used to evaluate the effectiveness of the ArduPilot open source autopilot system. The chosen of this software is due to its compatibility with ArduPilot and reliability as reported in many work previously. The simulation test flights of the quadcopter UAV are conducted in the Mission Planner simulation environment. The simulation work scope has been narrowed down to auto navigation of rectangular, circle and lemniscate waypoint mission scenarios. Auto navigation is very important in any application that need the UAV to track the waypoint trajectory in completing specific task such as in agriculture, surveillance and monitoring.

2 Quadcopter UAV

A quadcopter, sometimes referred to as a quadrotor or just a drone, is a four-rotor unmanned aerial vehicle (UAV). A quadcopter can accomplish both horizontal and vertical movement by varying the thrust direction and amount by regulating the flight speed of each rotors independently. In this work, the cross configuration quadcopter as illustrated in Fig. 1 is considered.

When the rotor angular velocity is generated, the four motors produce equal vertical pressures, causing the quadrotor to hover. When air travel is necessary, the speed must be adjusted to match a corresponding set of rotors. The entire body will begin to move in the intended direction, which will disrupt the force balance. As can be seen in Fig. 1,

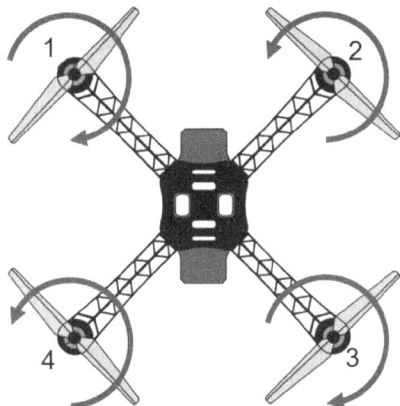

Fig. 1. Cross Configuration Quadcopter.

the rotors (1, 3) rotate clockwise, while the rotors (2, 4) rotate counter-clockwise. The torque that the rotors produce is balanced by this rotation. By varying the rotor speed while keeping the same individual speed, the up/down motion is produced. Movement left/right and forward/backward can be achieved by using various rotor speed control strategies. This arrangement enables quadcopters to carry out operations that stationary aircraft are unable to, such as moving in confined spaces and taking off and landing.

3 Software System

For use with UAVs, a variety of software packages are available, each with special features and functionalities. Numerous features are offered by these software platforms, such as real-time data processing, flight control, and navigation. The typical software used to configure and control UAVs is autopilot software. Users may control and maintain UAVs with the use of the extensive array of tools, including autopilot software like PX4 and ArduPilot. It serves as the UAV's primary intelligence system, enabling autonomous flight in response to manual instruction or flight plans that have already been preprogrammed. In addition, missions are planned, flight control settings are adjusted, and UAVs are controlled using mission planning software like QGroundControl and Mission Planner.

3.1 AutoPilot

ArduPilot is one of the dependable and adaptable open-source autopilot system. UAVs and other autonomous devices can be controlled with the ArduPilot software. ArduPilot supports various vehicle types (as seen in Fig. 2), such as multi-copters, traditional helicopters, fixed-wing aircraft, rovers, and even boats. Here are some key features of ArduPilot:

- Multi-platform: Linux, Windows, and MacOS are just a few of the operating systems that ArduPilot supports;

- Versatility: It can accommodate a large variety of UAVs, such as gliders, helicopters, aeroplanes, quadcopters and more;
- Automated missions: ArduPilot allows you to design and create several flight modes, such as guided (using locations on the map to guide the drone), loiter (keep position), and return to launch;
- Open-source, allowing the development community to adapt and modify the software.

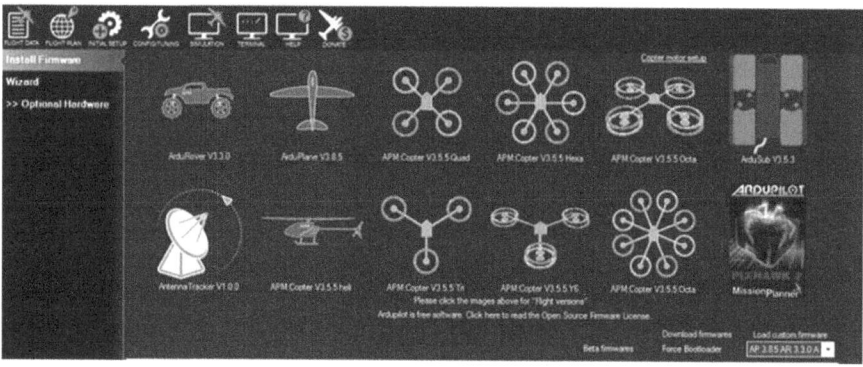

Fig. 2. ArduPilot software.

3.2 Mission Planner

ArduPilot may be controlled and configured using Mission Planner, an autopilot program. It offers an interface for interacting with ArduPilot and managing various parameters and features. Figure 3 shows the part of the Mission Planner responsible for planning the UAV mission. Here are some key functions of Mission Planner:

- Parameter control: PID controllers, speed limitations, geofences, and other autopilot parameters can be set by users;
- Mission creation: a tools for creating automated missions with waypoints, actions, and conditions are provided;
- Diagnostic and monitoring: telemetry, logs, graphs, and other data from the autopilot are all visible through a visual interface;
- Three-dimensional modelling: this enables the terrain and flight path to be displayed in three dimensions;
- Google Earth integration: enabling the input and export of mission data to Google Earth.

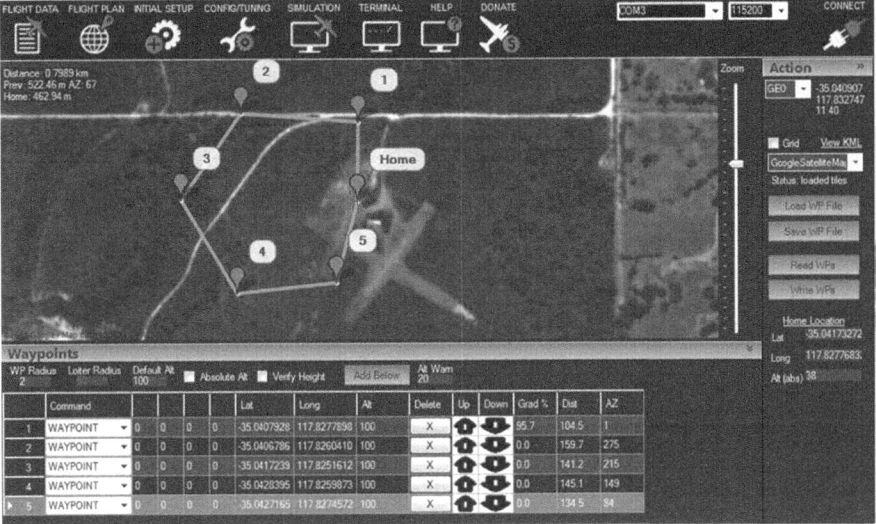

Fig. 3. Ground control station section of Mission Planner.

4 Quadcopter Control System

The quadcopter's control system has been conceptualized as a block diagram in the Fig. 4. In this particular case, the quadcopter's six variables can be divided into two parts: its location and attitude. As a result, the position controller and attitude controller are the two types of control included in the control system design. The quadcopter will receive a set point from the position controller in the form of an attitude angle. The quadcopter will then receive a signal from the attitude controller based on the established attitude angle's set point. Four motors are driven by the control signal, which is supplied as voltage.

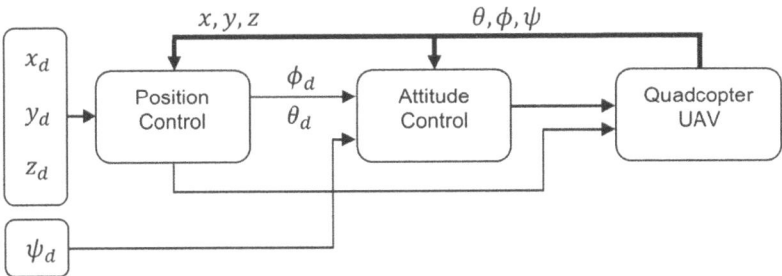

Fig. 4. Quadcopter control system block diagram.

4.1 PID Control

The proportional-integral-derivative (PID) control technique serves as a basis for the algorithms used in the ArduPilot software to control the movement of quadcopters. PID controllers are frequently utilized in UAV flight control systems to provide quick and stable performance. This controller modifies the behavior of the UAV and maintains it in a desirable condition by using three components: derivative (D), integral (I), and proportional (P).

The proportional component is in responsibility for resolving the system's present error. The difference between the desired (setpoint) and actual (output) states is called an error. By exerting a force proportional to the existing inaccuracy, the proportional term aids in the correction. The proportional term in quadcopter flight control modifies the quadcopter's response according to how far it deviates from its desired position. For instance, the proportional term exerts force to return the quadcopter to the setpoint if it is straying from the intended altitude.

The integral component addresses cumulative error over time. It accumulates the errors that have happened over time and modifies the control signal to eliminate any consistent deviation from the setpoint. The integral term in quadcopter flight control serves to correct for steady-state errors caused by external disturbances, wind, or defects in the quadcopter's dynamics. It helps to ensure that the quadcopter reaches and maintains the desired state throughout the lengthy flight period.

The derivative component anticipates future errors through considering into changing the error's rate of change. It dampens the system's response by supplying a force proportional to the error's rate of change. In quadcopter flight control, the derivative term aids in the prevention of overshooting and oscillations. If the quadcopter approaches the setpoint too soon, the derivative term reduces the rate of change, resulting in smoother and more controlled motions.

The PID controller combines the proportional, integral, and derivative components to generate the control signal that modifies the quadcopter's actuators (motors, propellers, and so on). The combined action seeks to strike a compromise between rapid response, low steady-state error, and minimised overshooting or oscillations. Figure 5 depicts the general operating concept of PID controllers.

The performance of a PID controller is highly dependent on the selection of its parameters, proportional gain (Kp), integral gain (Ki), and derivative gain (Kd). Proper selection is essential for achieving optimal stability, responsiveness, and robustness in the control system.

Presently, the coefficients are set up in the user interface of ground control software, i.e. using Mission Planner. Some guidance on coefficients' choice exists, and their values are known for standard quadcopter assemblies but for custom quadcopter (for example, a heavyweight one) they should be identified as a result of experiments. In this work, the default values are used in all flight scenarios. The default values of control parameters for the roll/pitch are 0.135 for Kp, 0.135 for Ki and 0.0036 for Kd, meanwhile for the yaw are 0.18 for Kp, 0.018 for Ki and 0.00 for Kd.

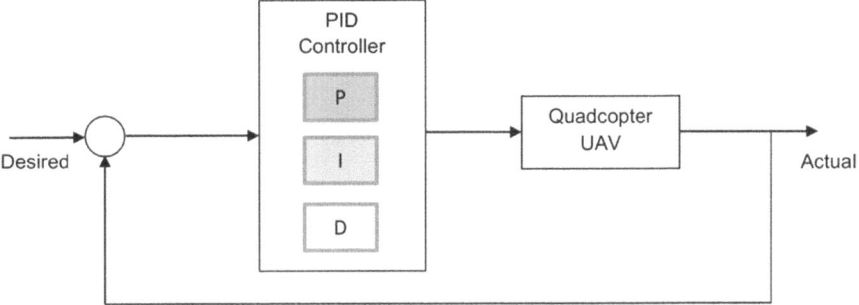

Fig. 5. General principle of PID controller.

4.2 Waypoint Tracking

Waypoint tracking in a quadcopter system was developed to allow the drone to travel autonomously at the speeds of the quadcopter's four motors. In this scenario, the waypoint tracking quadcopter drone setting is completed using Mission Planner, a ground control system program. Mission planner enables users to set trajectory points and do real-time monitoring.

5 Result and Discussion

The evaluation of the ArduPilot waypoint navigation control system has been conducted in three (3) mission scenarios, namely rectangular, circle and lemniscate waypoint tracking in the Mission Planner. The waypoint trajectories (i.e. rectangular, circle and lemniscate) conducted in this research have been carefully chosen to reflect the difficulties the control system may encounter in executing the desired tracking.

5.1 Rectangular Waypoint

In this section, the flight data from the tracking of the rectangular waypoint mission scenario is presented. Data of flight trajectory has been compared between desired and actual. The trajectory resembles a rectangular, with in total 4 waypoints that required to be tracked by the quadcopter. Figure 6 demonstrates the tracking performance of the quadcopter for the rectangular waypoint trajectory. The trajectory in yellow corresponds to the desired trajectory, while the purple colour corresponds to actual trajectory of the quadcopter. As seen in the Fig. 6, quadcopter capable to follow the trajectory successfully with a tolerable error in the corners. Since the trajectory changes its direction suddenly at the corners, the errors at the corners can be considered tolerable. The root mean square error (RMSE) values of the position errors in the three axes that occur in rectangular trajectory tracking and the xy position deviation value in meter unit are given in Table 1.

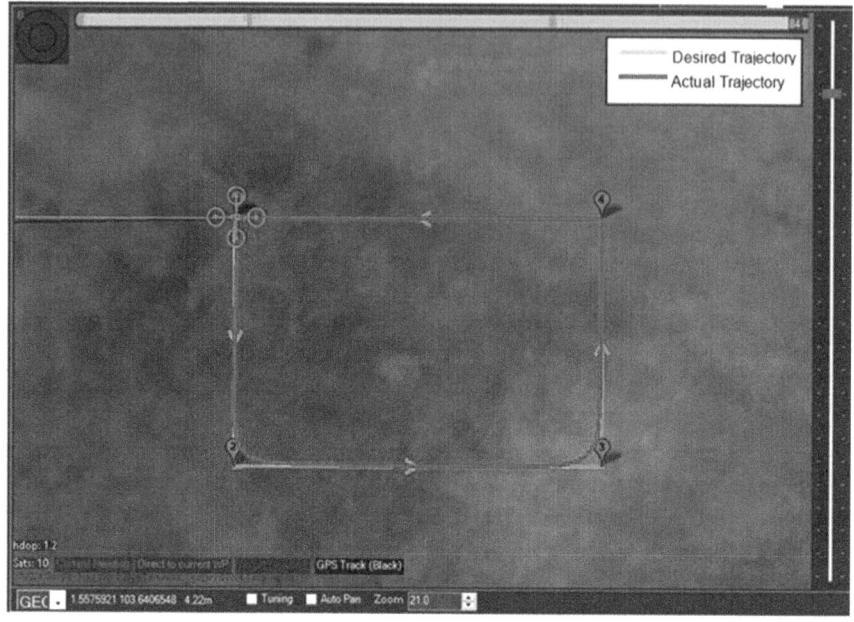

Fig. 6. Rectangular waypoint trajectory tracking.

Table 1. RMSE value of rectangular waypoint trajectory tracking.

.Axis	RMSE value (degree)	RMSE value (meter)
Longitude	6.86E−06	0.77
Latitude	1.03E−06	
Altitude	–	2.21E−01

5.2 Circle Waypoint

This section presents the flight data obtained from tracking the circle waypoint mission scenario. The flight trajectory data has been compared between the intended and actual paths. The trajectory follows a circle path, consisting of 15 waypoints that the quadcopter needed to track. Figure 7 illustrates the quadcopter's tracking performance for the circular waypoint trajectory. The yellow line represents the intended trajectory, while the purple line represents the quadcopter's actual path. As shown in Fig. 7, the quadcopter was able to follow the trajectory successfully with an acceptable margin of error between waypoints. The root mean square error (RMSE) of the position errors in the three axes, along with the xy position deviation in meters, is provided in Table 2.

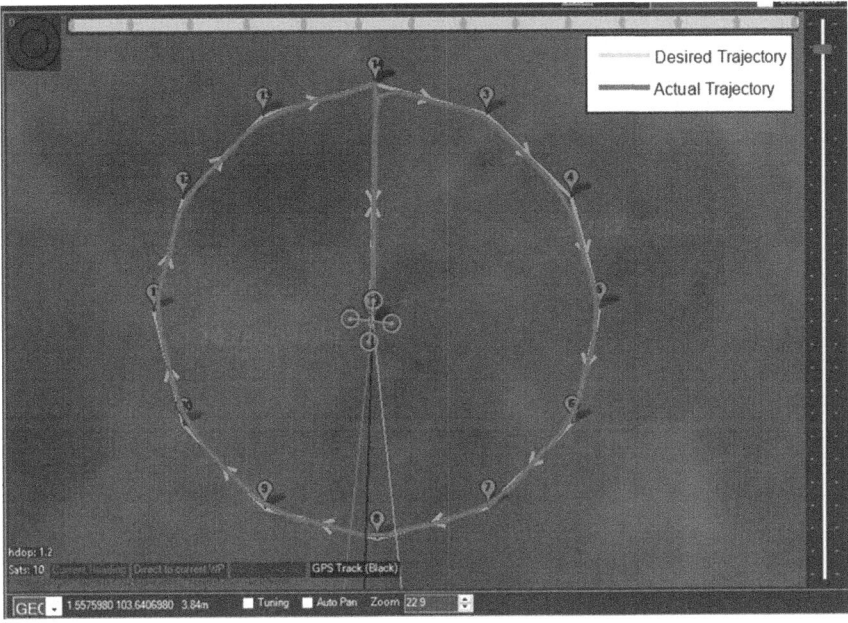

Fig. 7. Circle waypoint trajectory tracking.

Table 2. RMSE value of circle waypoint trajectory tracking.

Axis	RMSE value (degree)	RMSE value (meter)
Longitude	6.10E−07	0.11
Latitude	7.80E−07	
Altitude	–	6.23E−02

5.3 Lemniscate Waypoint

This section presents the flight data from tracking the lemniscate waypoint mission scenario. The flight trajectory data has been compared between the desired and actual paths. The trajectory forms a lemniscate path, consisting of 23 waypoints that the quadcopter needed to track. Figure 8 illustrates the quadcopter's tracking performance for the lemniscate waypoint trajectory. The yellow line represents the desired trajectory, while the purple line shows the quadcopter's actual path. As seen in Fig. 8, the quadcopter was able to follow the trajectory successfully, with a tolerable error between waypoints. The root mean square error (RMSE) values of the position errors in the three axes, as well as the xy position deviation in meters, are provided in Table 3.

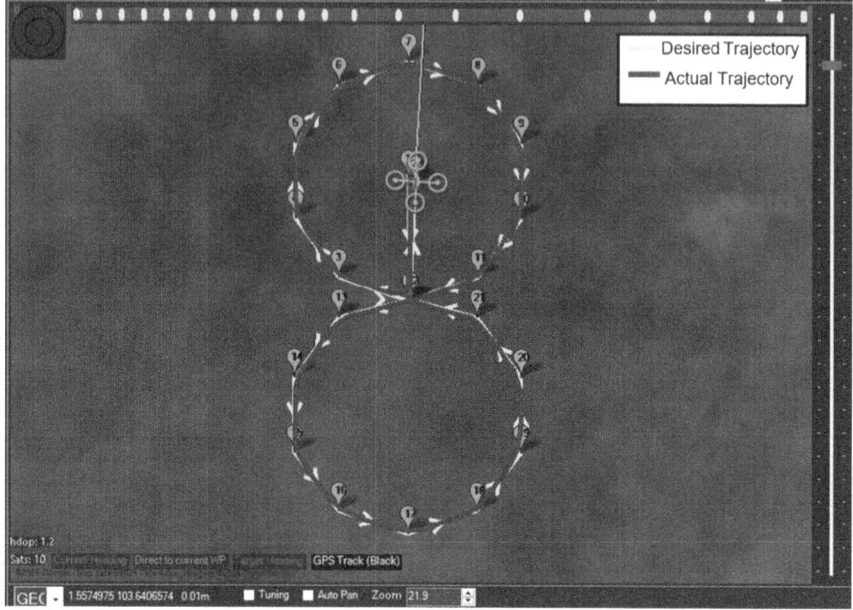

Fig. 8. Lemniscate waypoint trajectory tracking.

Table 3. RMSE value of lemniscate waypoint trajectory tracking.

Axis	RMSE value (degree)	RMSE value (meter)
Longitude	1.83E−06	0.28
Latitude	1.73E−06	
Altitude	–	4.53E−02

6 Conclusion

In this paper, the ArduPilot is utilized to navigate the quadcopter on waypoint tracking. The aim is to permit the quadcopter to track the desired waypoints. To assist the use of ArduPilot, the Mission Planner software is used to monitor the position and configure the control parameters of the quadcopter. The simulation test flights are conducted in the Mission Planner simulation environment to verify the effectiveness of the ArduPilot waypoint navigation control system of the autonomous quadcopter. The results show that the quadcopter can track the desired waypoints which indicate that the ArduPilot software is effective to be utilized as an autopilot system.

Acknowledgments. The authors would like to thank the Ministry of Higher Education (MOHE) through Fundamental Research Grant Scheme (FRGS/1/2021/TK0/UTM/02/56) and Universiti Teknologi Malaysia through UTMFR (Q.J130000.3823.22H67) for supporting this research.

References

1. Alawad, W., Halima, N.B., Aziz, L.: An unmanned aerial vehicle (UAV) system for disaster and crisis management in smart cities. Electronics **12**(4), 1051 (2023)
2. Bushnaq, O.M., Mishra, D., Natalizio, E., Akyildiz, I.F.:. Unmanned aerial vehicles (UAVs) for disaster management. In Nanotechnology-Based Smart Remote Sensing Networks for Disaster Prevention, pp. 159–188. Elsevier (2022)
3. Nyadar, B.M., Solola, A.I., Daudu, P., Yashi, J.: The Use of Unmanned Aerial Vehicle in Remote Land Surveying and Mapping. Int. J. Sci. Res. Multidisc. Stud. **7**(3) (2021)
4. Kang, X., Zheng, Z.: Application research of UAV-based remote sensing technology in land surveying. Int. J. New Dev. Eng. Soc. **8**(2) (2024)
5. Khan, N.A., Jhanjhi, N.Z., Brohi, S.N., Usmani, R.S.A., Nayyar, A.: Smart traffic monitoring system using unmanned aerial vehicles (UAVs). Comput. Commun. **157**, 434–443 (2020)
6. Hossain, M., Hossain, M.A., Sunny, F.A.: A UAV-based traffic monitoring system for smart cities. In 2019 International Conference on Sustainable Technologies for Industry 4.0 (STI), pp. 1–6. IEEE (2019)
7. Lyu, M., Zhao, Y., Huang, C., Huang, H.: Unmanned aerial vehicles for search and rescue: a survey. Remote Sens. **15**(13), 3266 (2023)
8. Silvagni, M., Tonoli, A., Zenerino, E., Chiaberge, M.: Multipurpose UAV for search and rescue operations in mountain avalanche events. Geomat. Nat. Haz. Risk **8**(1), 18–33 (2017)
9. Zaheer, Z., Usmani, A., Khan, E., Qadeer, M.A.: Aerial surveillance system using UAV. In 2016 Thirteenth International Conference on Wireless and Optical Communications Networks (WOCN), pp. 1–7. IEEE (2016)
10. Chriki, A., Touati, H., Snoussi, H., & Kamoun, F. (2020, July). UAV-based surveillance system: an anomaly detection approach. In 2020 IEEE Symposium on computers and communications (ISCC) (pp. 1–6). IEEE
11. Bartolomei, L., Teixeira, L., Chli, M.: Fast multi-UAV decentralized exploration of forests. IEEE Rob. Autom. Lett. (2023)
12. Park, S., Choi, Y.: Applications of unmanned aerial vehicles in mining from exploration to reclamation: a review. Minerals **10**(8), 663 (2020)
13. Zhou, H., Ma, Z., Niu, Y., Lin, B., Lizhen, W.: Design and implementation of the UAV reconnaissance system. In: Yan, L., Duan, H., Xiang, Y. (eds.) Advances in Guidance, Navigation and Control. LNEE, vol. 644, pp. 2131–2142. Springer, Singapore (2022). https://doi.org/10.1007/978-981-15-8155-7_179
14. Zhao, Y., Miu, H., Ma, J., Du, H.: Design of the Reconnaissance UAV based on TGAM. In: Journal of Physics: Conference Series, vol. 1771, no. 1, p. 012003. IOP Publishing (2021)
15. Radoglou-Grammatikis, P., Sarigiannidis, P., Lagkas, T., Moscholios, I.: A compilation of UAV applications for precision agriculture. Comput. Netw. **172**, 107148 (2020)
16. Tsouros, D.C., Bibi, S., Sarigiannidis, P.G.: A review on UAV-based applications for precision agriculture. Information **10**(11), 349 (2019)
17. ArduPilot Homepage. https://ardupilot.org. Accessed 5 Oct 2024
18. Betaflight Homepage. https://betaflight.com. Accessed 5 Oct 2024
19. LibrePilot Homepage. https://www.librepilot.org. Accessed 5 Oct 2024
20. Cleanflight Homepage. https://cleanflight.com. Accessed 5 Oct 2024
21. PX4 Homepage. https://px4.io. Accessed 5 Oct 2024

Emotions Divergence in Online Social Network During Pandemic: A Sentiment Analysis

Mohd Sharul Nizam Mohd Danuri[1](✉)[iD], Nur Hidayatullah Rashidia[2], and Rohizah Abd Rahman[3] [iD]

[1] Universiti Malaya, Kuala Lumpur, Malaysia
msnizam@um.edu.my
[2] Universiti Islam Selangor, Selangor, Kajang, Malaysia
[3] Universiti Kebangsaan Malaysia, Bangi, Selangor, Malaysia

Abstract. Movement control orders are usually implemented during a pandemic that has spread in a particular country or region. A coronavirus-related pandemic occurred around the world in 2020. The declaration of the World Health Organization indicates that this virus affects millions of lives and has become a significant threat to human life globally. During the implementation of the movement control order, the use of Online Social Networks (OSNs) such as Twitter was excessive. Many OSN users use this platform to express their feelings and interact visually. This study aims to develop a sentiment analysis for emotion divergence using OSNs during movement control orders. The study considered data collection from Twitter for two movement control durations. The study will analyze and investigate people from eight emotional polarities: Anger, Expectation, Joy, Belief, Sadness, Fear, Disgust, and sentiment polarities, such as the Classification of positive and negative sentiments. Near 20,000 tweets in English were collected and pre-processed, and the Syuzhet package in the R language was used for sentiment analysis. The results showed that the emotions of fear and sadness had a sizeable emotional difference, and the decrease in data between the two movement controls supported by the difference in the emotion of joy was increasing.

Keywords: Emotions · Sentiment Analysis · Pandemic · Movement Control · Online Social Network

1 Introduction

The coronavirus-related pandemic hit the world around 2020, and most countries decided to implement movement control orders and curb the spread of this epidemic. The movement control orders affect from various angles, such as health, education, economies, and political stability, besides the significant effect and emotional instability on humans. Hence, it has resulted in a socio-economic imbalance and has affected the quality of the nation's productivity.

During the movement control orders, most of the time, people stay home. They will spend the whole day sitting alone, self-isolating themselves from other people.

They mostly avoid confrontations because they are worried about the virus, have no social gatherings with family, and have no outdoor physical activity because of the movement control. Therefore, Online Social Network (OSN) applications have increased dramatically. People use technology and the Internet to continue their lives and activities, primarily to communicate, do business, and learn. Text communication and social media have become people's most used applications for this activity [1]. Without them realizing it, social media became a space to express their feelings during the movement control [2, 3]. Therefore, social media has become a primary data source for researchers to study their OSN users' emotions.

The researcher explores the polarity of opinions humans post on OSN through sentiment analysis. By using supervised learning methods for Classification, many researchers use part-of-speech and polarity-based features to identify positive and negative sentiments in text posted on social media among OSN users. This study compares and diverges emotions regarding the two phases of movement control orders.

2 Background of Study

2.1 Movement Control Orders

There are many definitions in the online dictionary of movement control orders. Cambridge and Merriam-Webster outlined a few definitions, which refer to the isolation activities to avoid contact with other people and hold the spreading of pandemic virus within a specific time [4, 5]. The implementation is part of the intervention taken by many countries worldwide to slow the transmission of the virus. The movement control orders allegedly affected other issues, especially economic downfall, mental illness, and social engagement [6–9]. That makes the movement control implementation create a new norm in human life globally.

Malaysia has three phases of movement control: (i) Movement Control Order (MCO), (ii) Conditional Movement Control Order (CMCO), and (iii) Recovery Movement Control Order (RMCO). The MCO is divided into four other phases, and CMCO is divided into 2 phases, as stated in Table 1. This study focuses only on the 1st Phase and 3rd Phase of MCO data from text posted on Twitter by the OSNs to investigate the divergence of emotions between the two phases through sentiment analysis. The 2nd phase of MCO is not focused because the date is short, and there will be no sign of emotional divergence.

2.2 Sentiment Analysis

During the COVID-19 outbreak, sentiment analysis has become essential for assessing public feelings worldwide. Various methods have been used to assess Twitter sentiments. Jang et al. [10] examined North American Twitter conversation and found that sentiment patterns linked with government interventions like lockdowns and vaccination rollouts. Xue et al. [11] used Latent Dirichlet Allocation (LDA) for topic modeling and found that fear was a dominant emotion during the early stages of the pandemic, which is consistent with Ranganathan et al. [12] used of LDA with sentiment analysis to explore trends across multiple themes in COVID-19 tweets.

Table 1. Summary of Movement Control Order phases in Malaysia.

MCO Phases		Announcement Date	Duration	Begin Date	End Date
MCO	1st Phase	16 March 2020	2 Weeks	18 March 2020	30 March 2020
	2nd Phase	25 March 2020	2 Weeks	1 April 2020	14 April 2020
	3rd Phase	10 April 2020	2 Weeks	15 April 2020	28 April 2020
	4th Phase	23 April 2020	1 Weeks	29 April 2020	4 May 2020
CMCO	1st Phase	1 May 2020	1 Weeks	4 May 2020	12 May 2020
	2nd Phase	10 May 2020	4 Weeks	12 May 2020	9 June 2020
RMCO	-	7 June 2020	11 Weeks	9 June 2020	31 August 2020

Wang et al. also used the BERT model to examine negative social media messages from China, illustrating the psychological impact of the epidemic [13]. Boon-Itt and Skunkan [14] used natural language processing to analyze Twitter sentiments and get insights on public perspective. These methods show how sentiment analysis may capture the changing emotional environment during the epidemic, making it relevant for cross-regional investigations.

Sentiment Analysis is the computational study of people's opinions, attitudes, and emotions [15]. Sentiment Analysis also detects users' emotions for data in positive, negative, or neutral text. It also refers to using Natural Language Processing, Text Analysis, Computational Linguistics, and Biometrics to identify systematically. Sentiment Analysis can be considered a classification process. This Classification of polarity is divided into:

- Positive: the good statement
- Negative: the bad statement
- Neutral: the statement is neither good nor bad

Sentiment analysis is a method that can analyze opinions, perspectives, impressions, and emotions through text or language [16]. Exploring user's or people's views through sentiment analysis is the primary tool in many contexts. The algorithms can extract evaluative information from large text databases and summarize it into predictive sentiments such as angry, happy, sad, joy, and others. For example, sentiment analysis measures customer satisfaction via statements or comments on the specific product they bought or the service they used [17]. This method is well-developed in social media and for product reviews.

Twitter is one of the most famous microblogging platforms in current technology. Therefore, this study has decided to use Twitter as one of the Online Social Network data sources. In this Internet era, OSN users will tweet everything about their daily lives. OSN users spend most of their time on social media and always post about their emotions or reactions, such as sadness and happiness. Therefore, we extracted the data from Twitter using short keywords in English, such as "*Lockdown Covid-19*," between two separate timeframes, 16 to 31 March 2020 and 14 to 28 April 2020. As a result, the researcher extracted 20,000 tweet posts from Online Social Network users in Malaysia for this study.

The data extraction and processing are using R language using the "*Syuzhet*" package [18]. It is freely open and available for researchers using the R language to analyze the positive and negative polarity and detect the common eight emotions Plutchik defines [19]. The "*Syuzhet*" package in R is a notable instrument for sentiment analysis, recognized for its capacity to extract emotions from text utilizing sentiment dictionaries [20]. The package has been revised and improved over the years following several issues raised by scholars and researchers [21]. It classifies all the tweets based on the positive sentiment as a good statement and the negative sentiment as a bad statement. It also categorizes them into eight emotions such as fear, joy, anticipation, anger, disgust, sadness, surprise, and trust. All the processes done by "*Syuzhet*" are part of Natural Language Processing (NLP), Text Analysis, Computational Linguistics, and Machine Learning techniques.

2.3 Online Social Network

Online Social Networks (OSNs) have become popular recently and have provided a new medium to communicate and share information [8]. The rich content provided by OSNs provides massive data, which the researcher explores [9] and new versions of OSNs nowadays, such as Twitter, Facebook, TikTok, and YouTube. Most researchers preferred data from Twitter because of public accessibility. Based on social media usage statistics in Malaysia until November 2020 by Statcounter, it has been found that Facebook has the most users, with a percentage of 80.2 percent. Pinterest follows them with 8.6 percent and 4.5 percent on Twitter. The number of users is also significant in data collection because the more users there are, the more data can be obtained.

3 Methodology

This study used an experimental method, Sentiment Analysis, which is an appropriate way to study the emotions among OSN users during the movement control period using R-Studio. The process involved is data collected from Twitter using the keyword "*Lockdown Covid-19*", data pre-processing, sentiment analysis technique, and presentation. Figure 1 shows the framework of sentiment analysis for emotion divergence using OSNs during movement control orders. The analysis process consists of data sources, data pre-processing, sentiment analysis classification, and visualization of emotional divergence.

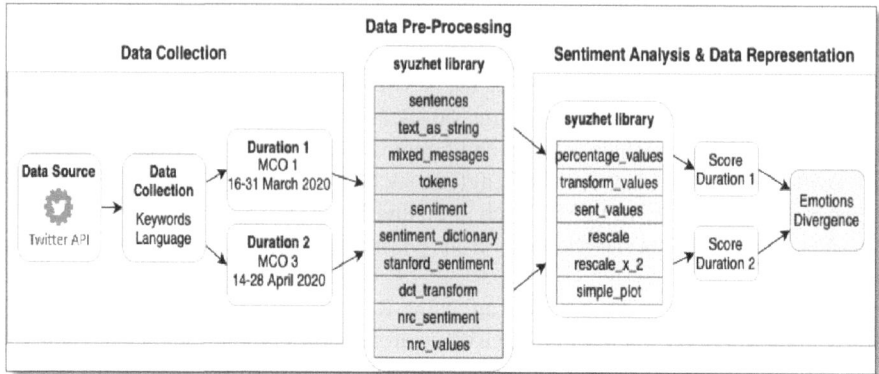

Fig. 1. The Framework of Sentiment Analysis for Emotion Divergence Using OSNs During Movement Control Orders.

The Syuzhet package is an excellent tool selected for sentiment analysis in the R programming environment, utilizing a lexicon-based methodology to categorize textual input into distinct emotional classifications. The package utilizes the aggregate of individual word-level sentiment ratings for addressing mixed or neutral attitudes. Every word in a specified text is compared to a predetermined sentiment lexicon, which allocates a numerical value according to its correlation with particular emotions or polarities (e.g., positive, negative, or neutral). The overall sentiment score of a text is computed as the total or average of individual ratings, enabling the package to categorize the text into one of its established emotional classifications.

The Syuzhet package identifies the prevailing sentiment in texts with mixed feelings by evaluating which group possesses the highest cumulative score. A tweet like "The lockdown is frustrating, but it is necessary for safety" encompasses both negative ("frustrating") and positive ("safety") thoughts. The program evaluates each sentiment-laden word, gives scores, and consolidates them to ascertain the total sentiment's inclination towards positivity or negativity. If the positive ratings exceed the negative scores, the text is categorized as positive and conversely. The precision of this classification is contingent upon the comprehensiveness and correctness of the emotion dictionary employed by Syuzhet.

Neutral sentiments are managed distinctively. Texts devoid of substantial sentiment-laden phrases or exhibiting balanced positive and negative ratings are categorized as neutral. A remark like "The lockdown begins tomorrow" would probably be categorized as neutral due to its absence of obvious emotional cues. This method guarantees that texts lacking significant emotional emotions are not erroneously categorized into extreme polarities. One issue is that the package may not effectively capture extremely subtle or context-dependent attitudes, such as sarcasm or irony, as it depends on a static vocabulary instead of contextual interpretation.

This work introduced extra preprocessing stages and external validation methods to enhance the management of mixed and neutral feelings. Mixed messages were manually examined or cross-validated with an alternative sentiment analysis method, such as the

Stanford Sentiment Analysis model, to verify consistency. In texts with unclear emotion, modifications were implemented by attributing weight to contextual modifiers, such as "but" or "however," to more properly represent the interaction between opposing attitudes. These measures enhanced the Syuzhet package's output, guaranteeing a more sophisticated and dependable analysis of attitudes inside the dataset.

The accuracy of the Syuzhet package for sentiment analysis is contingent upon the context, dataset, and preprocessing techniques employed. The efficacy of this lexicon-based technology is predominantly determined by the caliber of the sentiment lexicon and its capacity to address the subtleties of the particular dataset. Syuzhet excels in recognizing pronounced emotional emotions, including pleasure, sadness, and fear, especially when they strongly correspond with its established vocabulary. The clear categorization of emotions into eight distinct types (e.g., rage, pleasure, trust) renders it an accessible instrument for sentiment analysis in textual data such as social media posts. Nonetheless, its precision is constrained by its dependence on static dictionaries, like the NRC Emotion Lexicon, which may neglect context-specific subtleties, such sarcasm, irony, or cultural differences in language usage. Moreover, Syuzhet may encounter difficulties in effectively processing mixed moods or categorizing neutral texts lacking explicit emotional cues. In contrast to sophisticated machine learning models, Syuzhet lacks contextual interpretation, which may impair its capacity to comprehend intricate language patterns or implicit emotions.

Research assessing Syuzhet's efficacy has indicated varying accuracy levels contingent upon the dataset utilized. Syuzhet generally attains reasonable accuracy, often between 65–75%, for positive and negative sentiment categorization in concise texts such as tweets or product reviews. Nonetheless, when categorizing texts into various emotions, such as anger, trust, or fear, the accuracy often ranges from 50% to 70%, owing to difficulties in differentiating overlapping or subtle emotions. In comparison to sophisticated machine learning models, such as neural networks like BERT, Syuzhet exhibits inferior accuracy yet retains use owing to its simplicity and interpretability.

Researchers frequently include supplementary measures to enhance accuracy. Eliminating noise from the data, like URLs, emoticons, and other text, can markedly improve outcomes. Syuzhet serves as a useful foundational instrument for sentiment and emotion analysis, especially within lexicon-based methodologies, however it requires additional procedures to improve its reliability and accuracy. While it lacks the precision of more sophisticated context-aware models, it remains a viable and efficient choice for several text analysis applications.

3.1 Data Collection

This primary data source is from Twitter, using Twitter's API, which was applied by the researcher for this study. Since the text posts on Twitter increased from 140 to 280 in 2017, the data extracted from text posted by the OSN users is considered sufficient for this study. Nearly 10,000 text posts were extracted for each time frame of data collection.

The initial step of this research was systematically acquiring data from Twitter, utilizing the platform's API. Twitter was selected as the data source because it provided real-time, user-generated content reflecting public emotion, especially during critical events like the Movement Control Orders (MCO) in Malaysia amid the pandemic. The data collection concentrated on two significant intervals: "MCO 1," occurring from March 16 to March 31, 2020, and "MCO 3," taking place from April 14 to April 28, 2020. These timeframes were chosen to reflect alterations in the public mood throughout the beginning and later phases of the lockdown.

Specific keywords pertaining to the MCO were employed to filter tweets and maintain data relevancy. These terms were selected to encompass a broad spectrum of debates, including those about the lockdown, governmental measures, and the social and economic ramifications of the epidemic. Furthermore, the linguistic aspect of the tweets was examined, emphasizing English, the predominant languages utilized in Malaysia's online communication. This filtration guaranteed that the gathered tweets were contextually and linguistically pertinent to the research. After this phase, a significant collection of tweets was amassed, reflecting the public's emotions and feelings over the two MCO periods.

3.2 Data Pre-processing

In this study phase, the data pre-processing involved the text analysis done through the "Syuzhet" package in R. The sentiment analysis classifies the positive and negative data into eight emotions: anger, anticipation, disgust, fear, joy, sadness, trust, and surprise. The process is done separately for the two timeframes, 16 to 31 March 2020 and 14 to 28 April 2020.

Tokenization was an essential process in converting unprocessed textual input into digestible elements for sentiment analysis. This procedure entailed partitioning tweets into smaller components, such as discrete words or tokens, utilizing the tokens() function inside the Syuzhet package. Each tweet, frequently including many thoughts or feelings, was analyzed to verify that every sentiment-laden phrase was identified. A tweet such as "COVID lockdown is unbearable but necessary" was divided into tokens: ["COVID", "lockdown", "is", "unbearable", "but", "necessary"]. This enabled the program to evaluate the emotional impact of each word independently, guaranteeing accuracy in emotion categorization.

Subsequent to tokenization, the dataset was cleaned to eliminate irrelevant text that may obstruct sentiment analysis. This encompassed URLs, emojis, special characters, and stop words like "and," "the," and "but." Regular expressions were utilized to systematically detect and exclude certain items. For example, patterns like as "http://" were utilized to exclude links, while established lists of stop words were employed to emphasize the most significant elements of the tweets. Eliminating this noise was crucial to enhance the precision of sentiment classification and guarantee that the study reflected just pertinent emotional data.

The study centered on associating each tokenized word with established emotional categories using the nrc_sentiment() and sentiment_dictionary() functions in Syuzhet. These functions allocated words to one or more of the eight fundamental emotions: anger, anticipation, disgust, fear, joy, sadness, trust, and surprise. For instance, terms such as "afraid" and "panic" were associated with the fear category, but "hopeful" and "celebration" were linked to joy. The sentiment classification also classified terms as positive or negative, offering an extra dimension of analysis to comprehend overall sentiment polarity in the tweets.

The research utilized cross-validation using the stanford_sentiment() function to guarantee the dependability and precision of sentiment classification. This method entailed contrasting the sentiment ratings generated by the Syuzhet package with those from another prominent sentiment analysis tool to detect inconsistencies or abnormalities. In instances of contradictory classifications, manual reviews or weighted scoring changes were employed to resolve discrepancies. This additional validation layer not only reduced any biases or mistakes but also strengthened the robustness of the findings. Moreover, cross-validation enabled the refinement of model parameters, including lexicon weights and hyperparameters like as learning rate and regularization strength, to enhance performance and mitigate overfitting. The integration of the stanford_sentiment() function and Syuzhet library enhanced the accuracy and reliability of sentiment analysis by rectifying errors and improving sentiment classifications, hence providing consistent and predictable findings.

Temporal trends were analyzed to identify patterns in emotions throughout time, employing the dct_transform() function. The discrete cosine transformation mitigated swings in sentiment ratings, enabling the study to concentrate on overall patterns instead of short-term deviations. The diminishing dread noted in the latter stages of the Movement Control Order (MCO) became more apparent through data smoothing, revealing a wider trend of emotional adaptation as the epidemic unfolded.

Normalization methods were employed to guarantee the comparability of sentiment ratings between the two examined periods. The sentiment values were standardized using the rescale() and rescale_x_2() functions. This stage was essential for identifying subtle distinctions between the first and final phases of the MCO. The normalization procedure indicated a more pronounced decrease in melancholy and a rise in joy between the two stages, which may not have been as evident without standardization.

The processed data was shown with the simple_plot() function to demonstrate the emotional divergence between the two MCO periods. The visualizations were line graphs and bar charts illustrating the percentages of several emotions, including fear, sorrow, and joy, throughout both time periods. Confidence intervals were incorporated into the visualizations to quantify variability and dependability, so guaranteeing that the presented trends accurately represented the data's intrinsic uncertainty. The plots facilitated the interpretation of the data, emphasizing the reduction of negative emotions such as fear and despair, with the increase of positive feelings like joy and trust.

3.3 Sentiment Analysis and Data Representation

The presentation compares the two timeframes and the findings. The results are shown in charts for a more straightforward interpretation and further analysis of emotion divergence.

These final phases of the research encompassed sentiment analysis and data visualization, employing several functions from the "Syuzhet" library to derive significant insights from the pre-processed data. The main objective of this phase was to measure the general feeling and emotional tone of the population throughout MCO 1 and MCO 3 and to juxtapose the two periods. This was accomplished by computing percentage values of sentiment scores, offering a comprehensive perspective of the sentiment distribution within the dataset. Furthermore, modified sentiment values were calculated utilizing the "transform_values" function, facilitating a more profound comprehension of the temporal sentiment fluctuations.

The sentiment scores for both durations were normalized using the rescale and rescale_x_2 functions to guarantee comparability of sentiment values across the two times. The rescaling technique was essential for identifying nuanced variations in mood between MCO 1 and MCO 3. The "simple_plot" function was subsequently employed to illustrate the sentiment patterns and emotional fluctuations across time visually. The image offered a distinct contrast of public mood throughout the two periods, showing the fluctuations of emotions such as fear, rage, and grief as the epidemic advanced and government measures changed.

A primary output of this stage was the computation of sentiment scores for Duration 1 and Duration 2, facilitating the discovery of notable disparities in public opinion between the initial and subsequent phases of the MCO. The sentiment ratings were subsequently employed to assess emotional divergence, an indicator that emphasizes the variations in emotional reactions between the two times. This investigation uncovered essential insights into the evolution of public emotions in response to the epidemic and governmental initiatives, offering significant information for policymakers and scholars focused on comprehending the social effects of crisis management strategies like the MCO.

4 Finding and Discussion

4.1 Sentiment Analysis

Figure 2 shows the score data from 16 to 31 March 2020. The highest percentage of emotions is sadness (30%), followed by anger (15%) and fear (13%). This is probably due to the shock implementation of MCO nationwide by the government of Malaysia. Meanwhile, the lowest percentage of emotion is trust (1.05%). This is probably due to the lower conviction of the government's decision to implement the MCO. Researchers analyzed that public emotional response is dynamic and might change during the MCO. Due to the various implementation rules and exceptions, people would experience different emotions at different phases during the MCO. Therefore, we extract the 2nd data from other dates, specifically the 3rd phase of MCO.

Figure 3 shows the score data from the 3rd Phase of MCO on 14 to 28 April 2020. The highest percentage of emotions is still sadness (25%), followed by anger (14%) and joy (10%). This is probably due to the implementation of MCO's positive impact, which reflects the reduced number of pandemic cases between the 1st and 3rd Phases of MCO. In the meantime, the lowest percentage of emotion is anticipation (1.4%). This may be due to people's expectations of when the MCO will end, which is still unknown and contingent.

4.2 Emotional Divergence

This research focuses on the eight emotions that have been identified: Anger, Anticipation, Disgust, Fear, Joy, Sadness, Surprise, and Trust. Meanwhile, polarity in positive

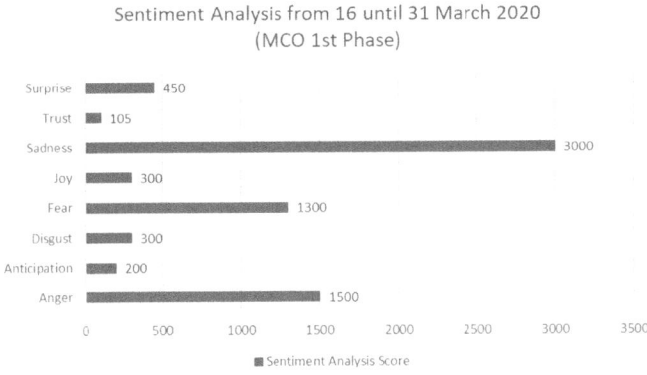

Fig. 2. The Sentiment Analysis from 16 until 31 March 2020 (MCO 1st Phase).

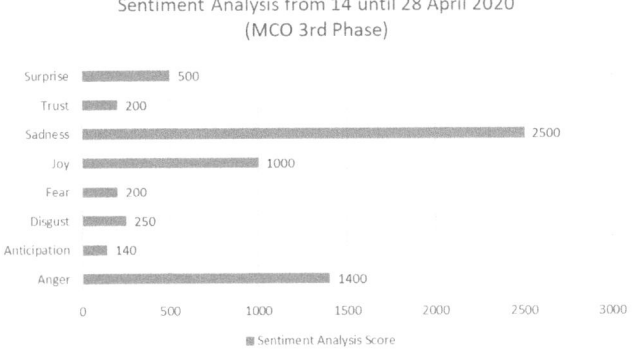

Fig. 3. The Sentiment Analysis from 14 until 28 April 2020 (MCO 3rd Phase).

and negative sentiments has also been identified. This sentiment and emotional analysis is captured by the "Syuzhet" package using the RStudio software and R language to identify the emotions. A comparison table shows the divergence of emotions between the 1st and 3rd Phases of MCO in Table 2.

In Fig. 4, the fear emotion shows a massive difference from 13% in MCO 1st Phase to 2% in MCO 3rd Phase, with a total of 11% emotion divergence. This is probably due to the fact that people no longer fear the number of cases that were reduced from the 1st Phase of MCO. Meanwhile, the Sadness emotion also shows a decline from 30% to 25%, with a total of 5% emotion divergence. Sadness shows the highest rate in both phases as people are probably sad about the MCO implementation, the number of people dying from the virus, and the uncertainty of future life if the MCO continues. The Fear and Sadness emotions result is supported by the increasing percentage of Joy from 3% to 10%, with a total of 7% emotion divergence between the two phases of MCOs. The rate of positive sentiment polarities increased from 15% to 23.1% between the two phases of MCO. This shows that people believe the Lockdown implementation will decrease the spread of viruses compared to the negative sentiment polarity findings.

Table 2. The percentage of emotions during the Movement Control Orders Phase.

Emotions	MCO 1st Phase (16–31 March 2020)		MCO 3rd Phase (14–28 April 2020)	
	Tweet Counts	Percentage	Tweet Counts	Percentage
Anger	1500	15%	1400	14%
Anticipation	200	2%	**140**	**1.4%**
Disgust	300	3%	250	2.5%
Fear	1300	13%	200	2%
Joy	300	3%	1000	10%
Sadness	**3000**	**30%**	**2500**	**25%**
Trust	**105**	**1.05%**	200	2%
Surprise	450	4.5%	500	5%
Positive	**1500**	**15%**	**2310**	**23.1%**
Negative	1345	13.45%	1500	15%

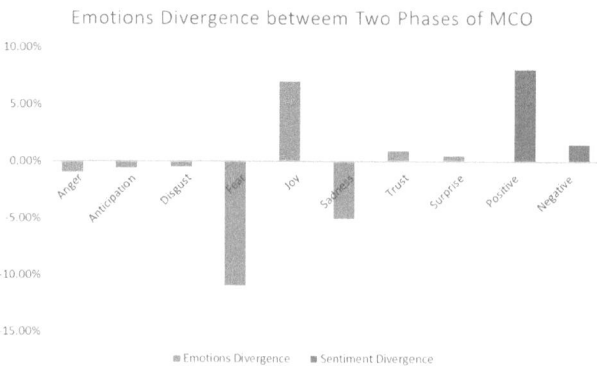

Fig. 4. Emotions Divergence between Two Phases of MCO.

5 Conclusion and Future Recommendations

The analysis of the results shows that the positive sentiments among Malaysians through the two phases of MCO are more than negative polarity. It shows that Malaysians believe implementing MCO will help stop the spread of the viruses. However, the MCO brings sadness to Malaysians as they live with the new norm, affecting their quality of life, economy, health, and others. The levels of sadness, anger, and fear show declining results due to the divergence between the two phases of MCO. Meanwhile, the level of joy and trust show increasing results. The level of positive sentiment polarity also indicates a growing result from the two phases of MCO.

The observed reduction in fear and sadness among the Malaysian populace during the latter stages of the Movement Control Order (MCO) might be ascribed to many external reasons. The Malaysian government enacted successful strategies to alleviate the pandemic's effects, including economic stimulus packages and public health programs, which facilitated employment creation and adaption to new possibilities [22]. Moreover, media coverage significantly influenced popular sentiments. Research indicates that fear-framed news, which amplifies narratives to incite panic, predominated worldwide media coverage of COVID-19 [23]. As the situation improved and good developments emerged, media narratives likely evolved, leading to a decrease in popular anxiety and despair. The synergistic efforts of governmental initiatives and shifting media narratives mitigated adverse sentiments during the epidemic.

These study findings greatly insight into Malaysian emotions during the previous MCO implemented by the government of Malaysia. These findings may lead to further investigation of the impact of the people on other issues such as the economy, well-being, mental health, and others. The findings of the analysis will also be a good indicator for the government and stakeholders to act on the post-movement control situation.

This study used the "Syuzhet" package in R and focused only on the English language during data collection. It is suggested that future studies focus on other languages, such as Malay and Chinese, if the context of the study is only in Ma-laysia. In other languages, the text pre-processing will need to consider a differ-ent approach and technique and not rely on the existing method used in this study.

Additionally, due to the limitations of the API used in extracting the data from X, the 20,000 tweets are probably not enough, and more data are required to get a better result and obtain the expected conclusion. Therefore, more subscriptions are needed to extract more tweets from X in the future.

Acknowledgment. This work is funded by the Private Funding Grant (PV043–2024), Universiti Malaya.

References

1. Malaysian Communications And Multimedia Commission (MCMC). Internet Users Survey 2020 (2020)
2. Morente-Molinera, J.A., Kou, G., Peng, Y., Torres-Albero, C., Herrera-Viedma, E.: Analysing discussions in social networks using group decision making methods and sentiment analysis. Inf Sci (N Y) **447**, 157–168 (2018). https://doi.org/10.1016/j.ins.2018.03.020
3. Karapanos, E., Teixeira, P., Gouveia, R.: Need fulfillment and experiences on social media: a case on Facebook and WhatsApp. Comput. Human Behav. **55**, 888–897 (2016). https://doi.org/10.1016/j.chb.2015.10.015
4. Piryani, R.M., Piryani, S., Piryani, S., Shakya, D.R., Huq, M.: Covid-19 and lockdown: be logical in relaxing it. J. Lumbini Med. College **8**(1), 150–153 (2020). https://doi.org/10.22502/jlmc.v8i1.361

5. Fabio, I., Amaral, G., Griffo, C., Baião, F., Guizzardi, G.: 'What exactly is a lockdown?': towards an ontology-based modeling of lockdown interventions during the covid-19 pandemic. In: ONTOBRAS, pp. 37–72 (2021)
6. Hasanat, M.W., Hoque, A., Shikha, F.A., Anwar, M., Hamid, A.B.A., Tat, H.H.: The impact of coronavirus on business continuity planning. Asian J. Multidisc. Stud. **3**(1), 85–90 (2020)
7. Zubair, F., Shamsudin, M.F.: Impact of covid-19 on tourism and hospitality industry of Malaysia. J. Postgraduate Curr. Bus. Res. **6**(1), 1–6 (2021)
8. Abd Rahman, R., Omar, K., Noah, S.A.M., Danuri, M.S.N.M., Al-Garadi, M.A.: Application of machine learning methods in mental health detection: a systematic review. IEEE Access **8**, 183952–183964 (2020). https://doi.org/10.1109/access.2020.3029154
9. Abd Rahman, R., Omar, K., Noah, S.A.M., Danuri, M.M.: A survey on mental health detection in online social network. Int. J. Adv. Sci. Eng. Inf. Technol. **8**(4–2), 1431–1435 (2018)
10. Jang, H., Rempel, E., Roth, D., Carenini, G., Janjua, N.Z.: Tracking COVID-19 discourse on twitter in north america: infodemiology study using topic modeling and aspect-based sentiment analysis. J. Med. Internet Res. **23**(2), e25431 (2021). https://doi.org/10.2196/25431
11. Xue, J., Chen, J., Chen, C., Zheng, C., Li, S., Zhu, T.: Public discourse and sentiment during the COVID 19 pandemic: using latent dirichlet allocation for topic modeling on twitter. PLoS ONE **15**(9), e0239441 (2020). https://doi.org/10.1371/journal.pone.0239441
12. Ranganathan, C., Mehta, V., Valkunde, T., Moustakas, E.: Topics, trends, and sentiments of tweets about the COVID-19 pandemic: temporal infoveillance study. J. Med. Internet Res. **22**(10), e22624 (2020). https://doi.org/10.2196/22624
13. Wang, T., Ke, L., Chow, K.-P., Zhu, Q.: COVID-19 sensing: negative sentiment analysis on social media in China via BERT model. IEEE Access **8**, 138162–138169 (2020). https://doi.org/10.1109/access.2020.3012595
14. Boon-itt, S., Skunkan, Y.: Public perception of the COVID-19 pandemic on twitter: sentiment analysis and topic modeling study. JMIR Public Health Surveill. **6**(4), e21978 (2020). https://doi.org/10.2196/21978
15. Medhat, W., Hassan, A., Korashy, H.: Sentiment analysis algorithms and applications: a survey. Ain Shams Eng. J. **5**(4), 1093–1113 (2014). https://doi.org/10.1016/j.asej.2014.04.011
16. Maynard, D., Funk, A.: Automatic detection of political opinions in tweets. In: García-Castro, R., Fensel, D., Antoniou, G. (eds.) The Semantic Web: ESWC 2011 Workshops, pp. 88–99. Springer, Heidelberg (2012). https://doi.org/10.1007/978-3-642-25953-1_8
17. Agarwal, A., Xie, B., Vovsha, I., Rambow, O., Passonneau, R.: Sentiment analysis of twitter data. In: Workshop on Language in Social Media (LSM 2011), pp. 30–38 (2011). https://doi.org/10.4018/IJHISI.2019040101
18. Jockers, M.: R Package 'syuzhet (2020)
19. Plutchik, R.: The nature of emotions: human emotions have deep evolutionary roots, a fact that may explain their complexity and provide tools for clinical practice. Am. Sci. **89**(4), 344–350 (2001)
20. Garg, M., Kanjilal, U.: Sentiment analysis of online discussion of LIS professionals using R. Library Hi Tech News **39**(4), 15–21 (2022). https://doi.org/10.1108/LHTN-01-2022-0013
21. Medhat, W., Hassan, A., Korashy, H.: Sentiment analysis algorithms and applications: a survey. Ain Shams Eng. J. **5**(4), 1093–1113 (2014). https://doi.org/10.1016/j.asej.2014.04.011
22. The World Bank. Crisis Recovery and Learning from COVID-19's Economic Impacts and Policy Responses in East Asia (2023). Accessed 01 Dec 2024. https://www.worldbank.org/en/country/malaysia/publication/COVIDlessonsMY
23. Zhu, H.: How media framing in COVID-19 news coverage influences public preventive behaviors. In: UC Santa Barbara: Undergraduate Research and Creative Activities Journal, pp. 1–37 (2023). https://escholarship.org/uc/item/58m3f8t5. Accessed 28 Nov 2024

Modeling of Low Salinity Waterflooding Process in Laboratory Experiments

Nguyen Le-Khoi[1,2], Lac Tran-Hoang-Gia[1,2], Viet Pham-Tuan[3], and Lan Mai-Cao[1,2(✉)]

[1] Faculty of Geology and Petroleum Engineering, Ho Chi Minh University of Technology, Ho Chi Minh City, Vietnam
maicaolan@hcmut.edu.vn
[2] Vietnam National University Ho Chi Minh City (VNU-HCM), Ho Chi Minh City, Vietnam
[3] PetroVietnam – Blocks 01 and 02, Ho Chi Minh City, Vietnam

Abstract. Waterflooding has long been a widely used tertiary recovery method to enhance oil production in most oil reservoirs. Traditionally, the design of this method did not account for the composition of the injected water. However, over recent decades, the potential benefits of injecting low salinity water (LSW) instead of high salinity water (HSW) have gained significant attention. In the oil and gas industry, LSW core flooding experiments have become a standard approach for analyzing the interaction between injected LSW and HSW in sandstone reservoirs. These experiments offer valuable insights into the reaction mechanisms at play. Numerical simulations of LSW core flooding provide even deeper understanding, revealing how various ions in the injected water and FW impact oil recovery under different conditions. The primary aim of this paper is to simulate the LSWF process in sandstone plug cores, refining the accuracy of the model to closely match experimental results and extending its applicability. Additionally, the study seeks to verify the key mechanisms behind LSWF in sandstone reservoirs. The simulation results closely aligned with the experimental data, with a relative error of less than 5%. Moreover, the simulations confirmed the validity of the Multi-component Ion Exchange mechanism and refuted the Fine Migration mechanism through detailed analysis of ion concentrations in the aqueous phase and mineral composition. This high level of accuracy demonstrates the reliability of the model and its potential for effective application in future scale-up processes.

Keywords: LSW · LSWF · Low salinity waterflooding · plug core model

1 Introduction

The method of LSWF has emerged as an effective and sustainable solution. With thorough investigation, smart water injection has the potential to enhance the production from both heavy-oil reservoirs, which make up nearly 80% of the world's petroleum reserves, and conventional light-to-medium oil reservoirs [1]. This helps optimize the extraction process, reducing residual oil and increasing overall production. Notably, the investment costs for LSWF are generally lower than those for other enhanced oil recovery methods, like chemical or thermal methods. Additionally, this method has better environmental protection capabilities. The use of LSW helps minimize negative impacts on ecosystems and groundwater sources while reducing waste generated during extraction. This is particularly important in the context of increasingly strict environmental regulations and growing community awareness of environmental protection.

However, experimental studies and simulations related to the LSWF process are still inadequate. Among these, the experimentation and simulation of plug core samples is a crucial task for evaluating the potential reliability of the LSWF method. Since geochemical reactions in sandstone reservoirs have not been fully understood, the predictive results from previous methods have been uncertain, leading to significant errors. A more comprehensive approach to predicting the extract ability of an oil field involves first simulating plug core samples, which can then be scaled up to larger oil fields. While this method provides more accurate predictive results, it demands significantly more data, effort, and advanced expertise to create a valid reservoir model [2].

This study aims to simulate plug core samples obtained from laboratory experiments conducted on two oil production wells in Field X (the actual field name is withheld per operator requirements) located in Southern Vietnam, to aid in future enhancement efforts. Furthermore, the mechanisms when LSWF are also considered to explain the oil recovery process, although the exact mechanisms behind the low-salinity effect remain unclear, the most widely accepted explanation for LSWF is wettability alteration. Two primary theories have been proposed to explain this alteration: multicomponent ion exchange [3] and fine migration [4].

The remainder of this paper is organized as follows: Sect. 2 presents a detailed explanation of the mechanisms and outlines the workflow used for laboratory procedures and the modeling of plug core samples. Section 3 details and analyzes the findings of this study. Lastly, Sect. 4 highlights the key conclusions, emphasizing the accuracy of the modeling results in comparison with experiments.

2 Methodology

In this work, the research investigates the effect of LSWF on oil recovery in a sandstone reservoir. Both laboratory experiments and numerical modeling were conducted to examine the fundamental mechanisms involved. Core samples were subjected to FW flooding followed by LSWF, while one-dimensional models were developed to simulate these processes. The analysis focused on geochemical reactions and mechanisms. The results suggest that LSWF can improve oil recovery by altering wettability of rock and potentially through multicomponent ion exchange. The workflow shown in Fig. 1 and Fig. 4 shows the methodology in laboratory procedures and simulation followed for this work.

2.1 Laboratory Procedures

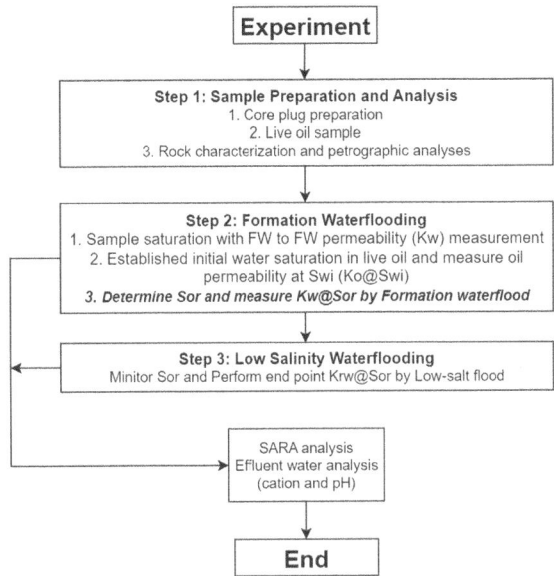

Fig. 1. Laboratory Procedures for LSWF

Sample Preparation, and Analysis

Firstly, conventional cores were collected, checked at field, and core plugs were cut, trimmed, and wrapped in nickel sleeves. Residual hydrocarbons were extracted using Soxhlet apparatus with toluene, while salt and water were removed with methanol. Samples were dried in a humidity oven, and their dry weight was measured. The grain volume was measured using the UltraPore-300TM through Boyle's Law gas expansion method, and calculate the grain density was subsequently based on these measurements. Porosity was determined using helium injection, and pore volume (PV) was measured using the CMS-300TM by helium expansion. Klinkenberg permeability was assessed using the CMS-300TM Unsteady-State Permeameter, where helium flow and pressure decay were analyzed to calculate permeability. Analysis results show that of the 4 plug core samples used for LSWF, approximately 3.7 cm in diameter and 6 cm in thickness, were prepared for waterflood tests. It has porosity ranging from 14.0% to 25%, permeability from 5.5 mD to 1000 mD, and pore volume from 11,169 cm^3 to 14,603 cm^3, as we can see in Fig. 2.

Secondly, PVT analysis was conducted to determine live oil samples for LSWF tests by recombining wellhead oil and gas samples using the reported gas-oil ratio. The well-stream composition was calculated based on the oil-gas ratio and separator gas and oil compositions. The reservoir fluid composition was determined using spike flash and gas-liquid chromatography. At 120°F and atmospheric pressure, the reservoir fluid separated into oil and gas phases, and the evolved gas and residual liquid were analyzed to obtain the $C_{36+}\%$ weight composition of the reservoir fluid. PVT analysis result provides oil properties as hydrocarbon composition up to C_{36+} for oil and C_{12+} for gas. With saturation pressure adjustment to 1700 psig, it produces GOR of 416 scf/stb, oil volume factor (B_o) of 1.214 bbl/bbl, oil viscosity of 0.785 cP at reservoir condition.

Finally, rock characterization and petrographic analyses to describe formation rocks both qualitatively and quantitatively to assess the potential for LSW injection based on petrographic characteristics and pore geometry. Pore size analysis was conducted using high pressure mercury injection capillary pressure (HPMI), as shown in Fig. 3. Each pore facies were categorized based on depositional fabrics and diagenetic histories, which influence pore geometry. Petrographic analysis identified controlling parameters in the pore system, and reservoir rock samples were classified according to these properties affecting pore system quality. The results show that high pressure mercury injection (HPMI) tests on reservoir rock samples revealed pore throat radii ranging from 4.8 to 27.4 μm at threshold pressures of 3.9 to 22.4 psi. Pore throat radii $\geq$ 1 μm ac-counted for 40% to 83% of the pore space in the samples, with oil-water threshold pressures ranging from 0.44 to 2.51 psi. Additionally, petrographic analysis of arkose sandstones from well (interval 1800 m) shows that the rocks are primarily composed of quartz (about 40%), plagioclase (about 45%), and lesser amounts of K-feldspar (about 2%), with diagenetic clay minerals including kaolinite (about 50%), Illite, chlorite, and smectite. Quartz overgrowths and a small amount of pore-filling/replacing clay minerals such as kaolinite (about 6%) are present, while carbonate minerals and pyrite are absent.

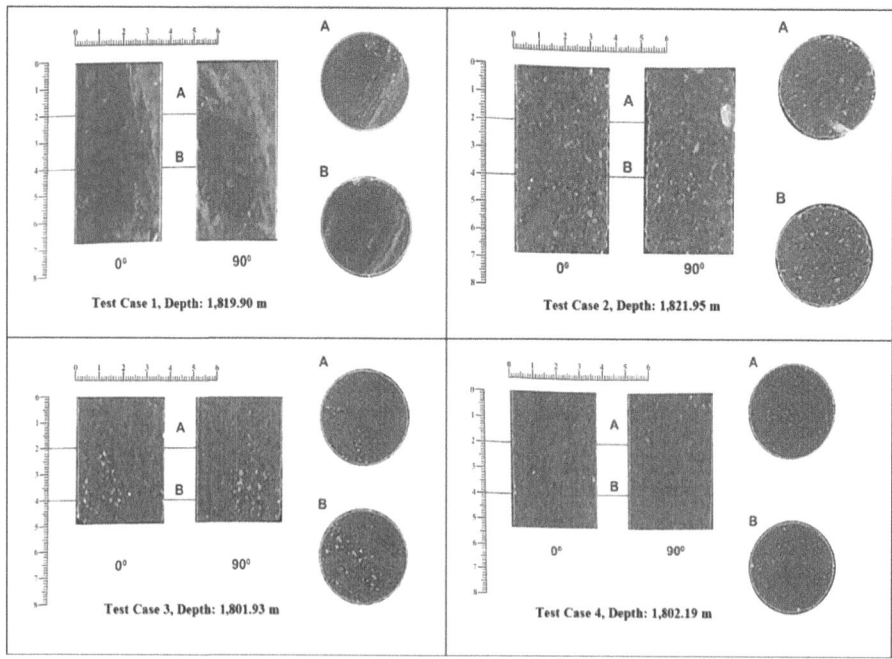

Fig. 2. Plug CT scanning images (Test Case 1–4)

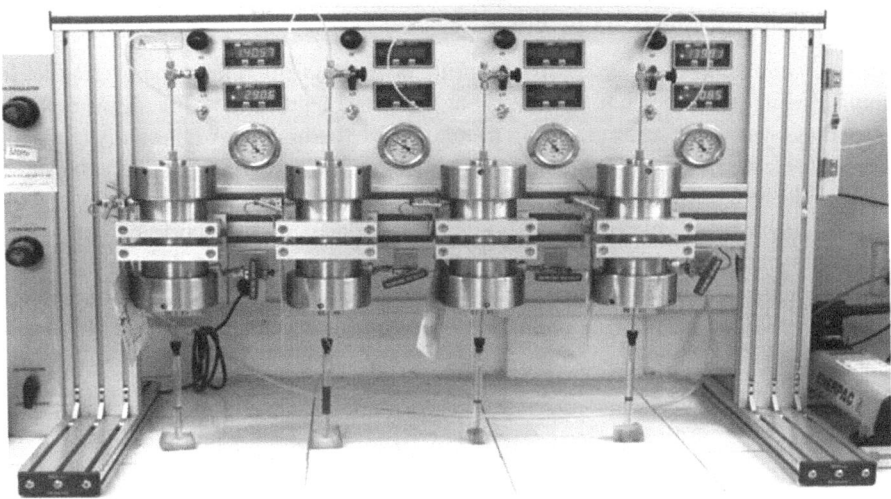

Fig. 3. Pore size analysis by capillary pressure measurements using high pressure mercury injection capillary pressure (HPMI)

Formation Water (FW) Injection and Low Salinity Water (LSW) Injection

The process of LSW injection begins by drying and weighing core samples, then placing them in a vacuum chamber filled with synthetic FW to achieve full water saturation under applied pressure. Once saturated, the sample is loaded into a core holder under reservoir conditions, and FW is flushed through it to measure absolute water permeability (K_w). Afterward, the initial water saturation of the core is determined using the porous plate method, followed by flushing live oil through the sample until stabilization. The sample is then aged for two weeks, and the effective oil permeability ($K_{o@Swi}$) is measured.

Following this, FW flooding is performed on the plug core at a constant flow rate. Produced oil and water volumes, pressure differential, and cumulative time are recorded to determine the oil recovery versus PV injected. The end-point water relative permeability at residual oil saturation ($K_{rw@Sor}$) is also calculated by flowing FW through the core sample under reservoir conditions.

Lastly, LSWF was conducted after FW flooding, with the plug core injected at a constant flow rate, monitoring oil recovery factor, residual oil saturation when water cut (WCT) until reaching $> 99.95\%$. The oil recovery factor was determined using the injected water pore volume and the data on oil recovery in relation to water cut. End-point water permeability ($K_{rw@Sor}$) was determined by flowing LSW through the core sample at reservoir conditions, measuring differential pressure and temperature, and calculating effective water permeability. Effluent oil from 1 PV and 5 PV was collected for SARA analysis, while effluent water was tested for cation concentration and pH. Residual fluid content was analyzed using the Dean-Stark method, where vaporized toluene was used to extract oil and water, and water saturation was calculated from collected water volume and pore volume.

2.2 Plug Core Flooding Simulation

This core simulation aims to clarify the primary mechanism behind the effect of LS combined with geochemical reactions. Using the process outlined in Fig. 4, we created a core model with properties matching those of our experiments. Additionally, we performed plug core numerical simulations by injecting FW and LSW in two stages. To ensure model reliability, the simulation outcomes were history matching with the recovery factor and residual oil saturation. While the heterogeneity of history matching was not explicitly analyzed, the resulting relative permeability curve enables an investigation into how multi-component ion exchange impacts the recovery factor. This was achieved by applying two sets of relative permeability curves to represent the effects of HSW and LSW.

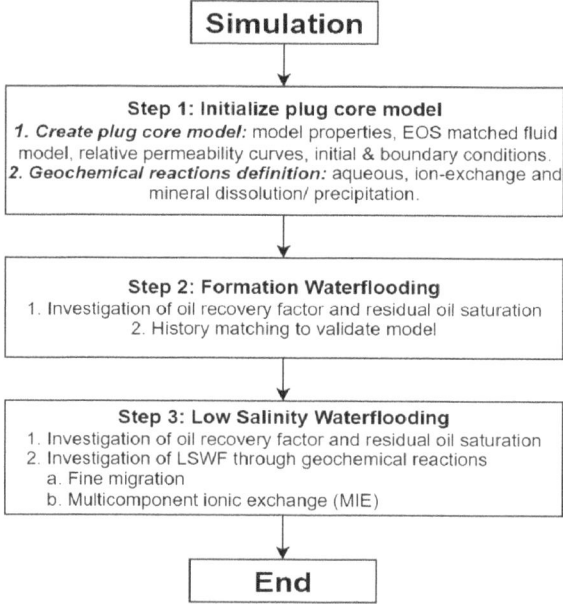

Fig. 4. Simulation workflow for LSWF

Initialize Plug Core Model

Four one-dimensional models were created to simulate the plug core sample experiments described above. Each model is composed of three grid blocks, and the second block represent the plug core model with approximately 3.7 cm in diameter and 6 cm in thickness, were prepared for waterflood tests. It has porosity ranging from 14.0% to 25%, permeability from 5.5 mD to 1000 mD, and pore volume from 11.169 cm^3 to 14.603 cm^3.

The geochemical equations that govern mineral dissolution and precipitation are referred to as rate-dependent reactions. These reactions describe changes in the rock minerals and assess their impact on the carbonate core during low salinity water injection.

Additionally, a second type of geochemical reaction, known as aqueous reactions, takes place. These reactions are selected based on the mineral composition of the rock [5]. Aqueous reactions involve the interaction between water and SiO_2, occur spontaneously, and are represented as equilibrium reactions. For the main lithology of the samples is light grey sandstone with siltstone lamination, veins. Sandstone is mainly quartz, plagioclase with much smaller amount K-feldspars, minor amount of mica, rock fragments, the geochemical equations used are shown in Eqs. (1)–(10).

Mineral Reactions:

$$\text{Quartz} \rightleftarrows SiO_2 \tag{1}$$

$$\text{K} - \text{feldspar} + 4 H^+ \rightleftarrows Al^{3+} + 2 H_2O + K^+ + 3SiO_2 \tag{2}$$

$$\text{Calcite} + H^+ \rightleftarrows Ca^{2+} + HCO_3^- \tag{3}$$

$$\text{Dolomite} + 2 H^+ \rightleftarrows Ca^{2+} + 2 HCO_3^- + Mg^{2+} \tag{4}$$

$$\text{Illite} + 8 H^+ \rightleftarrows 2.3 Al^{3+} + 5 H_2O + 0.6 K^+ + 0.25 Mg^{2+} + 3.5SiO_2 \tag{5}$$

$$\text{Kaolinite} + 6 H^+ \rightleftarrows 2 Al^{3+} + 5 H_2O + 2SiO_2 \tag{6}$$

Aqueous Reactions:

$$CO_2(aq) + H_2O \rightleftarrows H^+ + HCO_3^- \tag{7}$$

$$H^+ + OH^- \rightleftarrows H_2O \tag{8}$$

Ion Exchange Reactions [3]:

$$Na^+ + 0.5\, Ca - X_2 \rightleftarrows Na - X + 0.5\, Ca^{2+} \tag{9}$$

$$Na^+ + 0.5\, Mg - X_2 \rightleftarrows Na - X + 0.5\, Mg^{2+} \tag{10}$$

where X represents the clay present in the reservoir rock.

Investigation of the Mechanisms and Geochemical Reactions During Formation Waterflooding and Low Salinity Waterflooding

FW first was injected into the plug core model to investigate residual oil saturation and cumulative oil recovery the geochemical reactions and verify the mechanisms of the sandstone reservoir as conventional waterflooding by analyzing the ion concentrations in both injection water and production water.

After FW injection, LSW will be injected to investigate the potential for further in-creases in oil recovery, further reductions in residual oil saturation and, importantly, how geochemical reaction mechanisms affect the production process by considering two mechanisms, Fine Migration and Multicomponent Ion Exchange.

3 Results and Discussion

In this work, four test cases were modeled using different plug core samples in the laboratory, originating from oilfield X in Southern Vietnam. These test cases aimed to evaluate the end-point residual oil saturation and water permeability following oil displacement, as well as the maximum potential cumulative oil recovery using FW and LSW (around 3004 ppm). The historical and parameters used in the modeling process were adjusted to closely resemble the experimental process as closely as possible.

The cumulative oil recovery, oil residual saturation and the changes of ions concentration results in aqueous and minerals from LSWF experiments and simulations provide a detailed view of the performance of this method when applied to different plug core samples.

3.1 Cumulative Oil Recovery and Oil Residual Saturation Results

As illustrated in Table 1 and Figs. 5, 6, 7 and 8, the LSWF simulation process is detailed. First FW experiments yielded an average residual oil saturation (S_{or}) of 27.8% PV and an average cumulative oil recovery of 42.98% PV (60.5% of OOIP). The simulation models closely matched these results, with an average Sor of 29.65% and an average cumulative oil recovery of 46.498%. The relative error between the experimental and simulation results was only 0.2% for residual oil saturation (S_{or}) and 1.338% for cumulative oil recovery, demonstrating the precision and reliability of the simulations. During high salinity waterflooding, two common ion exchange reactions occur, involving Na^+, Ca^{2+}, and Mg^{2+} ions, as shown in Eq. (9) and Eq. (10). These reversible reactions indicate that Na^+ ions are absorbed by the exchanger, Mg^{2+} and Ca^{2+} ions are released when HSW is injected into a reservoir. In other words, injecting HSW into a sandstone reservoir increases the concentrations of Mg^{2+} and Ca^{2+} in the produced water due to their replacement by Na^+ ions from the HSW.

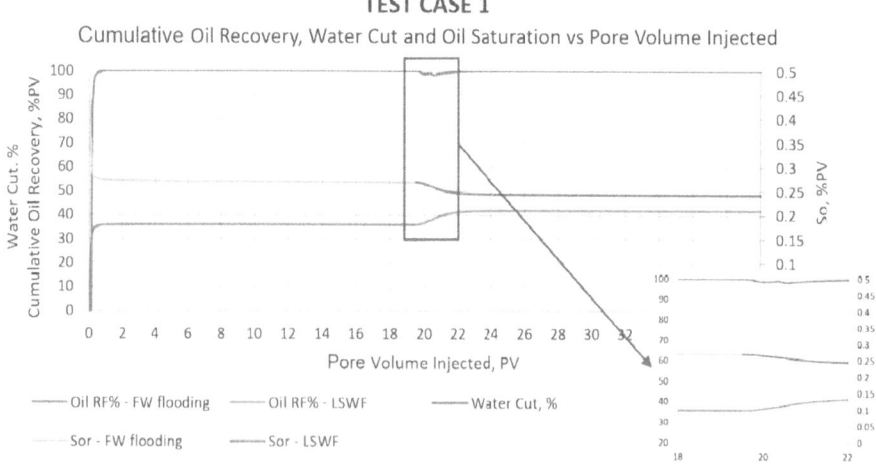

Fig. 5. Cumulative Oil Recovery and Residual Oil Saturation versus PV Injected – Test Case 1

For LSWF, the experiments showed a reduction in the final average S_{or} to 22.22% PV and a corresponding increase in average cumulative oil recovery by 5.55%, equivalent to 8.1% of OIIP. The simulations reflected this trend as well, with an average cumulative oil recovery increase of 5.28% and improvements ranging from 3.29% to 6.34% for 4 samples, resulting in an overall average recovery of 51.778%. The relative error between the experiment and simulation results for cumulative oil recovery was low, at 1.955%. Additionally, in the simulations, S_{or} decreased to 24.1% in Test case 1 and 28.7% in Test case 4, with an average reduction of 3.175%, further validating the consistency of both experimental and simulation data. Moreover, the use of LSWF led to a slightly reduction in WCT, helping to sustain higher oil pro-duction and improve overall cumulative oil recovery.

3.2 Changes in Ions Concentration in Aqueous and Mineral Moles

As can be seen in Figs. 9, 10, 11 and 12 for LSWF, effluent analysis results show that Ca^{2+} concentrations decreased by an average of 84.21%, while Mg^{2+} concentrations dropped by approximately 91.44% compared to the initial concentration in aqueous in 4 plug core models. The above phenomenon can be explained as follows. Wettability alteration towards a more water-wet condition during LSWF is commonly proposed as the reason for enhanced oil recovery [6]. Many studies have reported the impact of low salinity brine on wettability changes [7–9]. Experimental and simulation findings have shown that LSW significantly impacts the shape and end points of the relative permeability curves, leading to a reduction in relative permeability of water and an increase in relative permeability of oil[10–12]. This phenomenon can be physically explained by the ionic exchange between the injected water and FW, and mineral dissolution/precipitation [6]. In other words, multicomponent ionic exchange showed that during LSWF, Ca^{2+} and Mg^{2+} ions strongly adsorb onto the rock matrix, replacing polar compounds and organometallic complexes on the surface. In this work, our simulation results of four test cases also show great agreement with the above experiments through the sharp decrease of Ca^{2+} and Mg^{2+} during injection, emphasizing the strong adsorption of these ions onto the reservoir rock surface, thus releasing oil in plug core, thus improving oil recovery.

Table 1. Residual oil saturation and Oil recovery factor by LSWF and average relative error in both experiments and simulations

Test case number	Low salinity waterflooding (LSWF)			
	Residual oil saturation, %		Cumulative oil recovery factor, %	
	Modeling	Average relative error, %	Modeling	Average relative error, %
1	24.100	8.151	42.030	1.955
2	26.00		54.080	
3	27.100		55.550	
4	28.700		55.450	
Average	26.475		51.778	

In addition, the fine migration mechanism was also considered in this work, the simulation results of 4 plug core models for Kaolinite moles changes are presented in Fig. 13. The simulation results show that when injecting LSWF Kaolinite moles does not change significantly, at most only about 0.035 gmole. Previous experiments observed that fines, primarily kaolinite clay fragments, were released from rock surfaces, enhancing spontaneous imbibition recovery as salinity decreased in various sand-stones [4]. Fines migration was believed to improve mobility control by plugging pore throats, reducing water permeability in swept zones, and diverting flow to un-swept zones [6]. However, researchers have noted higher recovery from LSWF without observing fines migration, leading to questions about whether fines migration is directly responsible for the enhanced oil recovery [3, 13, 14]. In this study, the simulation results indicate that the reduction in kaolinite moles may be attributed to the low salinity conditions, characterized by decreased ionic strength. This reduction lowers the screening potential due to a decrease in multivalent cation concentration, leading to an increase in the ζ-potential. Consequently, the expansion of the electrical double layer enhances the electrostatic repulsion between clay and crude oil. When these repulsive forces surpass the binding forces mediated by the multivalent cation bridge, the oil particles can detach from the clay surface and it affects the Kaolinite moles on the surface of the plug core [15]. It is therefore concluded in this study that the dissolution of Kaolinite during LSWF injection is not one of the oil recovery mechanisms.

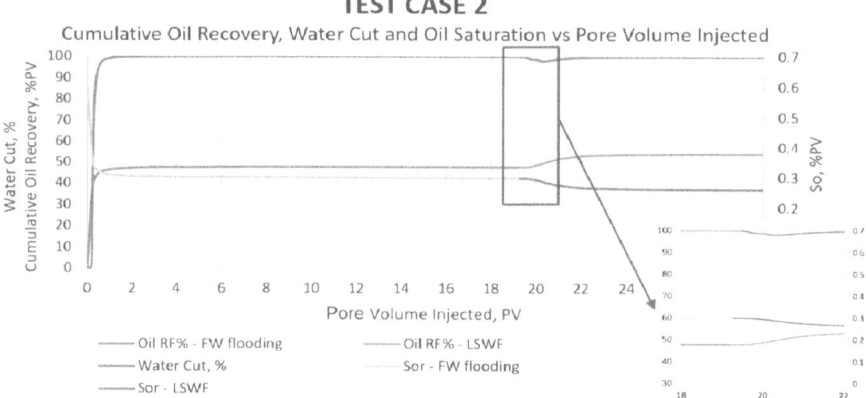

Fig. 6. Cumulative Oil Recovery and Residual Oil Saturation versus PV Injected – Test Case 2

Modeling of Low Salinity Waterflooding Process 205

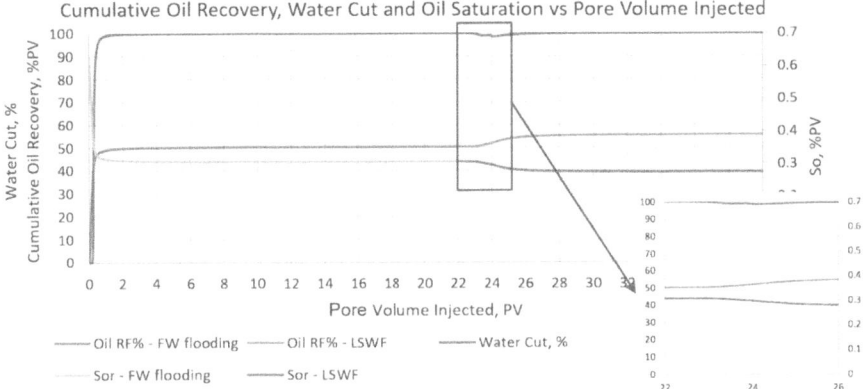

Fig. 7. Cumulative Oil Recovery and Residual Oil Saturation versus PV Injected – Test Case 3

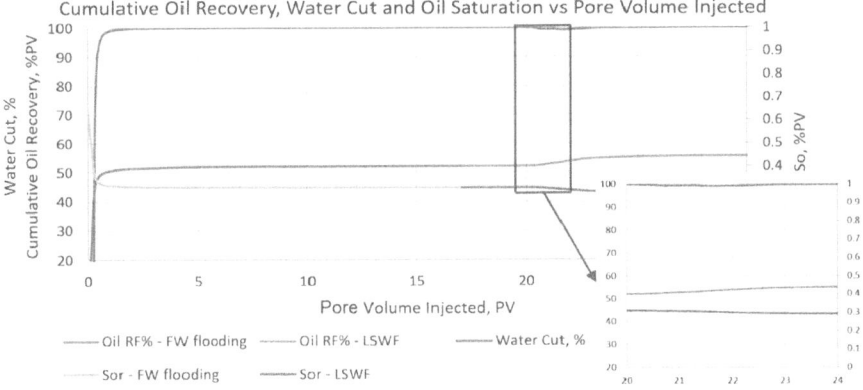

Fig. 8. Cumulative Oil Recovery and Residual Oil Saturation versus PV Injected – Test Case 4

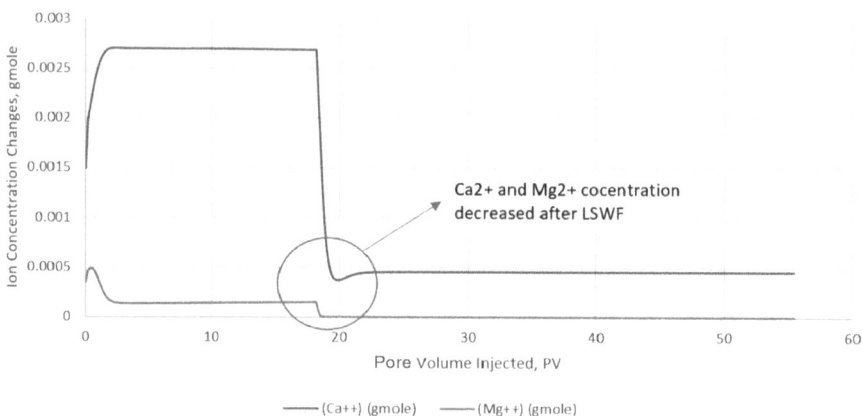

Fig. 9. Mg^{2+} and Ca^{2+} Concentration versus Pore Volume Injected – Test Case 1

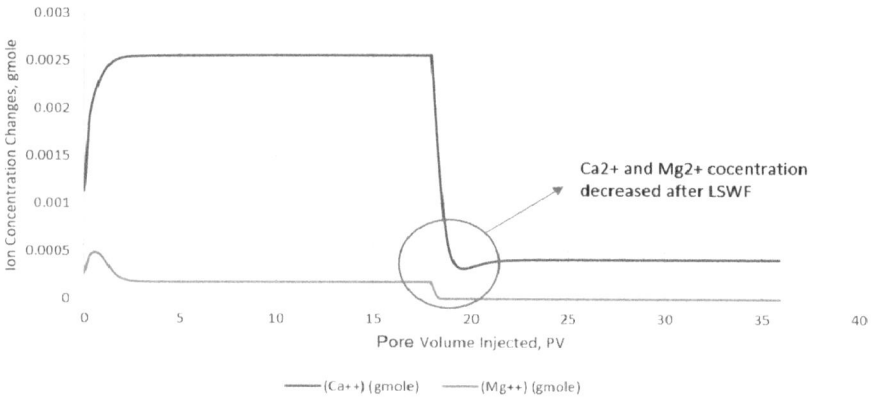

Fig. 10. Mg^{2+} and Ca^{2+} Concentration versus Pore Volume Injected – Test Case 2

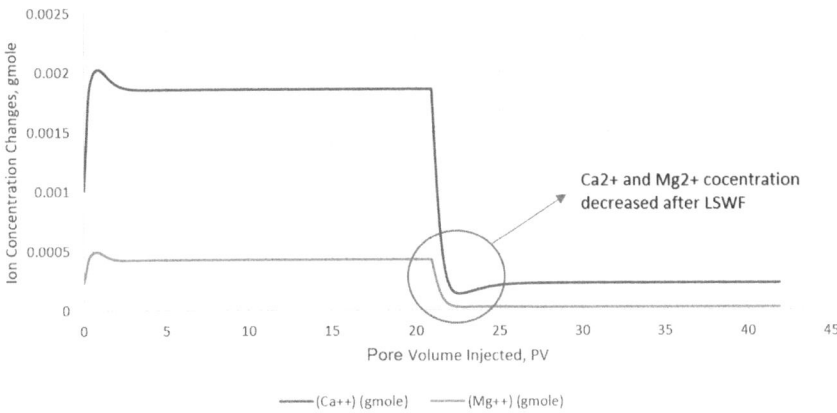

Fig. 11. Mg^{2+} and Ca^{2+} Concentration versus Pore Volume Injected – Test Case 3

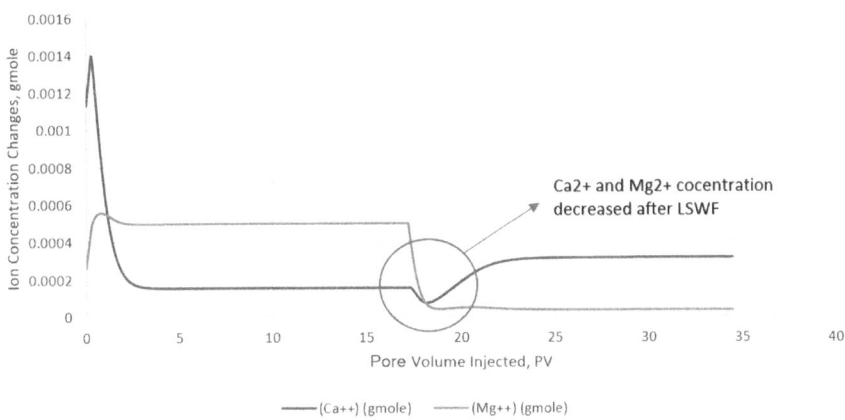

Fig. 12. Mg^{2+} and Ca^{2+} Concentration versus Pore Volume Injected – Test Case 4

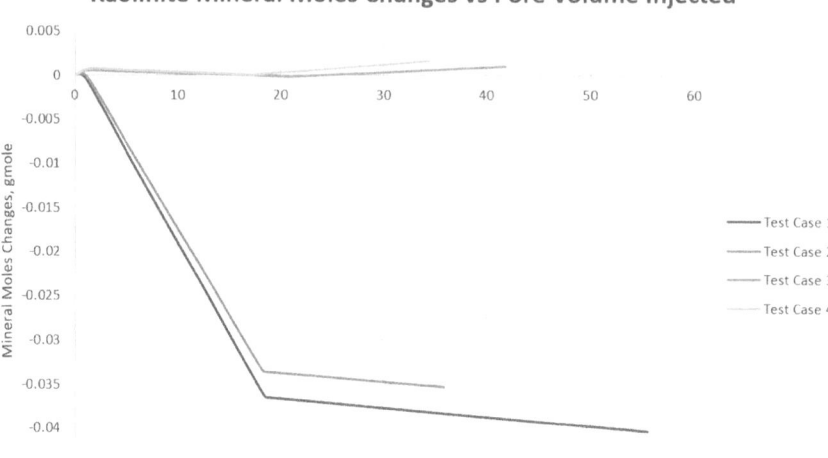

Fig. 13. Kaolinite Mineral Moles Changes vs Pore volume Injected in Four Test Cases

4 Conclusion

This paper presents a study on the accuracy of simulation models calibrated with experimental data in predicting tertiary oil recovery potential and changes in ion concentrations in water and minerals through the LSWF method. Four core samples with diverse lithology compositions, permeability, and porosity were successfully simulated, demonstrating high accuracy, and affirming the reliability of the model.

The study's findings prove that a complete core sample experiment can be accurately simulated using advanced simulation software. Notably, the analysis from this research not only provides precise predictions of the ion concentrations of Mg^{2+} and Ca^{2+} in the produced water, confirming the validity of the Multicomponent Ion Ex-change mechanism, but also refutes the Fine Migration mechanism by analyzing the concentration of Kaolinite minerals in the aqueous environment.

Furthermore, both the experimental and simulation findings highlight the significant potential of the LSWF technique in improving oil recovery, especially in reservoirs with pre-existing low salinity aquifers. Additionally, this method is environmentally sustainable, as it greatly lowers CO_2 emissions in comparison to other chemical oil recovery techniques, such as those using surfactants or alkali.

Acknowledgements. We express our gratitude to Ho Chi Minh City University of Technology (HCMUT), VNU-HCM, and PetroVietnam – Blocks 01 & 02, for their generous support in providing time and facilities for this research.

References

1. Afekare, D.A., Radonjic, M.: From mineral surfaces and coreflood experiments to reservoir implementations: comprehensive review of low-salinity water flooding (LSWF). Energy Fuels **31**, 13043 (2017)
2. Islam, M.R., Hossain, M.E., Mousavizadegan, S.H., Mustafiz, S., Abou-Kassem, J.H.: Advanced Petroleum Reservoir Simulation: Towards Developing Reservoir Emulators (2016)
3. Lager, A., Webb, K.J., Black, C.J.J., Singleton, M., Sorbie, K.S.: Low salinity oil recovery - an experimental investigation1. Petrophysics - SPWLA J. Form. Eval. Reserv. Descr. (2008)
4. Mechanisms of Improved Oil Recovery From Sandstone by Low Salinity Flooding | Petrophysics | OnePetro (2024)
5. Maguri, A.O.A.: Fluid/ Rock interaction effects on Oil Recovery for Low Salinity Water Flooding (2021)
6. Dang, C., Nghiem, L., Nguyen, N., Chen, Z., Nguyen, Q.: Mechanistic modeling of low salinity water flooding. J. Pet. Sci. Eng. (2016)
7. Tang, G.-Q., Morrow, N.R.: Influence of brine composition and fines migration on crude oil/brine/rock interactions and oil recovery. J. Pet. Sci. Eng. (1999)
8. Effect of Wettability on Waterflood Recovery for Crude-Oil/Brine/Rock Systems | SPE Reservoir Engineering | OnePetro (2024)
9. Surface forces and wettability – ScienceDirect (2024)
10. Webb, K.J., Black, C.J.J., Al-Ajeel, H.: Low Salinity Oil Recovery – Log-Inject-Log (2024)
11. Coreflooding Oil Displacements With Low Salinity Brine (2024)
12. Fjelde, I., Asen, S.M., Omekeh, A.: Low Salinity Water Flooding Experiments and In-terpretation by Simulations (2024)
13. Impact of Brine Chemistry on Oil Recovery | Earthdoc (2024)
14. Lager, A., Webb, K.J., Collins, I.R., Richmond, D.M.: LoSalTM enhanced oil recovery: evidence of enhanced oil recovery at the reservoir scale (2024)
15. Lee, K.S., Lee, J.H.: Hybrid Enhanced Oil Recovery Using Smart Waterflooding. Gulf Professional Publishing, Houston (2019)

Multimodal Biometric Recognition Using Fuzzified BTC with a Novel Hybrid JSO-CSO Algorithm

Indu Singh[1], Siddartha Aggarwal[2], Shrey Gupta[2], Vansh Khandelwal[2], and Abhishek Verma[3(✉)]

[1] Department of Computer Science and Engineering, Delhi Technological University, Delhi, India
[2] Department of Electrical Engineering, Delhi Technological University, Delhi, India
[3] Department of Computer Science and Engineering, National Institute of Technology Srinagar, Srinagar, India
vabhishek9621@gmail.com

Abstract. Biometric authentication is critical for secure access in diverse domains, emphasizing the need for high accuracy to prevent security breaches. Multimodal biometrics fuse traits to fortify security, vital across diverse domains, surpassing unimodal systems' vulnerability. This paper introduces a multi-modal biometric recognition using palm print and iris datasets. In our approach, Gabor filters are used for feature extraction with an innovative fuzzy bit transition coding (BTC). Followed by, a novel hybrid Jellyfish Search Optimization (JSO) and Cat Swarm Optimization (CSO) algorithm for feature selection of iris and palm print modalities. Firstly, the palm print images undergo feature extraction using the fuzzified BTC and the preprocessing of iris images is done using a Convolutional Neural Network(CNN) to refine the dataset and identify regions of interest (ROI). After that, extracted features are then selected using our hybrid JSO-CSO algorithm, which are then utilized for classification in a deep learning based ResNet model. The two biometric modalities process in parallel, and their classification results are fused using weighted score-level fusion for the final recognition result. This method improves feature interpretability as well as classification performance; it has been experimentally validated against MMU iris database, CASIA-palm print dataset and a custom multimodal database resulting in peak accuracy of 97.6%. The proposed hybrid JSO-CSO algorithm outperforms other techniques in terms of precision and convergence, thus providing a robust solution for multi-modal biometric recognition.

Keywords: Multimodal Biometric Authentication · Feature Extraction · Metaheuristic Optimization · Deep Learning · Score Level Fusion

1 Introduction

Authentication methods like passwords and tokens are more vulnerable to cyber threats. Biometric recognition represents a robust option that employs special human traits in identity determination. Biometrics can be broadly grouped into physiological and behavioral modals [3]. Physiological modalities such as fingerprints, iris scans and facial geometry examine unique physical attributes whereas behavioral modalities like signature recognition, keystroke dynamics and gait analysis focus on individual patterns of behavior. To prove this point, biometric systems identify people by taking their personal biometric information and comparing it with what is stored about them [18].

Various factors can compromise system accuracy [14]. Multimodal biometric systems address limitations by combining two or more modalities, harnessing each trait's unique strengths for enhanced authentication robustness and security [17]. Deep learning algorithms are pivotal in extracting intricate patterns from multimodal biometric data, enabling continual learning and adaptation to improve accuracy and resilience [19].

The proposed model introduces a novel framework for multimodal biometric authentication. The preprocessing of iris images to refine the dataset and identify regions of interest (ROI) is done using a convolutional neural network (CNN) and feature extraction from palm-print images involves an innovative binary transition code (BTC) infused with fuzzy logic. Then, a novel hybrid JSO-CSO algorithm is used for effective feature selection. Additionally, a deep learning based ResNet architecture is employed for the classification of features from both iris and palm-print datasets. Finally, the classification results from both modalities are fused using a weighted score-level fusion approach to achieve the final recognition outcome. The proposed methodology is evaluated based on the various performance metrics.

The following summarizes the key findings of this study:

1. *The model incorporates a novel feature extraction method using fuzzified bit transition code (BTC) combined with Gabor filtering.*
2. *A novel combination of Jellyfish Search Optimization (JSO) and Cat Swarm Optimization (CSO) is proposed for efficient and accurate feature selection.*
3. *Efficient deep learning based ResNet Lightweight Convolutional Neural Network (CNN) architectures are used for classification.*

This paper has been structured as follows: In Sect. 2, prior studies conducted on biometric authentication are reviewed. The methodology proposed in the present study is outlined in Sect. 3, which deals with pre-processing, classification and weighted score level fusion. Experimental outcomes and their analyses is present in Sect. 4. Section 5 presents the conclusion and future work.

2 Literature Review

Multimodal biometric authentication represents a critical frontier in the domain of cybersecurity. The fusion of various biometric traits prevents unauthorized access, ensuring robust protection in digital transactions (Fig. 1).

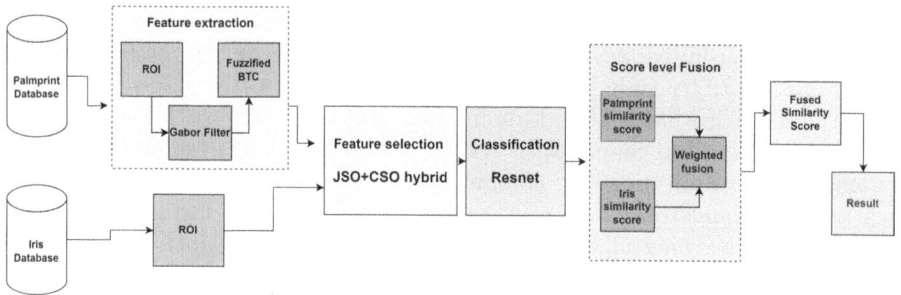

Fig. 1. Model Architecture

In the early years, *Fierrez-Aguilar et al. (2003)* [6] explored various score fusion strategies for multimodal biometric verification. A novel SVM-based approach for combining match scores from face recognition, fingerprint verification, and signature verification was proposed in their study.

The study presented by *Raghavendra et al.(2006)* [15] addressed the problem of developing efficient and effective fusion schemes of corresponding biometric modalities, that are effectively coded by employing Log-Gabor transformations that results in high dimensional feature spaces. Their research introduced innovative fusion techniques at match score and feature levels, showing a 6% performance improvement using Log-Gabor feature fusion over traditional methods of match score fusion.

Kihal et al. (2014) [11] proposed a multimodal biometric Fusion combining iris and palmprint data, using Wavelet packet decomposition for feature extraction. Various fusion strategies such as concatenation, weighted sum, and Hamacher t-norm were applied at feature, score, and error levels. This approach achieved enhanced recognition accuracy on Casia and PolyU databases.

Habibeh Naderi et al. (2017) [13] a bimodal iris-palmprint system employing various wavelet filters including log Gabor, DCT, Walsh & Haar for feature extraction. To tackle the high dimensionality resulting from wavelet transforms, Laplacian eigenmaps and SVD were employed for dimensionality reduction. The system outperformed unimodal systems significantly. Particularly, log Gabor filters provided the most discriminative features, reducing the number of features by 99.49%.

Hariprasath and santhi (2019) [8] proposed a recognition system using iris and palmprint modalities, employing a multi-resolution approach. Wavelet packets extracted iris features, while Gabor filters and Grey Level Co-occurrence

Matrices (GLCM) captured palmprint characteristics. This study showcased the effectiveness of combining texture-based and decomposition techniques, achieving a 94.5% recognition rate with iris and palm fusion in just 15.25 microseconds.

Gona and subramoniam (2022) [7] used Gaussian filters to preprocess photos of faces, fingerprints, and irises before extracting features using Grey Level Co-occurrence Matrices (GLCM). Partially Component Analysis (PCA) lowered the number of dimensions. Lastly, the data is categorized using a DLCNN. This method potentially increased recognition accuracy by utilizing deep learning for feature extraction. Addressing limitations of unimodal biometrics, *Balraj and Abirami (2022)* [5] presented a multimodal system using iris and face recognition with Ant Colony Optimization (ACO) for score level fusion. The system uses ACO to optimize weight selection within various fusion rules, such as sum, tanh, mean, etc., after extracting features from each modality.

The study by *Ipeayeda et al. (2023)* [10] proposed a fusion approach using the Gravitational Search Algorithm (GSA) for multimodal biometrics. Their method achieved superior recognition accuracy compared to traditional techniques, demonstrating the potential of GSA for efficient feature fusion in multimodal systems.

C. Vensila et al. (2023) [20] proposed a multimodal biomteric authentication system using face, fingerprint, and finger vein traits. Local Binary Patterns (LBP) were employed for feature extraction, and an Adaptive Particle Swarm Optimization (APSO) algorithm optimized these features. Classification was achieved through an Extreme Learning Machine (ELM), achieving high accuracy (95.69% sensitivity and 94.29% specificity) with specific parameter settings.

3 Proposed Methodology

The palmprint biometrics are initially preprocessed using Gabor filter and fuzzified BTC to obtain bit transition coding planes as multi dimensional arrays. The iris biometrics are also initially preprocessed for ROI using a CNN based deep learning model. Figure 2 shows original and preprocessed iris images. The obtained biometrics are transferred for feature selection that utilizes novel metaheuristic algorithm combining JSO and CSO. The selected features are also send for classification using the ResNet model. These resulting arrays are then send for classification using the ResNet model. The palmprint and iris biometrics are processed in parallel with the foretold mechanisms to obtain their classification scores. The scores are combined at the end using weighted normalized sum rule fusion to obtain the resultant classification.

3.1 Feature Extraction Using Fuzzified BTC

The palm-print and iris biometrics use different feature extraction modules as the datasets contain varying inputs. The palm-print biometrics are extracted using a novel combination of bit-transition-code planes (BTC) and fuzzy logic implemented on preprocessed dataset using many techniques. First, the dataset

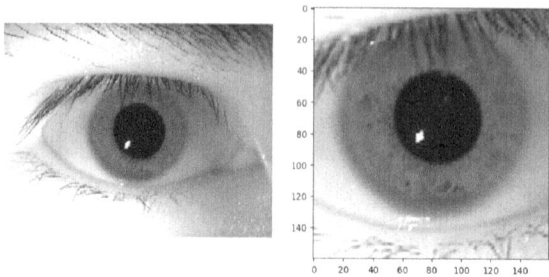

Fig. 2. (a) Original and (b) Preprocessed iris image

is exposed to ML based techniques to obtain ROI, then weiner filter is applied to remove blurs and finally, haar wavelet transformation is done for speeding up processing and ease computations. On these preprocessed dataset images gabor filter is applied followed by the fuzzified BTC to obtain the features.

Bit transition code is a feature extraction approach based on encoding binary transitions of asymmetric and symmetric parts of the gabor filtered dataset images at all pixel locations in the image. The two binarized images are concatenated using the Eq. 1.

At each pixel (x,y),

$$\bar{I}_{1,n,r}^{new}(x,y) = \begin{cases} 1, & \text{if } \bar{I}_{1,n,r}(x,y) > 0 \\ 0, & \text{otherwise} \end{cases}$$

$$\bar{I}_{1,n,i}^{new}(x,y) = \begin{cases} 1, & \text{if } \bar{I}_{1,n,i}(x,y) > 0 \\ 0, & \text{otherwise} \end{cases} \quad (1)$$

A count is made of the bit transitions in a binary string formed using combined vectors of binarized real and imaginary parts of 2D gabor filter of different frequencies. Then these two parts are binarized using zero-cross operation (thresholding at zero value) and stored in third dimension as they contain different information as explained in study *Kumar et al. (2002)* [12]. For efficient bit transition coding, the Eq. 2 is used to create a linear representation of the concatenated image.

$$M(x,y,2*n+1) = \bar{I}_{1,n,r}^{new}(x, y)$$

$$M(x,y,2*n+2) = \bar{I}_{1,n,i}^{new}(x,y) \quad (2)$$

The binary strings obtained by this process are further utilized in the encoding as given in Eq. 3. And the results are then given to Eq. 4 to count of bit-transitions (complimenting 0s and 1s) in the binary string are then utilized in texture representation to get the bit-transition matrix which acts as the extracted features. This methodology can be further understood from *Vyas et al. (2022)* [21]

$$\text{for } p = 1 \text{ to } \left\lfloor \frac{2N+1}{2} \right\rfloor \text{ do} \qquad (3)$$

$$btc(x, y, p) = \begin{cases} 1, & p \leq S(x,y) < p + \frac{N+1}{2} \\ 0, & \text{otherwise} \end{cases} \qquad (4)$$

Fuzzy logic is introduced in this process by replacing thresholding by zero. The thresholding can be done using different thresholds acting as membership functions for fuzzification. the fuzzified values are combined using fuzzy inference rule of whether the values are greater or less than a hyperparamter value lying between 0 and 1 based on the dataset. the value used in implementation was 0.67. Finally a value is obtained using largest of max defuzzification technique in the texture representation section. Figure 3 shows the image after bit transition coding. The end result provides a modified BTC utilizing fuzzy logic according to the needs of the dataset.

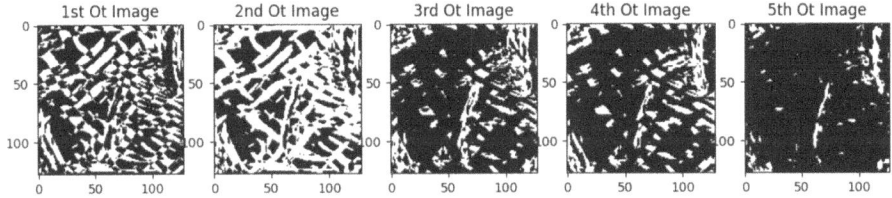

Fig. 3. Palmprint after bit transition coding

This feature extraction was not performed on the iris dataset as they were only preprocessed for getting the ROI. The ROI for iris images were found by creating a simple CNN model. This can also be done using an existing module. The iris dataset images after ROI were detailed and provided optimum results hence, not requiring any specific feature extraction process to be implemented.

3.2 Feature Selection Using JSO-CSO Hybrid

Feature selection is done for both datasets in parallel using this same method. A novel meta-heuristic optimization algorithm is developed as hybrid of jellyfish search optimization and cat swarm optimization. The novel algorithm is utilized with K-Nearest Neighbour (KNN) algorithm to optimize the score of the KNN. The features are exposed to KNN for the use its KNN score as its objective function. This method provides the nearest neighbours as the selected features for any data point.

The developed JSO-CSO Algorithm utilizes a dynamic approach combining jellyfish behavior to explore problem spaces and Cat behaviour to exploit the problem spaces. The method also adapts JSO's exploration strategy employing passive and active motion's exploitation strategy. This aims to discover the best solution within the specified search space and time frame (iterations).

The algorithm begins by defining the objective function and setting crucial parameters, like space boundaries, population, number of iterations. Objective function defines a particles current position's virtue. A particles aims to optimize its position value and eventually reach optimum value across the whole search space.

Algorithm 1 shows the pseudo code of the novel Algorithm. A particle's motion is majorly defined using 3 equations implemented based on a dynamic strategy. Initially a population of particles is defined, each particle's position represents a solution. Fitness values for the positions of all the particles are calculated. The most distinctive feature of this algorithm is its dynamic strategy based on time control (c(t)) given in Eq. 5.

$$C(t) = \left| \left(1 - \frac{t}{T}\right)(2 \times \text{rand} - 1) \right| \quad (5)$$

where, **t** is the current iteration and **T** are the total number of iterations. When c(t) is greater than or equal to 0.5, the algorithm imitates jellyfish behavior by having individuals follow ocean currents as shown in Eq. 6. This equation entails one of the exploration strategy of the algorithm. the current direction is determined and the particle's position is updated accordingly mimicking the way jellyfish explore their environment.

$$\mathbf{p}_i^{(t+1)} = \mathbf{p}_i^{(t)} + \text{rand}_1 \left(\mathbf{g} - \beta \times \text{rand}_2 \times \boldsymbol{\mu}\right) \quad (6)$$

where, $p_i^{(t)}$ is the particle's position i at iteration t, g denotes global best position, β is a hyper-parameter constant, $\boldsymbol{\mu}$ is the mean position of all particles, $rand_1$ and $rand_2$ are random numbers between 0 and 1. However, when c(t) is less than 0.5, the algorithm employs a different approach. It divides the jellyfish behavior into two types: passive and active motion. The choice between the two is made using a randomization process. If the random value falls within a certain range, the particle exhibits passive motion and updates its position using Eq. 7.

$$\mathbf{p}_i^{(t+1)} = \mathbf{p}_i^{(t)} + \gamma \times \text{rand} \times (\mathbf{max} - \mathbf{min}) \quad (7)$$

where, γ is a hyper-parameter constant. $rand$ is a random number between 0 and 1, max and min are the maximum and minimum bounds of the search space respectively. When active motion strategy is followed the values are determined using a direction. The particle's velocity in every dimension is calculated using Eq. 8 and used for updating its position according to Eq. 9. this motion is derived from the tracing equation of cat swarm optimization. These equations showcase the exploitation part of the algorithm.

$$\mathbf{v}_i^{(t+1)} = \mathbf{v}_i^{(t)} + rand \times c_0 \times (\mathbf{g} - \mathbf{p}_i^{(t)}) \quad (8)$$

$$\mathbf{p}_i^{(t+1)} = \mathbf{p}_i^{(t)} + \mathbf{v}_i^{(t+1)} \tag{9}$$

Where, $\mathbf{p}_i^{(t)}$ is the particle's position i at iteration t, $\mathbf{v}_i^{(t)}$ is the particle's velocity i at iteration t, $rand$ is a random number between 0 and 1, c_0 is a hyper-parameter constant, and $\mathbf{g}$ represents the global best position.

Algorithm 1: Hybrid Jellyfish - Cat Swarm Optimization Algorithm

Input: Population size N, maximum number of iterations T, problem dimension D, Objective function Φ, Search Space S, All the Hyper-parameters
Output: Best solution $\mathbf{x}^*$
Initialize population of particles X with their initial position and velocity.
Calculate the initial fitness values of all particles and find the initial global position X_{best}.
for $t = 1$ to T **do**
    Calculate the time control as $C(t) = \left|\left(1 - \frac{t}{T}\right)(2 \times \text{rand} - 1)\right|$;
    for $i = 1$ to N **do**
        if $C(t) >= 0.5$ **then**
            particle goes with the ocean current ;
            μ = average position of all particles;
$$\mathbf{p}_i^{(t+1)} = \mathbf{p}_i^{(t)} + \text{rand}_1\left(X_{best} - \beta \times \text{rand}_2 \times \boldsymbol{\mu}\right);$$
        end
        else if $rand > 1 - C(t)$ **then**
            particle shows passive motion ;
$$\mathbf{p}_i^{(t+1)} = \mathbf{p}_i^{(t)} + \gamma \times \text{rand} \times (\mathbf{max} - \mathbf{min});$$
        end
        else
            particle shows active motion ; velocity of particle ;
$$\mathbf{v}_i^{(t+1)} = \mathbf{v}_i^{(t)} + rand \times c_0 \times (\mathbf{p}_{best}^{(t)} - \mathbf{p}_i^{(t)})$$
            position of particle ;
$$\mathbf{p}_i^{(t+1)} = \mathbf{p}_i^{(t)} + \mathbf{v}_i^{(t+1)}$$
        end
    end
    Calculate new global best position Calculate fitness of all particles
    $f_i = \Phi(\mathbf{x}_i)$;
end
return

The algorithm ensures that the particles stay within the defined search space and recalculates the fitness values of all the particles in every iteration. The iterations through these processes continues till a termination criteria is achieved or maximum number of iterations is reached.

At the end of the algorithm's execution, it outputs the best solution found. This typically corresponds to the particle with the highest fitness representing the optimal solution to the optimization problem.

3.3 Classification Using ResNet

In the proposed model architecture, a residual neural network model is used for the classification of selected features for prediction. The residual network is implemented as explained in the paper *Kaiming He et al. (2016)* [9] The foretold ResNet module is utilized for both the palm-print and iris datasets. Residual networks are deep learning models that are designed in such a way that weight layers can learn residual functions relative to the inputs of each layer. These networks are particularly good in image related task. These networks have an ease of optimization and achieve high accuracy by significantly increasing the depth of the network. We evaluate residual networks with depths up to 152 layers, eight times deeper than VGG networks, but with lower computational complexity.

3.4 Fusion

The Proposed model computes the palm-print and iris datasets in parallel and combines their individual classification result using a Novel weighted normalized sum rule fusion based approach to get the resulting final score. In many cases the matching of individual classifiers are not homogeneous creating a requirement for normalization and weight for the individual scores. The weights For different modalities are trained using long short term memory (LSTM) based deep learning model. This technique simply normalizes the modality scores using min-max normalization, multiplies them with their trained weights and adds them together resulting in the sum score as:

$$\text{Sum Rule} = \sum_{i=1}^{n} w_i \times s_i \tag{10}$$

where, w_i is the weight given to each modality i, s_i is the match score obtained from modality i and n is the number of modalities. The implemented LSTM deep learning model uses the combined dataset of iris and palmprint for training the 50 layers LSTM model network.

4 Experimentation and Results

4.1 Dataset

In this multimodal biometric system, two distinct databases were utilized to enhance identification accuracy: the MMU-Iris and the IIT Palmprint databases. The MMU-Iris database [1] comprises left and right iris images of 45 individuals, with each eye having 5 images in BMP format. Conversely, the IIT Palmprint

Table 1. Comparison of datasets used for training and testing

Dataset Used	Test	Train	No. of persons	Images of each person	Split for each person		Epochs Trained
					Test	Train	
Iris	90	360	45	10	2	8	100
Palm print	90	360	45	10	2	8	100

database [2] consists of 5–6 palmprint images per hand for 230 individuals in JPG format. Due to the limited number of iris images available, the use of the IIT Palmprint database was restricted to the first 45 individuals, thereby ensuring consistency in the multimodal dataset.

For the training and evaluation of the model, an 80:20 train-test split was adopted, where 80% of the data, comprising 360 images of iris and palmprint each, was allocated for training purposes, and the remaining 20%, comprising 90 images of iris and palmprint each was used for testing as shown in Table 1. The implementation was conducted in Python on the Google Colab platform, which provided a computational environment equipped with 12.7GB of CPU RAM. This setup facilitated efficient processing and analysis, thereby ensuring robust performance and accuracy in the multimodal biometric identification system.

4.2 Performance Evaluation

Benchmark Functions. The proposed hybrid optimization algorithm was thoroughly evaluated using several standard benchmark functions. These benchmark functions include the sphere, schwefel_2_22, schwefel functions each presenting unique challenges in terms of dimensionality, modality, and landscape complexity.

The Table 2 results show that the hybrid algorithm outperforms WOA, GWO, PSO, JSO & CSO in various benchmark functions, highlighting the superior precision and faster convergence rate of the hybrid algorithm. Specifically, the graph (Fig. 4) shows how model accuracy improves with the number of epochs. The hybrid algorithm demonstrates steady and reliable progress, with a strong correlation between increased accuracy and the number of epochs, confirming its effectiveness and robustness.

Accuracy v/s Epochs. The accuracy versus epochs graph of the machine learning model, which integrates iris and palmprint similarity scores using a weighted average method, illustrates a steady increase in accuracy throughout the training period. As training advances, accuracy consistently improves, indicating the model's iterative refinement. The model eventually stabilizes, achieving a peak accuracy of 97.6%, demonstrating its effectiveness in leveraging both iris and palmprint features for precise classification. Additionally, the graph includes both training and validation accuracies, as shown in Fig. 5. Furthermore, when compared to other multimodal biometric models in the field, as seen

Table 2. Performance of proposed algorithm on benchmark functions

Function		WOA	GWO	PSO	JSO	CSO	Hybrid(JSO-CSO)
Sphere	Avg	3.12E-73	8.70E-28	1.85E-04	1.68E-02	3.27E-05	2.87E-03
	Std	1.65E-72	9.05E-28	1.98E-04	1.50E-02	3.66E-05	3.28E-03
Schwefel_2_22	Avg	2.31E-51	9.48E-17	4.82E-02	0.150423	0.012296	0.1854
	Std	8.10E-51	8.03E-17	6.38E-02	0.0735482	0.009892	5.3781
Schwefel	Avg	-1.07E+04	-5.94E+03	4.87E+03	710.706	721.12405	0.1671
	Std	1.79E+03	1.04E+03	1.27E+03	0.01835	17.667349	4.8496

Table 3. Comparison of presented approach with other previous approaches

Author	Modality	Technique Used	Accuracy
Razzak et al. (2010) [16]	Face + Finger Vein	CSLDA + Score Level Fusion	95%
Andersson et al. (2014) [4]	Skeleton Joint + Gait	KNN + Multi-layer Perceptron + SVM	88%
Hariprasath et al. (2019) [8]	Iris + Palmprint	LVQ Neural Network + Gabor + GLCM	94.50%
Balraj et al. (2022) [5]	Face + Iris	Score Level Fusion + ACO	96.42%
Vensila et al. (2023) [20]	Face + Fingerprint + Finger Vein	Local Binary Pattern + ELM + Adaptive PSO	97.14%
Our Proposed Model	**Iris + Palmprint**	**Fuzzified BTC + JSO-CSO + Weighted Score Level Fusion**	**97.6%**

in Table 3, our model's accuracy surpasses those of prior approaches, affirming the superiority of the proposed method for accurate biometric classification.

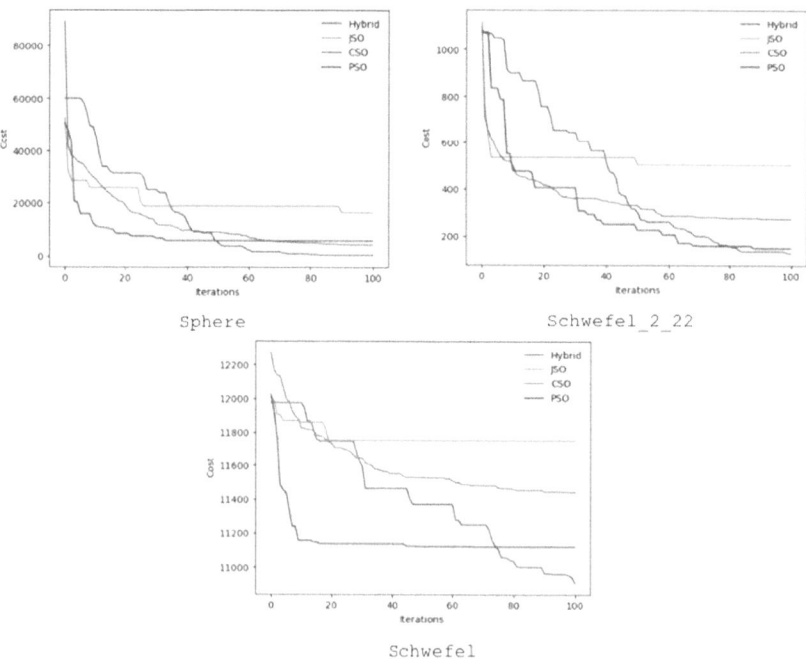

Fig. 4. Benchmark Functions (sphere, schwefel_2_22, schwefel)

Loss v/s Epochs. The loss versus epochs graph of the machine learning model depicts an exponential decrease in loss during the initial epochs, indicative of rapid learning and adaptation. Both training and validation losses exhibit this trend. However, after the initial decline, the rate of decrease gradually slows down, and the losses plateau, indicating diminishing returns in model improvement. Despite fluctuations, both training and validation losses generally follow a similar trajectory. Figure 6 suggests that the model effectively learns from the training data but may struggle to generalize to unseen data beyond a certain point.

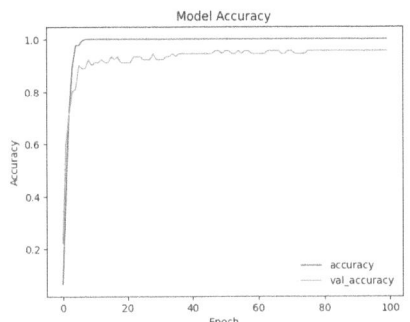
Fig. 5. Accuracy v/s Epochs

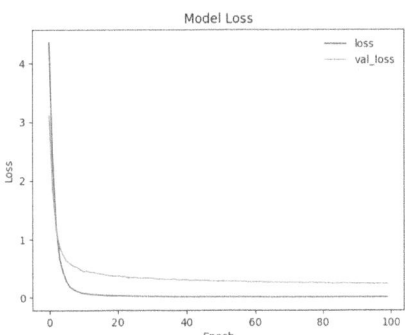
Fig. 6. Loss v/s Epochs

5 Conclusion and Future Work

In conclusion, this study introduces an advanced biometric system that combines palmprint and iris recognition to significantly improve identification accuracy. By using techniques like Gabor filtering and fuzzified Bit Transition Coding (BTC) for palmprints, and employing Convolutional Neural Networks (CNN) to focus on key areas of the iris, the system captures unique features from both biometrics. A hybrid optimization method, combining Jellyfish Swarm Optimization (JSO) and Cat Swarm Optimization (CSO), refines feature selection, making the classification process more efficient using K-Nearest Neighbour (KNN). Deep learning, through the ResNet architecture, alongside a weighted fusion technique, enhances the system's accuracy further, with Long Short-Term Memory (LSTM) helping to optimize these weights. The system achieves an impressive 97.6% accuracy, demonstrating its potential to be highly effective, adaptable, and reliable for real-world biometric identification needs.

Future work could explore integrating additional biometric modalities, such as fingerprints or voice and testing on larger, more diverse datasets with real-world conditions and implementing advanced neural architectures to further enhance the robustness of the system. Real-time performance optimization is also a potential area of development.

References

1. MMU iris dataset. https://www.kaggle.com/datasets/naureenmohammad/mmu-iris-dataset
2. IIT Delhi palmprint dataset. https://www4.comp.polyu.edu.hk/~csajaykr/IITD/Database_Palm.htm. https://www.kaggle.com/datasets/naureenmohammad/mmu-iris-dataset
3. Abdulrahman, S.A., Alhayani, B.: A comprehensive survey on the biometric systems based on physiological and behavioural characteristics. Mater. Today Proc. **80**, 2642–2646 (2023). https://doi.org/10.1016/j.matpr.2021.07.005. https://www.sciencedirect.com/science/article/pii/S2214785321048513. sI:5 NANO 2021
4. Andersson, V.O., Araujo, R.M.: Full body person identification using the kinect sensor, pp. 627–633 (2014). https://doi.org/10.1109/ICTAI.2014.99
5. Balraj, E., Abirami, T.: Performance improvement of multibiometric authentication system using score level fusion with ant colony optimization. Wirel. Commun. Mob. Comput. **2022**, 4145785 (2022). https://doi.org/10.1155/2022/4145785
6. Fierrez-Aguilar, J., Ortega-Garcia, J., Garcia-Romero, D., Gonzalez-Rodriguez, J.: A comparative evaluation of fusion strategies for multimodal biometric verification, pp. 830–837 (2003)
7. Gona, A.K., Subramoniam, M.: Multimodal biometric reorganization system using deep learning convolutional neural network, pp. 1282–1286 (2022). https://doi.org/10.1109/ICECAA55415.2022.9936398
8. Hariprasath, S., Santhi, M.: Bimodal biometric pattern recognition system based on fusion of iris and palmprint using multi-resolution approach. Signal Image Video Process. **14**(3), 519–527 (2020). https://doi.org/10.1007/s11760-019-01573-7
9. He, K., Zhang, X., Ren, S., Sun, J.: Deep residual learning for image recognition (2016)
10. Ipeayeda, F.W., Oyediran, M.O., Ajagbe, S.A., Jooda, J.O., Adigun, M.O.: Optimized gravitational search algorithm for feature fusion in a multimodal biometric system. Results Eng. **20**, 101572 (2023). https://doi.org/10.1016/j.rineng.2023.101572. https://www.sciencedirect.com/science/article/pii/S2590123023006990
11. Kihal, N., Chitroub, S., Meunier, J.: Fusion of iris and palmprint for multimodal biometric authentication, pp. 1–6 (2014). https://doi.org/10.1109/IPTA.2014.7001980
12. Kumar, A., Pang, G.K.: Defect detection in textured materials using gabor filters. IEEE Trans. Ind. Appl. **38**(2), 425–440 (2002)
13. Naderi, H., Soleimani, B.H., Matwin, S.: Manifold learning of overcomplete feature spaces in a multimodal biometric recognition system of iris and palmprint, pp. 191–196 (2017). https://doi.org/10.1109/CRV.2017.29
14. Oloyede, M.O., Hancke, G.P.: Unimodal and multimodal biometric sensing systems: a review. IEEE Access **4**, 7532–7555 (2016). https://api.semanticscholar.org/CorpusID:6981610
15. Raghavendra, R., Dorizzi, B., Rao, A., Hemantha Kumar, G.: Designing efficient fusion schemes for multimodal biometric systems using face and palmprint. Pattern Recognit. **44**(5), 1076–1088 (2011). https://doi.org/10.1016/j.patcog.2010.11.008. https://www.sciencedirect.com/science/article/pii/S0031320310005352
16. Razzak, M., Yusof, R., Khalid, M.: Multimodal face and finger veins biometric authentication. Sci. Res. Essays **5**, 2529–2534 (2010)
17. Ross, A., Jain, A.K.: Multimodal biometrics: an overview, pp. 1221–1224 (2004)

18. Rui, Z., Yan, Z.: A survey on biometric authentication: toward secure and privacy-preserving identification. IEEE Access **7**, 5994–6009 (2019). https://doi.org/10.1109/ACCESS.2018.2889996
19. Ryu, R., Yeom, S., Kim, S.H., Herbert, D.: Continuous multimodal biometric authentication schemes: a systematic review. IEEE Access **9**, 34541–34557 (2021). https://doi.org/10.1109/ACCESS.2021.3061589
20. Vensila, C., Boyed Wesley, A.: Multimodal biometrics authentication using extreme learning machine with feature reduction by adaptive particle swarm optimization (2023)
21. Vyas, R., Kanumuri, T., Sheoran, G., Dubey, P.: Accurate feature extraction for multimodal biometrics combining iris and palmprint. J. Ambient. Intell. Humaniz. Comput. **13**(12), 5581–5589 (2022)

Delayed Transitions in GPenSIM

Yuming Feng[1], Slawomir Samolej[2], and Reggie Davidrajuh[3]

[1] School of Computer Science and Engineering, Chongqing Three Gorges University, Wenzhou, China
[2] Department Computer and Control Engineering, Rzeszow University of Technology, Rzeszow, Poland
[3] Department Electrical Engineering and Computer Science, University of Stavanger, Stavanger, Norway
Reggie.Davidrajuh@uis.no

Abstract. The General-purpose Petri Net Simulator (GPenSIM) is a tool for modeling, simulation, and performance analysis of discrete event systems. GPenSIM is specially designed to model real-life industrial systems. Hence, the transitions in GPenSIM represent real-life events, such as machines. However, this induces a problem, namely that an enabled transition consumes tokens from its input places immediately. This act of an enabled transition immediately consuming tokens gives only monetarily visibility to the tokens, which can be a problem for certain cases where we need to see the tokens in their input places for a while. This paper proposes a method to delay an enabled transition ("delayed transitions") so that the tokens it is supposed to consume from its input places will be visible for a while. This paper, after presenting the theory and the method, also presents an application (the cat and mouse problem) of the delayed transitions.

Keywords: GPenSIM · Petri Nets · Time Petri Nets · Timed Petri Nets · Delayed Transition · Discrete Event Systems

1 Introduction

General-purpose Petri Net Simulator (GPenSIM) is a toolbox that runs on the MATLAB platform [9]. Some researchers use GPenSIM to model, simulate, and analyze the performance of discrete event systems. GPenSIM is popular because it is easy to learn, use, and extensible as it runs on the MATLAB platform [4].

This paper proposes "Delays" in enabled transitions, which is not available in GPenSIM. In GPenSIM, whenever a transition becomes enabled and all the conditions in the pre-processor files are satisfied, the transition is allowed to consume the input tokens and start firing. However, sometimes, we may want an enabled transition to wait for a while before consuming the input tokens so that the tokens in the input places are visible for a period of time; "Delayed Enabled Transitions" ("delayed transitions", for short) does that.

The structure of this paper is as follows: Sect. 2 is a literature study which briefly introduces Petri Nets and GPenSIM. A short discussion on how GPenSIM treats enabled transitions is given in Sect. 3. Section 4 proposes the delayed transitions. Finally, Sect. 5 shows a sample application.

2 Literature Review

Petri nets are well-known for modeling discrete systems because of their properties like self-documentation (graphical front), simple mathematics (linear algebra), and explicit state information (known as "reachability tree" in Petri Net literature) [3,15,17].

References [1,8,14,16] and [5] classify the association of time and Petri Nets under three classes of time-dependent Petri nets:

1. Time Petri Nets.
2. Timed Petri Nets.
3. Petri Nets with Timed Places.

Time Petri nets are classic Petri nets where each transition t_j is associated with a time interval [*eft*, *lft*]. When t_j becomes enabled, it cannot fire before *eft* (earliest firing time), and it has to fire no later than *lft* (latest firing time) time units after being enabled; note that after enabled, when a transition waits for firing, it can become disabled by the firing of another transition. Here, *eft* and *lft* are relative to the point in time when t_j was last enabled. As in the case of classical (P/T) Petri nets, the firing of a transition itself does not take up any time (in other words, unlike transitions in GPenSIM, t_j is primitive, and its firing time is zero). The interval bounds (*eft* and *lft*) are non-negative rational numbers or ∞.

In **Timed Petri nets**, a firing time (duration) is incorporated with each transition (transitions in GPenSIM are either make Timed Petri net or Untimed Petri net). In a Timed Petri net, GPenSIM does not allow assigning zero time duration to transitions. However, [14] allows zero firing times.

Petri Nets with Timed Places is to allow delays between events by assigning retention times for tokens in the places. It is noteworthy that, unlike the previous two classes in which time is attached to transitions, in Petri nets with Time Windows, time is attached to tokens in places. This Petri net class has been used to model applications in system diagnostics.

General-purpose Petri Net Simulator (**GPenSIM**) is a new Petri Net simulator developed by the third author of this paper. Some researchers use GPenSIM for modeling, simulation, and performance analysis of discrete-event systems. GPenSIM runs on the MATLAB platform.

GPenSIM was developed as computer software to work with Petri Nets. Due to its enormity, GPenSIM was developed in three stages:

- Stage-1: Supporting P/T Petri nets and Timed Petri Nets [4].
- Stage-2: Supporting Colored Petri Nets [6].

- Stage-3: Supporting modeling of large real-life industrial discrete systems with modular approach [5].

GPenSIM also offers a rich set of built-in functions so that modelers can implement compact models using these built-in functions [7].

2.1 Formal Definition of Petri Nets

A **Petri Net** (*aka* Marked Petri Nets) is defined as a five-tuple [10,12,13]:

$$PN = (P, T, A, K, M),$$

where (N is the natural number),

- P is a set of places, $P = \{p_1, p_2, \ldots, p_{n_p}\}$.
  $\forall p_i \in P$, $K(p_i) \to N \cup \{\infty\}$, is the maximum capacity of the place p_i.
- T is a set of transitions, $T = \{t_1, t_2, \ldots, t_{n_t}\}$.
  $P \cap T = \emptyset$.
- A is the set of arcs (from places to transitions and from transitions to places). $A \subseteq (P \times T) \cup (T \times P)$. The arc weight $w(a_{ij}) \in N \cup \{0\}$, $\forall a_{ij} \in A$.
- M is the row vector of markings (tokens) on the set of places.
  $M = [M(p_1), M(p_2), \ldots, M(p_{n_p})] \in N \cup \{0\}$, M_0 is the initial marking.

A transition in a Petri Net is enabled *iff* all of its input places have tokens more than the input arc weights. Formally, $\forall p \in P$, $M(p) \geq w(p,t)$.

3 Enabled and Firing Transitions in GPenSIM

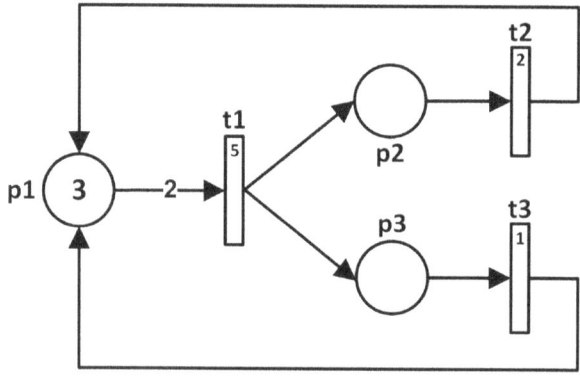

Fig. 1. Simple Petri net.

Figure 1 shows a simple Petri Net with three places, **p1** to **p3**, and three transitions, **t1** to **t3**. Only **t1** is enabled as its input place **p1** has three tokens (more than the input arc weight of two).

For simplicity, let us assume the firing times of transitions as deterministic, 5 TU, 2 TU, and 1 TU, for **t1**, **t2**, and **t3**, respectively.

Let us run this Petri net. Since **t1** is enabled, it will immediately consume two tokens from **p1** (as the input arc weight $w(p1, t1)$ is two). Also, the firing time of **t1** is 5 TU; hence, **t1** will hold the two consumed tokens as 'virtual tokens' for 5 TU [4]. When the clock strikes 5 TU, **t1** completes firing and deposits one token each in **p2** and **p3**. However, **p2** and **p3** only hold the deposited tokens momentarily, as the enabled transitions **t2** and **t3** consume these tokens immediately.

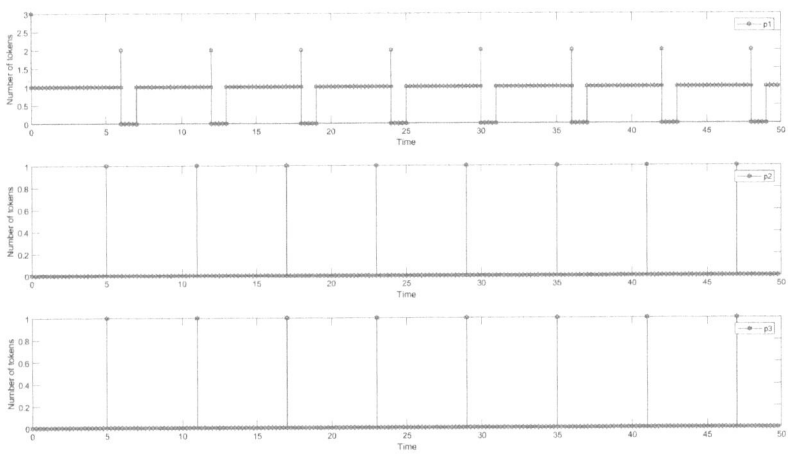

Fig. 2. Tokens in places **p1** to **p3**.

Figure 2 shows that when the Petri net is run, places **p2** and **p3** only hold one token *momentarily* as their output transitions **t2** and **t3** as become enabled and ready to consume the tokens. Only **p1** shows one token remaining in it most of the time; this is because **p1** possesses a surplus token which cannot be consumed by **t1**.

In GPenSIM, delaying an enabled transition from consuming tokens from its input places is not straightforward, as it demands a little bit of additional code (a delay mechanism) in the pre-processor files. Also, this extra code for delaying is not straightforward either, as it needs some deeper understanding of how GPenSIM works. The novelty of this paper - an algorithm and sample code for delaying the enabled transitions - is given in Sect. 4.

4 Delayed Transitions

GPenSIM realizes Timed Petri nets, as non-zero firing times (duration) are attached to transitions. However, the property of Time Petri nets, namely, an enabled transition that is forced to wait for a period of time, is a much-wanted property in some applications that lack in GPenSIM; an enabled transition that is forced to wait gives more visibility or traceability of tokens in places. As we have seen in section xx, an enabled transition in GPenSIM only provides momentary visibility of tokens in places.

Also, since the reason for proposing delayed transitions is to make tokens stay in the input places for a while, only a variant of Time Petri nets is needed, in which only *eft* (earliest firing time) is defined. There is no need for a *lft* (latest firing time).

4.1 Proposition (Delayed Transitions): $eft = lft \neq 0$

In the delayed transitions, when a transition is enabled, it can be delayed by a duration *eft*. Only after the time has passed and if the transition is still enabled (this transition may be disabled due to other firing transitions) will the preprocessor files be checked for additional conditions to satisfy. Hence, in delayed transitions, $eft = lft$.

4.2 Three Types of Delays

Also, the delayed transition should support three types of delays:

- Deterministic delay: E.g., 5 TU. All real-life events possess some elements of stochasticity. However, during the testing, developers feel comfortable with deterministic values so that the outputs of test inputs can be verified.
- Random delay: E.g., 'normrnd(5, 10)'. As stated above, all real-life events are non-deterministic, no matter how precise they are. Therefore, we should be able to assign random delays with random values already supported in Matlab.
- User-defined value: sometimes, the developers may need to feed some special values they generate for delays. Hence, we should be able to assign values output by user-defined functions as the delays.

4.3 Algorithm for Realizing Delayed Transitions

Algorithm 1 shows the three-part algorithm for realizing delayed transitions in GPenSIM. The algorithm proposes only adding code (extensions) to user files so that GPenSIM system files are not effected.

The first part is about the changes to make in the main simulation file. We need to declare the transitions that are to be delayed (by '`delays`': a constant or deterministic value, random value, or user-defined value). Also, we need to reset (initialize) the '`DELAY_UNTIL`' values with `NaN` ('not a number').

The second part is about the changes to make in the pre-processor COMMON_PRE. When a transition, let us call it t_j, is enabled, the COMMON_PRE will be automatically executed. In this file, first, we check whether t_j is a delayed transition (line 11). If it is not, we will continue with the rest of COMMON_PRE (line 21). If t_j is a delayed transition, then we check the value of `DELAY_UNTIL` (line 12). If this value is NaN, this means t_j is enabled newly; hence, we have computed the delay value and add this value to the current time and assign the sum (`DV`) to `DELAY_UNTIL`. Then, we check whether the current waiting time for the transition has passed (line 16). If the waiting period is not over, then the enabled transition should not allowed to start firing - hence, we return a logic false value for firing.

The third and final part is about the changes to make in the post-processor COMMON_POST. When a completes firing transition, let us call it t_j again, the COMMON_POST will be automatically executed. In this file, first, we check whether t_j is a delayed transition (line 11). If it is not, we will continue with the rest of COMMON_PRE (line 30). If t_j is a delayed transition, then we reset the value of `DELAY_UNTIL` to NaN (line 31).

5 Sample Application

Let us study the cat-and-mouse problem-a problem often discussed in books on "supervisory control." [18] originally proposed the problem, which was solved by Petri net-based supervisory controller in [2] and [11].

The Problem Description: A cat and a mouse coexist in a house comprising five rooms, from Room 1 to Room 5; see Fig. 3. Between the rooms, there are dedicated doors, meaning some doors only allow the cat to pass and the others - only the mouse. For example, door **C23** is for the cat to move from Room 2 to Room 3, whereas door **M32** is for the mouse to move from Room 3 to Room2.

The cat and the mouse are free to move to any room; however, they must not be in the same room at any time; hence, we have to design a controller ("supervisory controller") so that the doors can controlled to prevent any mishaps (that the cat and the mouse find in the same room).

5.1 Simulation Using GPenSIM

Figure 4 shows a Petri net model by [2]. This model consists of two parts, one for the cat's movement and the other for the mouse. Also, in this model, each physical room (e.g., Room 1) is shown as two logical rooms (e.g., **p1** for cat and **p6** for mouse) in the two models. Let us start the simulation by assuming the cat is in Room 3 (an initial token in **p3**) and the mouse is in Room 5 (and another initial token in **p8**).

Due to brevity, this paper does not show the files for implementing the Petri net using GPenSIM. The interested reader is encouraged to contact the authors for the code.

Algorithm 1 Algorithm to realize "Delayed Transitions" in GPenSIM

1: %%% **Part-1**: in **Main Simulation File (MSF)** %%%
2: Assign 'delays' to delayed transitions
3: Initialize 'DELAY_UNTIL' values with NaN
4: % (NaN: not a number)
5:
6:
7: %%% **Part-2**: in **COMMON_PRE** %%%
8: % let the enabled transition be t_j
9: **if** t_j is delayed **then**
10:     **if** t_j's DELAY_UNTIL value is NaN **then**
11:         compute delay value DV
12:         assign DV to t_j's DELAY_UNTIL
13:     **end if**
14:     **if** current_time < DELAY_UNTIL **then**
15:         firing is not allowed
16:         **return**
17:     **end if**
18: **end if**
19: % we are here because: Either t_j is not delayed
20: % or the delay time pass passed
21: % so, go on and check the other conditions,
22: % if there are any
23: ... % the rest of COMMON_PRE
24:
25: %%% **Part-3**: in **COMMON_POST** %%%
26: % let the fired transition be t_j
27: **if** t_j is delayed **then**
28:     reset t_j's DELAY_UNTIL to NaN
29: **end if**
30: ... % the rest of COMMON_POST

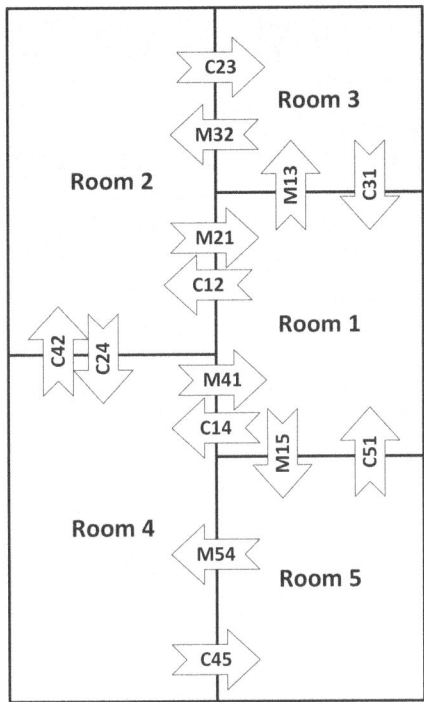

Fig. 3. Cat and mouse coexist in a house comprising five rooms.

The simulation result shown in Fig. 5 depicts the movement of the animals in different rooms. Figure 5 shows that during the time 6 TU and 9 TU, the cat and the mouse happen to be in the same room, Room 1, albeit momentarily. At time equals 6 TU, the cat and the mouse enter Room 1 at the same time. However, door **C12** immediately consumes the cat and starts transporting it to Room 2, whereas door **M13** immediately consumes the mouse and starts transporting it to Room 3. GPenSIM simulation only allows very brief meetings of these two animals, raising hopes that it is OK for them to meet (according to the original problem, this is not OK).

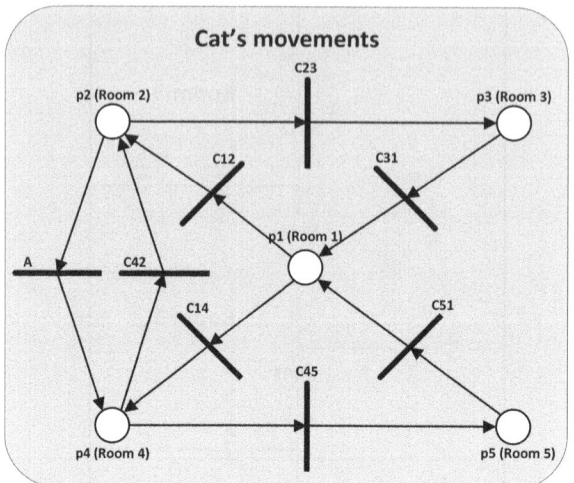

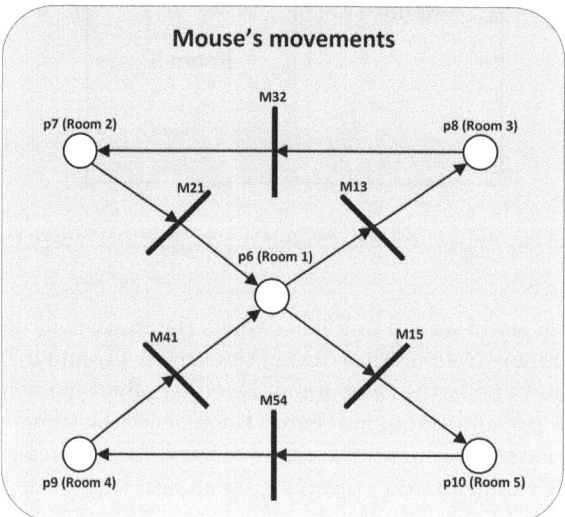

Fig. 4. Petri net model comprising two sub-graphs [2].

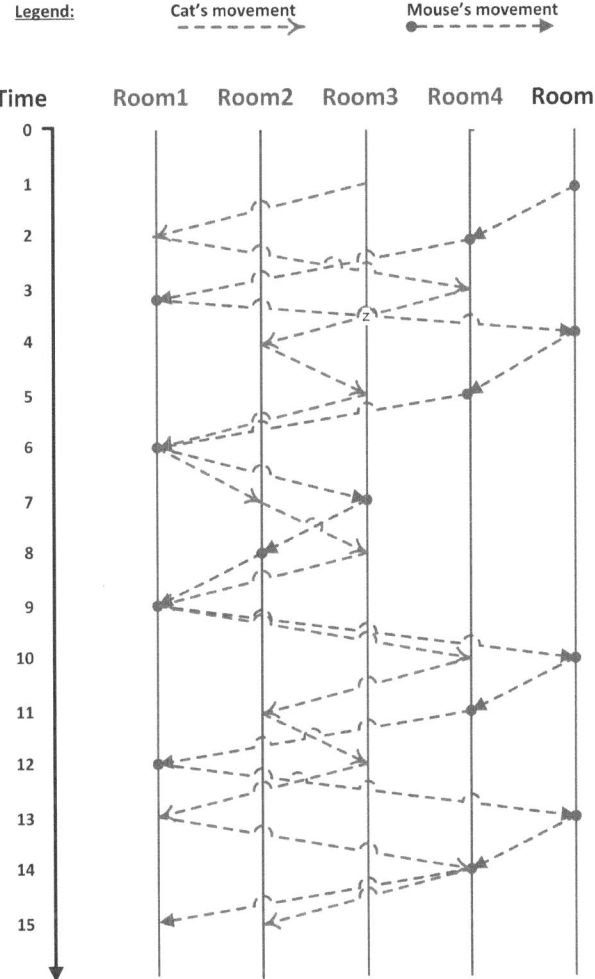

Fig. 5. Cat and mouse's movements for the first 15 TU.

5.2 Simulation Using Delayed Transitions in GPenSIM

Let us add delays to the doors so that they won't grab when the cat or mouse enters a room. The doors will be delayed by a random delay, a normal distribution with mean (μ) 5 and standard deviation (σ) 1. Listing 1.1 shows the main simulation file in which the doors are delayed.

Listing 1.1. Main Simulation File

```
%%%%%%%%%%%%%%%   The Plant   %%%%%%%%%%%%%%%
pns = pnstruct({'cat_pdf','mouse_pdf'});
% Initially: cat in Room 3, and mouse Room 5
dyn.m0 = {'Room3C',1, 'Room5M',1};

% firing time of all doors are 1 TU
dyn.ft = {'allothers', 1};

% all doors are delayed by a random delay
dyn.delayed = {'allothers', 'normrnd(5,1)'};

% simulations
plant = initialdynamics(pns, dyn);
sim = gpensim(plant);
subplot(5,1,1), plotp(sim, {'Room1C','Room1M'});
title('the plant');
subplot(5,1,2), plotp(sim, {'Room2C','Room2M'});
subplot(5,1,3), plotp(sim, {'Room3C','Room3M'});
subplot(5,1,4), plotp(sim, {'Room4C','Room4M'});
subplot(5,1,5), plotp(sim, {'Room5C','Room5M'});
```

Figure 6 shows the movement of the animals after delays are introduced to the doors. Figure 6 shows that now the cat and the mouse can land in the same room for a longer period of time (not momentarily):

- Room 3: (**Room3C** and **Room3M** are aliases for places **p4** and **p9**, respectively) Cat and mouse stay in the same Room 4 from 11.2 TU to 13 TU.
- Room 4: (**Room4C** and **Room4M** are aliases for places **p3** and **p8**, respectively) Cat and mouse stay in the same Room 3 from 23.8 TU to 26 TU.

In Listing 1.1 and Fig. 6, we can see the model of Cat and Mouse is labeled as 'the plant.' This is because the main objective of the cat and mouse problem is to design a controller to prevent these two animals from meeting each other. Hence, the design of the controller for the 'plant' is the next natural step. However, the focus of this paper is the delay of enabled transitions; hence, the synthesis of the supervisory control part is not discussed in this paper.

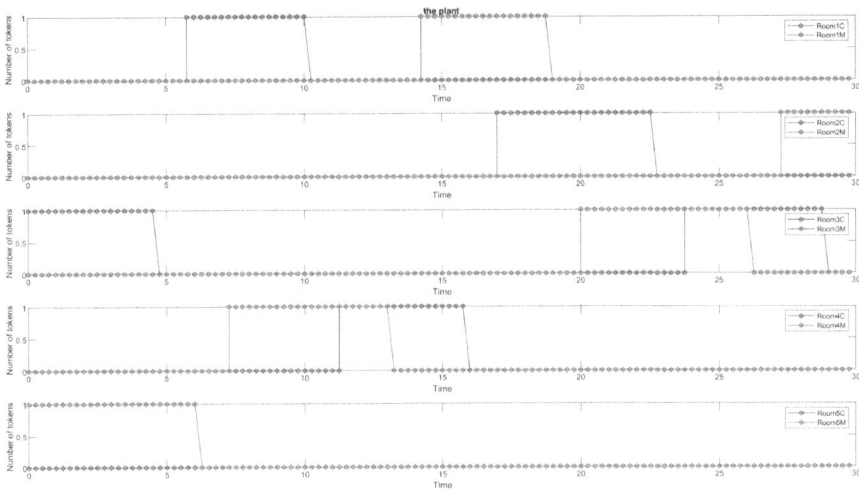

Fig. 6. Cat and mouse's movements after the delays are introduced to doors.

6 Concluding Remarks

GPenSIM is specially designed to model real-life industrial systems. Hence, the transitions in GPenSIM represent real-life events, such as production machines. However, this induces a problem, namely that an enabled transition consumes tokens from its input places immediately. An enabled transition (just like a production machine) consumes tokens from its input places ('raw material') and keeps these tokens as 'virtual tokens' during the firing time (production time).

This act of an enabled transition immediately consuming tokens gives only monetarily visibility to the tokens, which can be a problem for certain cases, as we saw in the previous section in which we need to see the cat and the dog in a room for a while, without being snatched away by the doors. The method proposed in this paper thus gives modelers an opportunity to choose GPenSIM for modeling problems that demand increased visibility of tokens in the input places.

References

1. Best, E., Devillers, R.: Petri net primer: a compendium on the core model, analysis, and synthesis. Springer (2024)
2. Boissel, O.R.: Optimal feedback control design for discrete-event process systems using simulated annealing. Ph.D. thesis, University of Notre Dame (1993)
3. David, R., Alla, H.: Petri nets for modeling of dynamic systems: a survey. Automatica **30**(2), 175–202 (1994)
4. Davidrajuh, R.: Modeling Discrete-Event Systems with GPenSIM. Springer, Cham (2018). https://doi.org/10.1007/978-3-319-73102-5

5. Davidrajuh, R.: Petri Nets for Modeling of Large Discrete Systems. Springer (2021)
6. Davidrajuh, R.: Colored Petri Nets for Modeling of Discrete Systems: A Practical Approach With GPenSIM. Springer (2023)
7. Davidrajuh, R.: GPenSIM Reference Manual. University of Stavanger (2024)
8. Desel, J., Reisig, W.: The concepts of petri nets. Softw. Syst. Model. **14**, 669–683 (2015)
9. GPenSIM: General-purpose Petri net simulator. Technical report (2019). http://www.davidrajuh.net/gpensim. Accessed 20 July 2020
10. Liu, G.: Petri Nets: Theoretical Models and Analysis Methods for Concurrent Systems. Springer (2022)
11. Moody, J., Antsaklis, P.J.: Supervisory control of discrete event systems using Petri nets, vol. 8. Springer (1998)
12. Murata, T.: Petri nets: Properties, analysis and applications. Proc. IEEE **77**(4), 541–580 (1989)
13. Peterson, J.L.: Petri net theory and the modeling of systems. Prentice Hall PTR (1981)
14. Popova-Zeugmann, L., Popova-Zeugmann, L.: Time petri nets. Springer (2013)
15. Reisig, W.: Understanding petri nets. Springer (2016)
16. Shi, H., Liu, H.C.: Fuzzy Petri Nets for Knowledge Representation, Acquisition and Reasoning. Springer (2023)
17. Silva, M.: Introducing petri nets. Practice of Petri Nets in manufacturing, pp. 1–62 (1993)
18. Wonham, W.M., Ramadge, P.J.: On the supremal controllable sublanguage of a given language. SIAM J. Control. Optim. **25**(3), 637–659 (1987)

Application Interface for GPenSIM

Yuming Feng[1], Slawomir Samolej[2], and Reggie Davidrajuh[3](✉)

[1] School of Computer Science and Engineering, Chongqing Three Gorges University, Wenzhou, China
[2] Department of Computer and Control Engineering, Rzeszow University of Technology, Rzeszow, Poland
[3] Department of Electrical Engineering and Computer Science, University of Stavanger, Stavanger, Norway
Reggie.Davidrajuh@uis.no

Abstract. The General-purpose Petri Net Simulator (GPenSIM) is a tool used by researchers to model, simulate, and analyze the performance of discrete event systems. Its popularity stems from its simple interface, extensibility, and compatibility with the MATLAB platform. This paper focuses on the crucial role of GPenSIM in Petri Net model-based software-to-software and software-to-hardware access and control, highlighting unique features and capabilities in its application interface.

Keywords: GPenSIM · Petri Nets · Model-based control · Application Interface · Discrete Event Systems

1 Introduction

General-purpose Petri Net Simulator (GPenSIM) is a toolbox that runs on the MATLAB platform [8]. Some researchers use GPenSIM to model, simulate, and analyze the performance of discrete event systems. GPenSIM is popular because it is easy to learn, use, and extensible as it runs on the MATLAB platform [13]. Literature studies reveal works that use GPenSIM for Petri Net model-based software-to-software and software-to-hardware control [1]. In these cases, the GPenSIM's application interface becomes decisive; Fig. 1 shows the application interface facilitating a Petri Net model-based access and control of external hardware and software. This paper presents the application interface of GPenSIM and how it enables Petri Net model-based software-to-software and software-to-hardware control.

The structure of this paper is as follows: Sect. 2 introduces Application Program Interface. Section 3 briefly introduces Petri Nets and GPenSIM. Section 4 presents the application interface of GPenSIM. Section 5 shows a sample application.

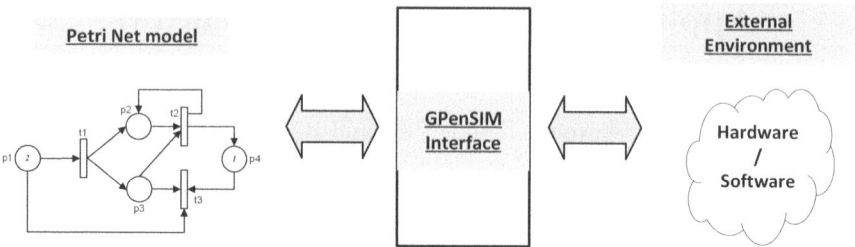

Fig. 1. GPenSIM's application interface facilitating a Petri Net model-based access and control of external hardware and software.

2 Application Program Interface

An Application Program Interface (API) for GPenSIM allows a Petri net model to interact with the outside (external environment), either software or hardware system. An API presented in this paper is about the interface a modeler can use to create interactions between a Petri net model and external software or hardware; also, An API presented in this paper is not a user interface (UI) or a graphical user interface (GUI), which are visible to users.

Since an API, as an interface, enables software programs to share data and functions, the API for GPenSIM does this followingly:

– Data to be shared.
– Functions that are available through the interface.
– Contact points.

Data to be Shared: Data can be shared between the Petri net model and the external software by declaring data as global variables (as elements of "global_info" packet.

Functions that are Available Through the Interface: all GPenSIM functions are available through the interface. The static and dynamic (run-time) details of the Petri net can be accessed using GPenSIM functions. Since GPenSIM is a MATLAB toolbox, the interface allows access to hundreds of MATLAB functions in virtually any domain of engineering.

Contact Points: GPenSIM provides two types of processor files that function as the control points. A specific pre-processor file is executed whenever the related transition is enabled. Only if the conditions coded in the pre-processor file are satisfied is the transition allowed to start firing. A specific post-processor file is executed whenever the related transition completes firing. GPenSIM also provides two more types of processor files: modular and common. Modular processor files and common processor files are relevant only to Modular Petri nets and Large Petri nets. The following sections shed more light on these processor files.

3 Petri Nets and GPenSIM

This section starts with a definition of Petri Nets and then gives a brief introduction to GPenSIM.

3.1 Petri Nets

Definition: A **Petri Net** (*aka* Marked Petri Nets) is defined as a five-tuple [12,15]:

$$PN = (P, T, A, K, M),$$

where (N is the natural number),

- P is a set of places, $P = \{p_1, p_2, \ldots, p_{n_p}\}$.
  $\forall p_i \in P$, $K(p_i) \to N \cup \{\infty\}$, is the maximum capacity of the place p_i.
- T is a set of transitions, $T = \{t_1, t_2, \ldots, t_{n_t}\}$.
  $P \cap T = \emptyset$.
- A is the set of arcs (from places to transitions and from transitions to places). $A \subseteq (P \times T) \cup (T \times P)$. The arc weight $w(a_{ij}) \in N \cup \{0\}$, $\forall a_{ij} \in A$.
- M is the row vector of markings (tokens) on the set of places.
  $M = [M(p_1), M(p_2), \ldots, M(p_{n_p})] \in N \cup \{0\}$, M_0 is the initial marking.

A transition in a Petri Net is enabled *iff* all of its input places have tokens more than the input arc weights. Formally, $\forall p \in P$, $M(p) \geq w(p, t)$.

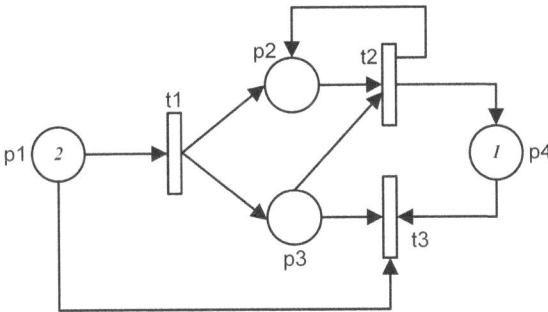

Fig. 2. Simple Petri net.

Figure 2 shows a simple Petri Net with four places, **p1** to **p4**, and three transitions, **t1** to **t3**. Only **t1** is enabled as its input place **p1** has two tokens (more than the arc weight of one).

3.2 GPenSIM

The third author of this paper developed GPenSIM as a tool for teaching a course on modeling, simulation, and performance analysis of discrete-event systems. Petri Net was chosen as the tool for this course because of its properties like self-documentation (graphical front), simple mathematics (linear algebra), and explicit state information (known as "reachability tree" in Petri Net literature) [2,16,17].

GPenSIM was developed as computer software to work with Petri Nets. Due to its enormity, GPenSIM was developed in three stages:

- Stage-1: Supporting P/T Petri nets and Timed Petri Nets [3].
- Stage-2: Supporting Colored Petri Nets [6].
- Stage-3: Supporting modeling of large real-life industrial discrete systems with modular approach [5].

GPenSIM offers a rich set of built-in functions so that modelers can make use of these functions to create compact models [7].

4 Application Interface of GPenSIM

An application interface is a tool and mechanism for a software developer to use as the entry point for the interaction between the particular software and other entities (software or hardware components). An application interface must be well-defined (the protocol) and easy to understand and use [14]. It must also protect the entities from both sides from any attempts to cause damage intentionally or unintentionally [10]. The well-known application interfaces are the operating systems and the internet protocols that allow applications to interact with each other.

4.1 Interfacing with a Small Petri Net

In order to make the GPenSIM's application interface as simple as possible, the following fundamental decisions were made during the design phase:

- In a Petri Net model implemented with GPenSIM, only transitions can be manipulated; places and arcs are static (cannot be influenced) once they are declared.
- A transition is composed of three parts: a specific pre-processor, a delay function, and a specific post-processor.
  - Specific Pre-processor: Whenever a transition becomes enabled, the compiler executes its specific pre-processor file (a MATLAB file); this pre-processor file is known as the 'specific' pre-processor file as only the transition is visible in this file; the other transitions are not visible. In this file, a user can code logical conditions that must be satisfied before the transition can start firing.

- Delay: If an enabled transition starts firing (because all the conditions in the pre-processor are satisfied), the compiler will put this transition in a zombie queue (the transition sleeps) for the duration of the firing time.
- Specific Post-processor: Whenever a transition completes firing, the compiler executes its specific post-processor file (a MATLAB file). This post-processor file is known as the 'specific' post-processor file, as only the transition is visible in this file, and the other transitions are not visible. In this file, a user can code actions that must be carried out after the transition completes firing.

Let's consider a practical example to understand the functioning of a transition better. Suppose we want a transition, which we'll call '**tLight**', to turn on a light for 30 s. In this case, in the specific pre-processor of **tLight**, we will switch on the light; then **tLight** will be put to sleep in the zombie queue for 30 s (the 'firing time'). Finally, when **tLight** completes firing (the compiler wakes up **tLight** from the zombie queue after 30 s), in its specific post-processor, we will switch off the light. (we are not discussing the hardware interfacing for switching on and off the light here; the hardware interfacing is discussed in Sect. 5).

Figure 3 shows the composition of **tLight** that includes the specific pre-processor, the delay function, and the specific post-processor. Listings 1.1 and 1.2 show the specific pre-processor and the specific post-processor of **tLight**.

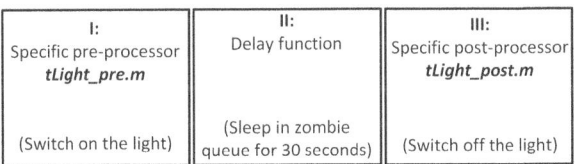

Fig. 3. Composition of the transition **tLight**.

Listing 1.1. Pre-processor of tLight (tLight_pre.m)

```
function [fire, transition] = tLight_pre(transition)

global global_info
disp('switching ON the light');
SwitchLamp(global_info.light, 'on');
fire = true;
```

Listing 1.2. Post-processor of tLight (tLight_post.m)

```
function [] = tLight_post(transition)

global global_info

disp('switching OFF the light');
SwitchLamp(global_info.light, 'off');
```

Summary: The Application Interface for Small Petri Nets:
In summary, the application interface for using a small Petri Net to access and control the other software and hardware consists of a set of processor files: a set of specific pre-processors (one pre-processor for each transition) and a set of specific post-processors (also one post-processor for each transition). Hence, Fig. 4 shows the application interface for the Petri Net shown in Fig. 2.

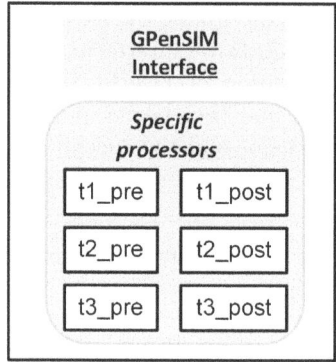

Fig. 4. The application interface for the small Petri Net shown in Fig. 2.

4.2 Interfacing with a Large Petri Net

There are only three transitions in the Petri Net shown in Fig. 2. Hence, the application interface of this Petri Net consists of three specific pre-processors and an equal number of specific post-processors. Some transitions may not need specific pre-processors as there are no additional conditions to satisfy before starting firing. Similarly, some transitions may not need specific post-processors as no post-firing actions need to be performed. Hence, the application interface of this Petri Net consists of a *maximum* of three specific pre-processors and a *maximum* of three specific post-processors.

However, when we confront **a large Petri Net**, the application interface presented in the previous section (Sect. 4.1) proves to be inadequate as it will demand a large number of specific pre-processors and specific post-processors. We need to have another solution:

– Many enabled transitions may have common conditions that they must satisfy before they start firing; hence, copying the same common code into multiple pre-processors seems redundant.
– Similarly, there may be common actions that are to be performed after the firing of some transitions; hence, copying the same common code for these actions into multiple post-processors also seems redundant.

Due to these reasons, two **common processors** are introduced: the "**COMMON_PRE**" is a pre-processor for placing the common conditions that are to be satisfied before for any enabled transition starts firing. The "**COMMON_POST**" is a post-processor for placing the common actions that are to be performed after any transition completes firing.

Figure 5 shows the application interface for dealing with large Petri Nets. The common processors (COMMON_PRE and COMMON_POST) facilitate reducing the application interface's complexity. However, some transitions may need special code (e.g., hardware-specific code) for their pre- and post-processors. Hence, the application interface also accommodates additional specific pre- and post-processors.

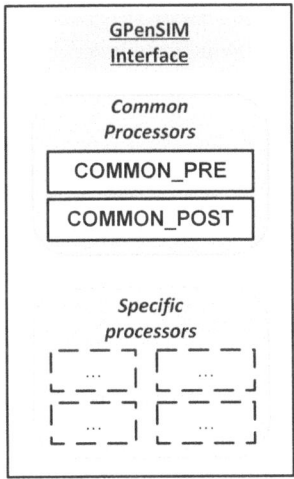

Fig. 5. The application interface for large Petri Nets.

4.3 Interfacing with a Modular Petri Net

Petri Nets' main weakness for modeling discrete systems is well-documented; namely, the Petri Net models become huge and very difficult to analyze [9,11]. Hence, [5] suggests a modular approach by introducing a new Modular Petri Nets, which is already implemented in GPenSIM. In this new modular approach, a Petri Net model is composed of Petri Modules, which are independent and can be run on different computers. The Petri Modules are interconnected by Inter-Modular Components (IMCs). Due to brevity, the formal definitions of Petri Modules and IMCs are not given in this paper as they are lengthy; the interested reader is referred to [4,5].

Figure 6 shows a Modular Petri Net composed of two Petri Modules (Module-**A** and Module-**B**) and two IMCs (IMC-**X** and IMC-**Y**).

A Petri Module follows strict guidelines:

- Tokens enter a Petri Module only through the Input Ports, which are transitions.
- Tokens leave a Petri Module only through the Output Ports, which are also transitions.
- Places cannot function as Input or Output Port of Modules.
- Local transitions and local places (the elements inside a module) cannot have a direct connection with elements outside the module.

Since a Petri Module has transitions as the Input and Output Ports, we need places to connect two Petri Modules; these places and other accompanying transitions make up the IMCs.

In a Modular Petri Net model, there are three types of processors: the first and second types have already been introduced as specific processors and common processors, respectively. The third and new type is the modular processor.

All the transitions in a Petri Module are visible in their Modular Processors; for example, Fig. 6 shows a Modular Petri Net composed of two Petri Modules (Module-**A** and Module-**B**) and two IMCs (IMC-**X** and IMC-**Y**). In this case, the transitions of Module-**A**, such as **tAI1**, **tAL1**, **tAL2**, **tAO1**, and **tAO2**, are visible in the Modular pre-processor and Modular post-processor of **A**. Similarly, the transitions of Module-**B**, such as **tBI1**, **tBI2**, **tBL1**, **tBL2**, and **tBO1**, are visible in the Modular pre-processor and Modular post-processor of **B**. However, the Input Ports and Output Ports of all modules (e.g., **tAI1**, **tAO1**, and **tAO2**, and **tBI1**, **tBI2**, and **tBO1**) and the IMC transitions (**tX1** and **tY1**) are also visible in the common processors. Note that local transitions of modules are not visible in the common processors.

Hence, the application interface of a modular Petri Net becomes more detailed as it consists of common processors, modular processors, and some specific processors. Figure 7 shows the application interface of a modular Petri Net.

5 Sample Application

This section presents a simple application of the Petri Net model-based control of external hardware. The first subsection describes the hardware, the next is on the Petri Net model, and finally, the application interface.

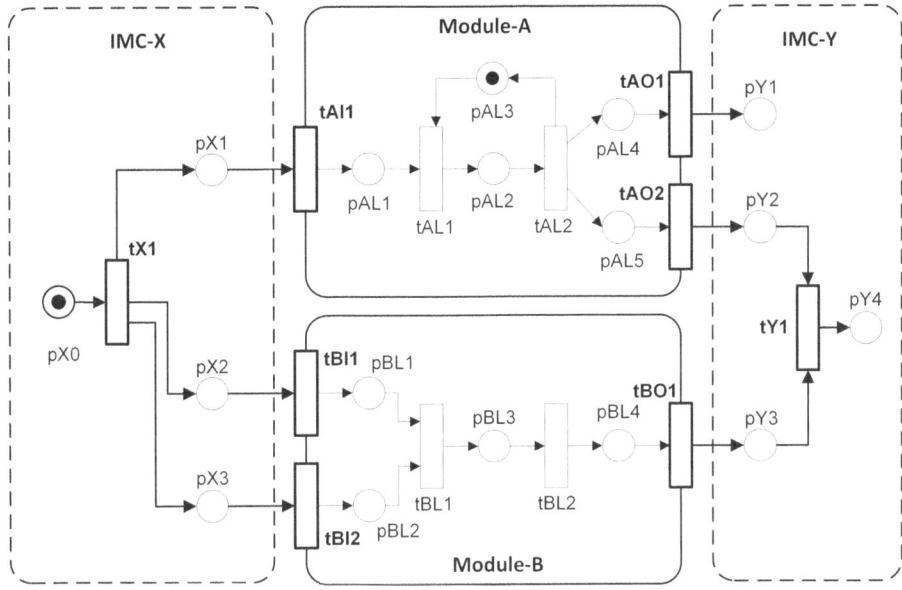

Fig. 6. A Modular Petri Net.

5.1 The Hardware

Figure 8 shows a simple LEGO robot gadget. The LEGO robot is equipped with three buttons as the inputs:

- The first button on the left is the "reset button". The reset button is for starting (or resetting) the robot.
- The button in the middle is the "light button". The light button is for switching on the green light for 30 s.
- The button on the right is the "stop button". The stop button is for shutting down the robot.

The LEGO robot is also equipped with three lights as the outputs; only the green light on the right edge will be used in this testing.

5.2 The Petri Net Model

Figure 9 shows a small Petri Net model for testing the LEGO robot. This Petri Net is composed of isolated subnets, which is normal for hardware-interfacing applications.

The first subnet is for switching the robot on and off. At the start, **tStart** begins firing, and its pre-processor ('tStart_pre') possesses all the (hardware-specific) code for the initialization of the LEGO robot; there is no need for a post-processor for **tStart**. When **tStart** completes firing (its firing time is of

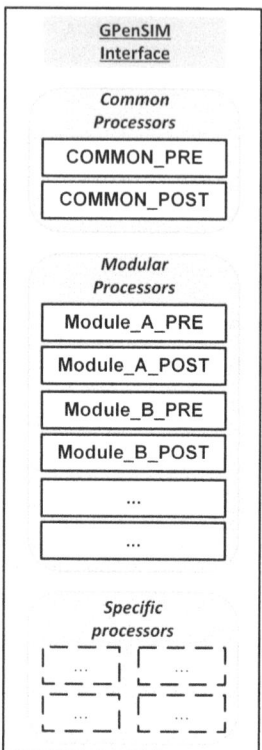

Fig. 7. The application interface for a Modular Petri Net.

little consequence), it will deposit a token into **pStop**, enabling **tStop**. In the pre-processor of **tStop** ('tStop_pre'), there is code for scanning the stop button; whenever the stop button is pressed, the code in **tStop**'s pre-processor will allow **tStop** to start firing. When **tStop** completes firing (its firing time is also of little consequence), its post-processor ('tStop_post') executes the code to clean up the memory and switch off the robot.

The second subnet is for switching on the green light. In the **tLightButton**'s pre-processor ('tLightButton_pre'), there is code for scanning the light button. Whenever the light button is pressed, **tLightButton** is allowed to fire. When **tLightButton** completes firing, the deposited token in **pLight** enables **tLight**.

As described in Sect. 4.1, the **tLight**'s pre-processor switches on the green light. Then **tLight** starts firing for 30 s (it is put to sleep in the zombie queue for 30 s), and when it completes firing, the green light is switched off in its post-processor. The end result is that the green light is switched on for 30 s whenever the light button is pressed.

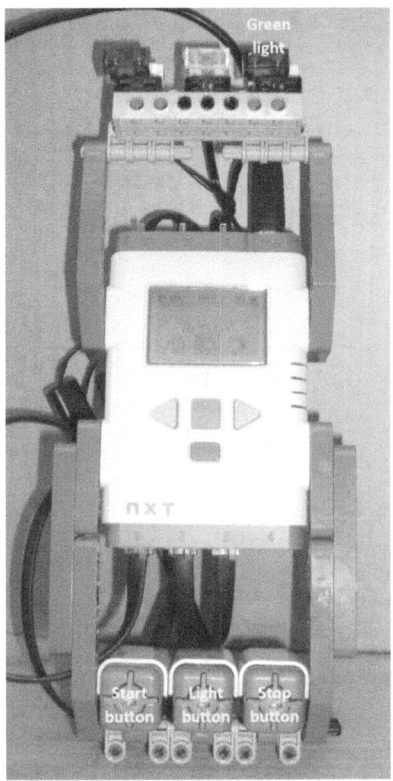

Fig. 8. The LEGO Robot.

5.3 The Application Interface

Figure 10 shows the application interface to make the Petri Net model control the LEGO Robot. Of course, since the Petri Net model is small, we are using the application interface shown in Fig. 4 (the first application interface for smaller Petri Net models).

Due to brevity, this section does not show any code. The interested reader is encouraged to contact the authors for complete code.

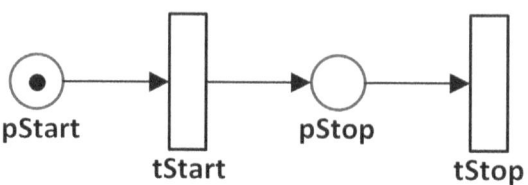

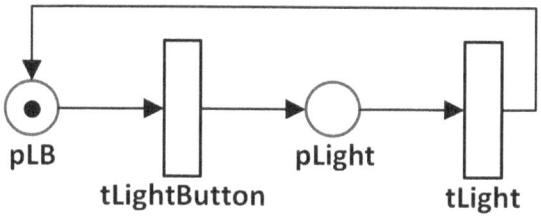

Fig. 9. The Petri Net model for controlling the LEGO NXT Robot.

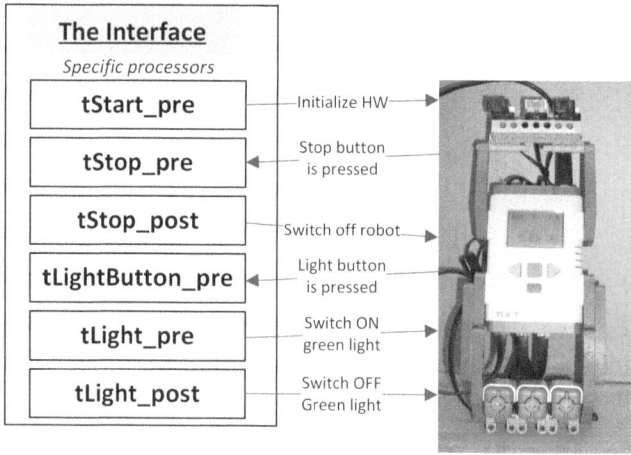

Fig. 10. The application interface between the Petri Net model and the LEGO NXT Robot.

6 Concluding Remarks

GPenSIM provides three application interfaces as shown in Fig. 4, 5, and 7. With these three application interfaces, a developer could make use of Petri Net models to access other functions in diverse MATLAB toolboxes and external software and control external hardware.

The application interface shown in Fig. 4 is for small Petri Net models, and the one shown in Fig. 5 is meant for larger models. However, both of these application interfaces are for monolithic (non-modular) Petri Nets. Though these two application interfaces are compatible with each other, it is advisable to use the application interface shown in Fig. 5 as it will provide a clear and simple set of contact points (by the use of common processors).

This application interface shown in Fig. 7 is for Modular Petri Net models only; this application interface is not interchangeable with the other interfaces. This is because the visibility of transitions in a Modular Petri Net model is not the same as in monolithic (non-modular) Petri Net models.

As a final note, it should be emphasized that this paper presents an application interface for Petri net models to control external hardware or software automatically. In other words, the application interface presented in this paper is not about "user interface" or GUI. Hence, this paper does not provide an evaluation of the application interface's effectiveness. Nor is it possible to involve user testing, performance benchmarks, or case studies demonstrating the application interface's impact on the modeling and simulation process.

References

1. Cameron, A., Stumptner, M., Nandagopal, N., Mayer, W., Mansell, T.: Rule-based peer-to-peer framework for decentralised real-time service oriented architectures. Sci. Comput. Program. **97**, 202–234 (2015)
2. David, R., Alla, H.: Petri nets for modeling of dynamic systems: a survey. Automatica **30**(2), 175–202 (1994)
3. Davidrajuh, R.: Modeling Discrete-Event Systems with GPenSIM. Springer, Cham (2018). https://doi.org/10.1007/978-3-319-73102-5
4. Davidrajuh, R.: A new modular petri net for modeling large discrete-event systems: a proposal based on the literature study. Computers **8**(4), 83 (2019)
5. Davidrajuh, R.: Petri Nets for Modeling of Large Discrete Systems. Springer (2021)
6. Davidrajuh, R.: Colored Petri Nets for Modeling of Discrete Systems: A Practical Approach With GPenSIM. Springer (2023)
7. Davidrajuh, R.: GPenSIM Reference Manual. University of Stavanger (2024). https://doi.org/10.31265/USPS.293
8. GPenSIM: General-purpose Petri net simulator. Technical report (2024). http://www.davidrajuh.net/gpensim. Accessed 20 July 2024
9. Grobelna, I., Karatkevich, A.: Challenges in application of petri nets in manufacturing systems. Electronics **10**(18), 2305 (2021)
10. Jerraya, A.A., Wolf, W.: Hardware/software interface codesign for embedded systems. Computer **38**(2), 63–69 (2005)
11. Jones, N.D., Landweber, L.H., Lien, Y.E.: Complexity of some problems in petri nets. Theoret. Comput. Sci. **4**(3), 277–299 (1977)
12. Murata, T.: Petri nets: properties, analysis and applications. Proc. IEEE **77**(4), 541–580 (1989)
13. Mutarraf, U., Barkaoui, K., Li, Z., Wu, N., Qu, T.: Transformation of business process model and notation models onto petri nets and their analysis. Adv. Mech. Eng. **10**(12), 1687814018808170 (2018)

14. Patterson, D.A., Hennessy, J.L.: Computer organization and design ARM edition: the hardware software interface. Morgan Kaufmann (2016)
15. Peterson, J.L.: Petri net theory and the modeling of systems. Prentice Hall PTR (1981)
16. Reisig, W.: Understanding petri nets. Springer (2016)
17. Silva, M.: Introducing petri nets. Practice of Petri Nets in Manufacturing, pp. 1–62 (1993)

Analyzing Production Line Schedules with Tecnomatix Plant Simulation

Jocelyn Yeoh[1], Li-Pei Wong[1](✉), Mohd. Zulkifli Che Wanik[2], and Siang Kok Chia[3]

[1] School of Computer Sciences, Universiti Sains Malaysia, USM, 11800 Pulau Pinang, Malaysia
lpwong@usm.my
[2] Program Management Engineering Advance Analytics Office, Western Digital (Sandisk Storage Malaysia Sdn. Bhd.), 14100 Pulau Pinang, Malaysia
[3] Automation Development Engineering, Western Digital (Sandisk Storage Malaysia Sdn. Bhd.), 14100 Pulau Pinang, Malaysia

Abstract. Manufacturing environments involve complex scheduling decisions with a high volume of data and daily order urgency. Every production stage holds its own defined schedule in optimizing single production objective or balancing a list of targeted objectives under specific constraints. Oftentimes, production planners find it challenging to effectively analyze and compare a sheer number of potential schedules with different criteria that a manufacturing operation could employ. While Gantt charts provide a basic timeline of production schedule, they are limited in their ability to interpret the intricacies of production performance. For better visualization and deeper insights of production operation on the shop floor, simulation approach emerges as a viable tool to get a full picture of the system behavior under various scenarios. This paper presents two simulations performed using Tecnomatix Plant Simulation. The first simulation analyzes a schedule optimized for a single objective, specifically minimizing makespan, while the second simulation examines a schedule optimized for two objectives namely, makespan and production line conversion frequency. By conducting these simulations, the impact of various schedules can be further obtained with interactive visuals and analytics. Manufacturers can make production strategies effectively in view of the statistical capabilities of the simulation software, thereby enhancing overall productivity and operational efficiency.

Keywords: discrete-event simulation · statistical analysis · machine utilization · throughput analysis · bottleneck analysis

1 Introduction

Production scheduling has always been a long-standing research topic across diverse domains. It holds paramount importance within manufacturing sectors in supplying on-demand products to customers, hence requires complex decision-making to allocate resources to perform a set of distinct tasks in a certain timeframe. The performance of the production lies in optimizing single or multiple objectives (e.g. minimization of makespan and reduction in line conversion frequency) subjected to a set of constraints (e.g. no pre-emption and job precedence).

Over the corridors of industrial time, the complexity of production scheduling expands with the diverse requirements in the industry, e.g. the need of balancing multiple objectives, deciding on efficient schedules, and visualizing the production flow. Every working day, production planners generate schedules in different forms, where each tailored to one or more targeted objectives: time-based (makespan and flow time), job-based (number of job orders and number of jobs execution) or line-based (total workload and line conversion frequency). This leads to a variety of schedules that planners need to examine for efficient production scheduling.

Existing production scheduling methods involve manual intervention, where all the machine line information is captured in tabular format (e.g. Excel-based application). While tabular data in spreadsheets offer basic planning capabilities in a simplified view, they lack the ability to visualize the impact of scheduling decisions or assess schedule robustness. Planners are not equipped to fully comprehend how different schedules will function in real-time. Besides, these methods are limited to visualizing the operation and behavior of the manufacturing environment in a global view when it comes to coordinating the entire production flow.

The scheduling issue not only involves identifying the optimal schedule among distinct forms of potential schedule, but also in determining which of these schedules will work effectively when applied in real-world production settings. To address this challenge, Discrete-event Simulation (DES) comes into play to simulate the execution of generated schedules by supporting visualization, performing what-if scenarios, and offering analytical insights to enhance production schedule planning [1, 8]. The research employs Tecnomatix Plant Simulation as the virtual environment to simulate various scheduling scenarios. Such an approach would enable a holistic view of scheduling, allow planners to evaluate different hypothetical scenarios [5, 6, 12], experiment with different configurations [7], and make strategic decisions in a simulated environment [4, 8]. Tecnomatix Plant Simulation will serve as the primary tool to streamline the output of the generated schedules and present the schedule analysis and their associated performance metrics at a glance.

The contributions of this paper are twofold. The first contribution shows the capability of the simulation approach in gathering better insights of scheduling performance by depicting production line schedules into a proper workflow model for potential improvements in the manufacturing process. Besides, the application of Tecnomatix Plant Simulation facilitates what-if analysis in further testing of various production scenarios in a risk-free virtual environment without incurring any implementation costs.

The remainder of the article is outlined as follows. Section 2 presents a literature review that covers the existing approaches and uses of Tecnomatix Plant Simulation software. Section 3 details the simulation methodology as the basis of the research framework. Section 4 discusses the experimental setup and results of the simulation tool. Section 5 concludes the article with a summary of the findings.

2 Related Work

A review of the relevant literature reveals the application of Tecnomatix Plant Simulation is employed across a wide range of domains. It is an indispensable tool for production planning while developing a new system or optimizing an ongoing workflow [3, 4, 12,

15]. Maidstone [9] identified DES tool as one of the most commonly applied practices in operational research. As the term implies, DES model represents a system by simulating a sequence of discrete processes into various states. The article examines the potential uses of Tecnomatix Plant Simulation software to simulate manufacturing processes, as well as the benefits of using such simulations. It also introduces analytic tools for optimizing the throughput and mechanical components for reducing the work-in-process of the manufacturing system.

Many researchers claimed that plant simulation tools should be viewed as a way to test and experiment with planned solutions in a virtual environment before implementing them in actual settings [6, 10, 13]. By simulating various forms of production schedules, simulation enables the optimization of time-based metrics (processing time, failure time, set-up time, and recovery time), line throughput, buffer sizing, machine allocation and job count with the configuration of operational parameters [13]. Musil, Laskovský and Fialek [12] demonstrated how Tecnomatix Plant Simulation can be applied to model and analyze logistics processes. They did a proof-of-concept by creating both 2D and 3D models of a sorting line system that handles municipal and hazardous waste. Moreover, a deeper understanding of the complicated manufacturing systems can be acquired by configuring the model and prompting statistical reports on the simulation [2]. This approach aids in examining the effectiveness of production schedules and initiating resource strategies before the real execution [4–6]. Additionally, it allows for the acquisition of valuable information without causing disruption to the operational equilibrium of the physical system [11].

Siderska and Perkowska [14] conducted an experiment with Tecnomatix Plant Simulation to investigate the production flow of sheet metal screws in the carpentry plant. They determined potential bottlenecks and analyzed the workloads of every machine on the simulation software. The simulation analysis shows that most machines are severely underutilized at merely 10% capacity and produce 91 screws per 16-h period. Through extensive testing on the simulation, the authors were able to make some useful decisions to optimize the overall production process. Among the ideal solutions are dividing the zinc plating and coating operations into separate workstations and adding two assembly stations. DES proved to be effective as a decision support tool for viewing the entire production flow and validating model designs [14].

Kikolski [6] simulated experiments by presenting three production scenarios using Tecnomatix Plant Simulation to analyze how different batch sizes (15, 30, and 45 items) of each component affects the production efficiency. The study examines a model system consisting of three processing stations and one assembly station, running over an 8-h shift. The results show that the smaller batches of 15 items performed the best, which produces 69 final products compared to 63 products with medium batches (30 items) and 56 products with large batches (45 items). The observation shows the impact of plant simulation in verifying the functionality of various processes.

Marasova, Saderova, and Ambrisko [10] utilized Tecnomatix Plant Simulation to create a production line model on material handling equipment and conveyance lines in an automotive company. The initial simulation reveals bottlenecks with some operations being blocked up to 63% of the time while others were either underutilized (17% active time) or overworked (99% active time). With the adjustment of production times, they observed some improvements but also new challenges. The performance of Operation A, B and E were improved to 65%, 49% and 70% of active time respectively [10]. However, the modification shows a negative impact on Operation G where its performance is reduced by 29%. The authors raised that while the experiments help in detecting line inefficiencies and reducing some production backlogs, they still require further research to find an ideal balance for the entire production line [10].

Kopec and coworkers [8] also showcased the effectiveness of using Tecnomatix Plant Simulation to improve manufacturing efficiency. Initially, the company faces several critical issues with inefficient transport between workplaces and severed downtime at three key workplaces - laser cutting, polishing, and grinding stations. Through simulation, researchers tested two main changes: replacing manual transporters with continuous conveyor belts and adding machines at bottleneck locations [8]. The results are boosted across all three workstations, with laser cutting inputs increasing from 72 to 105, polishing inputs from 211 to 384, and grinding inputs from 416 to 704. Additionally, the downtime decreases substantially at all stations. The simulation tool allows the company to test these improvements and enable them to accept larger orders.

Tecnomatix Plant Simulation software unfolds itself at the forefront of innovative simulation applications owing to the invention by a German company, Siemens PLM Software, which is a global leader in Product Lifecycle Management (PLM) and Manufacturing Operations Management (MOM) software [15]. This application enables the intuitive modeling of lifelike digital production systems, facilitating the thorough testing and optimization of system properties and performance, as well as visualizing complex production processes. Given the complex nature of manufacturing, Tecnomatix Plant Simulation can aid in scheduling optimization by conducting experimentation beforehand [7], for instance, considering factors such as machine capabilities, setup times, and batch sizes, each of these can significantly impact production efficiency. This is particularly useful in the production schedule planning stage, where potential strategies can be evaluated ahead [3, 5, 10].

3 Simulation Procedure

This section explains the step-by-step procedures of the entire research. A standard flowchart recommended by Siemens PLM for Tecnomatix Plant Simulation is presented in Fig. 1 [15]. The simulation study adopts a three-phase methodological approach: Preparation phase, Execution phase, and Evaluation phase.

According to the simulation study process as depicted in Fig. 1, it begins by initiating the problem and targeted objectives of a scenario (e.g. an existing system of a manufacturing plant). During the Preparation phase, a thorough analysis of the system under investigation is conducted to identify key factors and relationships. Data is acquired to understand system behavior and inform the model development stage. Next, the Execution phase proceeds with the formation of a model that represents the system's behavior and dynamics. The model is being validated to ensure its accuracy and reliability. Subsequently, experiments are performed and the model is analyzed to generate insights of the performance under various conditions. Finally, the Evaluation phase focuses on evaluating the results obtained from the experiments to assess their significance and implications. Recommendations are also suggested based on the findings of the study for decision-making and continual improvements. This iterative approach ensures a thorough simulation study and actionable outcomes of the investigated system.

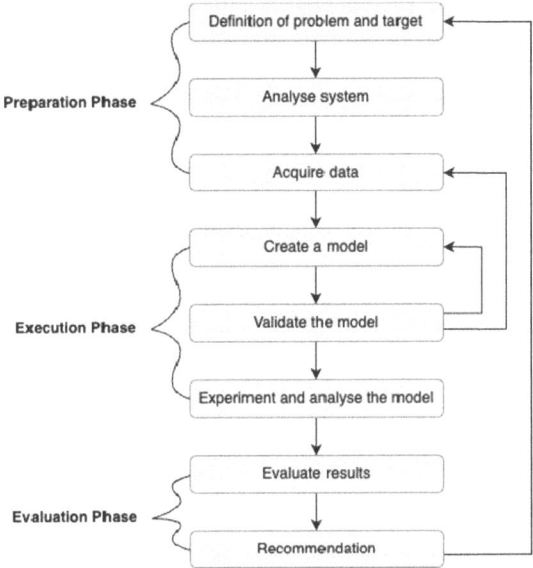

Fig. 1. Methodology of simulation study.

Moreover, a structured process of the research framework that shows all the crucial activities of the research work is outlined in Fig. 2. Each stage is presented to align with each of the crucial phases in the simulation study.

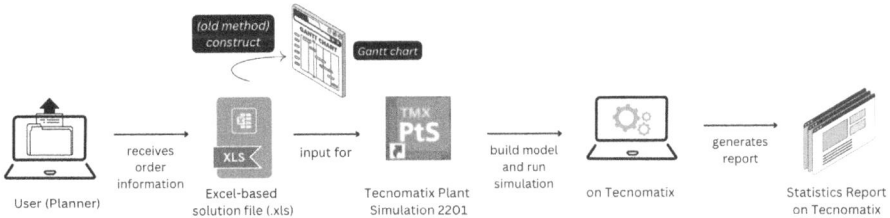

Fig. 2. Steps to simulate and analyze schedules with Tecnomatix Plant Simulation.

The implementation work of this simulation study delineates a stepwise approach as described in Fig. 2. Consider a scenario in a manufacturing plant, data will be gathered from manufacturing processes such as order status, job quantity, product types, processing times, production line characteristics, and resource constraints. These data are being captured and stored in Excel spreadsheet (.xlsx) format. Production scheduling traditionally relied upon Excel spreadsheets to construct static Gantt charts where tasks are represented by bars on a timeline to manage production operations. To address this limitation, the research uses Tecnomatix Plant Simulation such that the Excel-based solution files can be imported into the software as input data to be simulated and visualized at sight. All the data can be accessed and reflected on 'DataTable' in Tecnomatix Plant Simulation. Each of the timeline bars on the Gantt chart can be simplified into a uniform setup on Tecnomatix Plant Simulation. To assemble an entire production line model, basic elements or objects on Tecnomatix Plant Simulation such as 'Station', 'Source', 'Drain' and 'Connector' are inserted into the model frame as shown in Fig. 3. The functionalities of each of the element are described as follows: 'Station' acts as the working line to process workpieces; 'Source' acts as the starting point to produce product parts; 'Drain' acts as the endpoint where the finished products (processed job orders) are collected; and 'Connector' connects all the inserted objects on the frame.

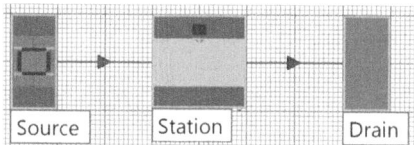

Fig. 3. Basic structure of a complete production line.

The simple model is designed to represent the conditions and scenarios of the problem based on the data received. A statistical report can be obtained after running the simulation model to analyze the performance of the model. The analytic capability of the simulation serves as a guide for informed decision-making by understanding the strengths and weaknesses of each schedule. On top of that, the generated report can be exported and saved as an HTML file, which is suitable for sharing and online viewing.

4 Experiments and Discussion

This section examines a real-world scheduling problem of a semiconductor manufacturing plant. The purpose of this study is to elevate the visualization of production flow on Tecnomatix Plant Simulation with different forms of schedules in the manufacturing industry. The simulation model provides a visual representation of schedules in different scenarios and validates the efficiency of production scheduling by analyzing simulation results for better manufacturing planning.

In this experiment, the input model of the manufacturing system has been reframed according to a one-day actual dataset from a semiconductor manufacturing company. The dataset involves 35 distinct job orders with 9 product types operated by 11 available machine lines. The problem instance provides the job processing times along with the operation sequence of the production lines. The scheduling problem is then tested on Tecnomatix Plant Simulation software for further experimentation. All the key information from the spreadsheet is extracted and labeled with their respective data type as follows: 'Order No.' (numerical, 1 to 35), 'Product Type' (alphabetical string, A to I), 'Machine No' (numerical, 1 to 11), and 'Processing Time' (decimal, in hours). Due to the specific naming conventions of the simulation model, a new column named 'OrderID' is created by having 'Product Type' combined with an underscore (_) and followed by 'Order No.' (e.g. A_1). It is observed that the different manufacturing parameters and production objectives applied during the production scheduling process can lead to variations in the processing times on identical machines.

The experimental work aims to increase productivity and reduce blocking time by following regular simulation procedures. The simulation approach is expected to facilitate the production planning and scheduling of manufacturers based on the analysis result. Experiments are carried out for the decision-maker in two different scenarios based on the use case instance, one in which to analyze a schedule which is generated by minimizing makespan (Schedule A) and another is to examine a schedule which is built by minimizing makespan and production line conversion frequency (Schedule B).

After retrieving all the job order information, 'DataTable' manages all the key information such as 'OrderID', 'OrderQTY', 'Processing Time' and 'Start Time' which will be translated into simulation parameters on Tecnomatix Plant Simulation. There are two different configurations for 'DataTable' named 'OrderInfo' and 'ProductInfo'. 'OrderInfo' configures the 'Source' and stores all the order-related information ('OrderID', 'OrderQTY', 'Product Part', 'Start Time') while 'ProductInfo' configures the 'Station' and references all the product-related information ('OrderID', 'OrderQTY', 'Processing Time'). A complete 3D model of the production schedule is built on the model frame with components such as 'Station', 'Source', 'Drain' and 'Connector'. Figure 3 showcases the foundation of designing a floor plan layout of the production schedule in a 2D planning view on Tecnomatix Plant Simulation. 11 production lines are created with each of its 'Source' and 'Drain'. All 'Stations' are renamed and labeled as 'Line1' up to 'Line11' for easy initialization. This simulation model is assumed to execute in a flexible job-shop scheduling environment, where all the machines are available at the time unit of zero. All the job orders are ready to be processed from the 'Source' to the following line 'Station' at time zero.

Analysis of the built system is generated after running the simulation in order to get an extensive picture of the relationship and performance of the model in production scheduling. The statistical report generated by Tecnomatix Plant Simulation can be categorized into three major parts based on the three main components ('Source', 'Station', and 'Drain') of the production line, namely overall resource statistics, machine utilization analysis for 'Station', and throughput analysis at 'Drain'. Overall resource statistics will provide the production performance of the model as a whole such as the total usage and efficiency of object components (e.g., 'Source', 'Line', and 'Drain') in different states. Machine line utilization analysis displays the capacity usage of each machine line which reflects the time when a product is being processed at the 'Station' in terms of their working time and waiting time. Throughput analysis gives information on the count of different product types that have done processing on the machine line and arrived at the 'Drain'. It shows the efficiency of the production line in handling each product type which has been assigned to the respective machine line at the granular view. The statistical data allows planners to refine the model for further improvements.

4.1 Schedule A

In the first scenario, a simulation model is built upon a single-objective optimization scheduling problem which focuses solely on the makespan minimization. The resulting optimal makespan value of the obtained schedule is 23.071 h. The Gantt chart for Schedule A illustrated in Fig. 4 is simulated on Tecnomatix Plant Simulation to attain further performance metrics for analysis.

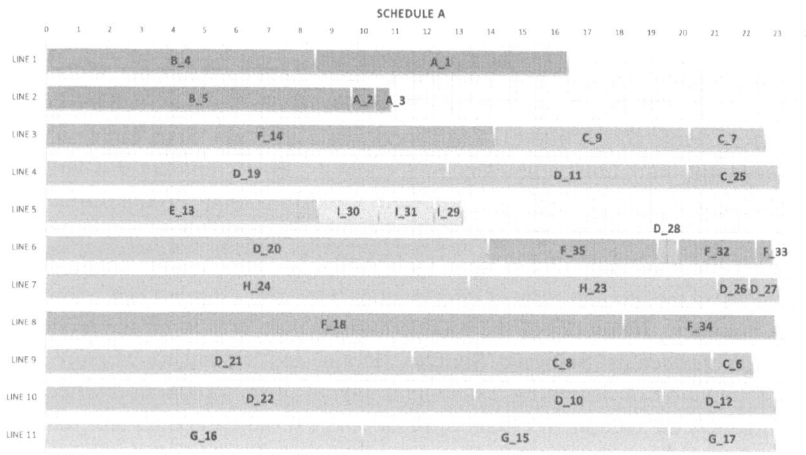

Fig. 4. Gantt chart for Schedule A.

After simulating the existing operation of Schedule A as shown in the created basic model (Fig. 3), the statistical report can be generated from the simulation. Based on the chart of the overall resource statistics in Fig. 5, the high percentage of the blocked time (yellow bar) indicates that the 'Sources' are not able to release product parts into the following station because of downstream congestion. The 'Drains' have a 100% waiting rate since they are always waiting to receive products from the upstream lines. The green bars at the 'Station' show that they are always active in processing products.

Table 1 presents the details of the machine line utilization at each 'Station' object (renamed as 'Line1' to 'Line11') with regard to the working time and waiting time based on Schedule A. The table explains the occupancy rate of the processing product on each production line which reflects its operating time (working time) and idle time (waiting time). The 'Sum' column in the 'Working Time' section (Table 1. (a)) represents the completion time for every single line. 'Line7' has the longest makespan of the entire model at 23:04:15.6 (HH:MM:SS) of time. 'Line1', 'Line2', and 'Line5' show relatively low utilization at 71.08%, 47.03% and 56.96% of the total simulated time compared to the rest of the 'Line' which are almost fully occupied. The remaining 28.92%, 52.97% and 43.04% of the time for 'Line1', 'Line2', and 'Line5' represent idle time (waiting time). The idle time of 'Line1', 'Line2', and 'Line5' can be observed by inferring from the data. These line stations occupy comparatively short makespan of 16:23:52.8, 10:51:0.0,

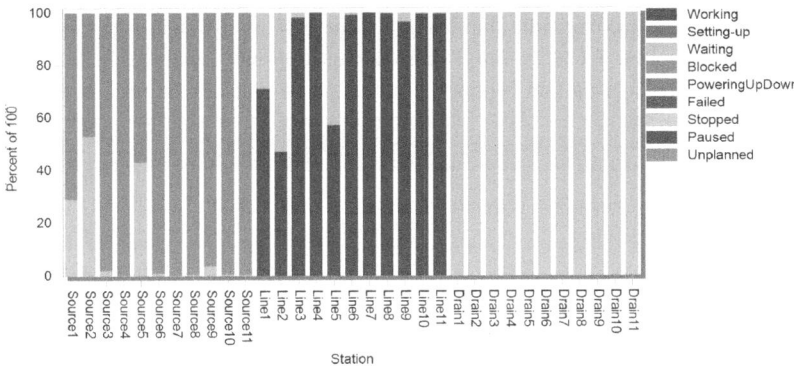

Fig. 5. Overall resource statistics of Schedule A.

and 13:08:31.2 respectively which eventually weigh up their idle time of not processing any products until the end time of the model at 23:04:15.6.

Table 1. Machine utilization analysis with (a) working time for Schedule A and (b) waiting time for Schedule A.

(a) Working time			(b) Waiting time		
Object	Portion	Sum	Object	Portion	Sum
Line1	71.08%	16:23:52.8	Line1	28.92%	6:40:22.8
Line2	47.03%	10:51:00.0	Line2	52.97%	12:13:15.6
Line3	98.05%	22:37:19.2	Line3	1.95%	26:56.4
Line4	99.95%	23:03:32.4	Line4	0.05%	43.2001
Line5	56.96%	13:08:31.2	Line5	43.04%	9:55:44.4
Line6	99.12%	22:52:04.8	Line6	0.88%	12:10.8
Line7	100.00%	23:04:15.6	Line7	0.00%	0.0
Line8	99.61%	22:58:51.6	Line8	0.39%	5:24.0
Line9	96.43%	22:14:49.2	Line9	3.57%	49:26.4
Line10	99.54%	22:57:54.0	Line10	0.46%	6:21.6
Line11	99.56%	22:58:08.4	Line11	0.44%	6:07.2

Table 2 brings about the summarized data of the throughput analysis generated by the Tecnomatix statistical report. The product type statistics in the report exhaustively break down the count of item units based on the product type that has been processed at each 'Station' and received at the 'Drain'. This table simplifies the information into a consolidated column of 'Product Types (item units)' with the name of all the product that have smoothly exited at the 'Drain' and the number of units in bracket. It also computes the total throughput which refers to the accumulated units at the 'Drain' as well as the product units processed over time that highlight the system productivity. For example, in 'Drain1', product A_1 (7125 units) gains a much higher throughput than product B_4 (2748 units), indicating that product A_1 has processed in a larger quantity across 'Line1' compared to B_4.

Table 2. Summarized throughput analysis for 'Drain' on Schedule A.

Object	Product Types (product units)	Total Throughput	Throughput per Hour	Throughput per Day
Drain1	A_1 (7125), B_4 (2748)	9873	427.94	10270.56
Drain2	A_2 (660), A_3 (450), B_5 (3116)	4226	183.17	4396.17
Drain3	C_7 (1680), C_9 (4275), F_14 (11998)	17953	778.16	18675.91
Drain4	C_25 (2025), D_11 (8246), D_19 (13937)	24208	1049.28	25182.78
Drain5	E_13 (1588), I_29 (148), I_30 (314), I_31 (290)	2340	101.43	2434.22
Drain6	D_20 (11856), D_28 (546), F_32 (2100), F_33 (448), F_35 (4488)	19438	842.53	20220.71
Drain7	D_26 (1120), D_27 (1036), H_23 (10500), H_24 (18000)	30656	1328.77	31890.43
Drain8	F_18 (15442), F_34 (4092)	19534	846.69	20320.58
Drain9	C_6 (900), C_8 (6570), D_21 (9839)	17309	750.25	18005.98
Drain10	D_10 (6460), D_12 (3902), D_22 (14900)	25262	1094.97	26279.23
Drain11	G_15 (13000), G_16 (13500), G_17 (4508)	31008	1344.03	32256.60

4.2 Schedule B

The second scenario introduces an analysis of a more complex schedule in balancing multiple objectives that involves a concurrent minimization of makespan and production line conversion frequency. The obtained Schedule B is expected to explore more advanced analysis techniques on the simulation software. Based on the resulting schedule, the optimal makespan value is 23.448 h with 4 occurrences of line conversion. The Gantt chart for Schedule B is shown in Fig. 6 and the respective model can be simulated as displayed in Fig. 3.

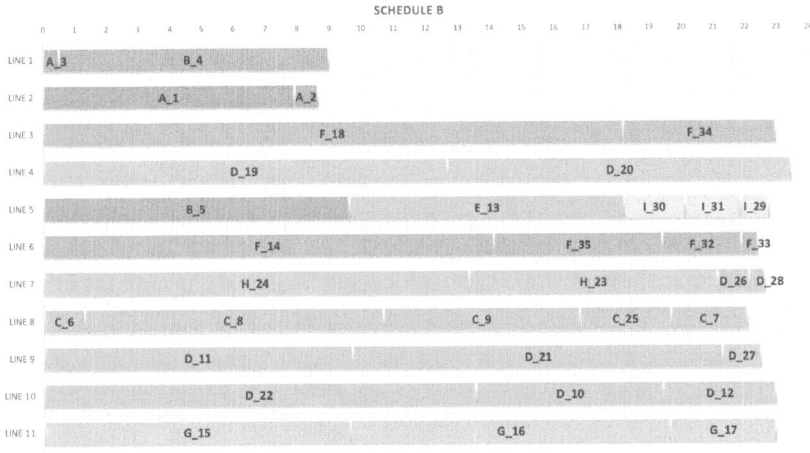

Fig. 6. Gantt chart for Schedule B.

Schedule B possesses distinct characteristics in terms of the individual performance of each component compared to the Schedule A. Figure 7 depicts the bar chart of the overall resource statistics that represents the performance of each object component on the model frame. All the 'Source' of the current scenario remains a high percentage of blocked time which is identical to the condition in Schedule A. Despite a longer makespan of Schedule B (23.448 h) as compared to Schedule A (23.071 h), Schedule B which tackles an additional scheduling objective (production line conversion frequency) demonstrates a better trade-off in performance. There is a clear reduction in the blocked time for 'Source', for example, 'Source1' and 'Source2' decrease from 71% and 47% in Schedule A to 38.28% and 36.88% in Schedule B, hence, it indicates a better flow in the system and less congestion.

The machine utilization analysis for each manufacturing line activity is shown in depth in Table 3. As referenced in Table 3. (a) on the working time analysis for Schedule B, the total time it takes to complete all the jobs in each line is observed in 'Line 4' with the longest makespan of the entire model at 23:26:52.8. With the multi-objective nature of the scheduling problem, it is seen that 'Line1' and 'Line2' are underutilized, accounting for 38.30% and 36.89% of total simulated time as compared to the remainder of the 'Lines' which are almost fully utilized. Subsequently, the waiting time analysis

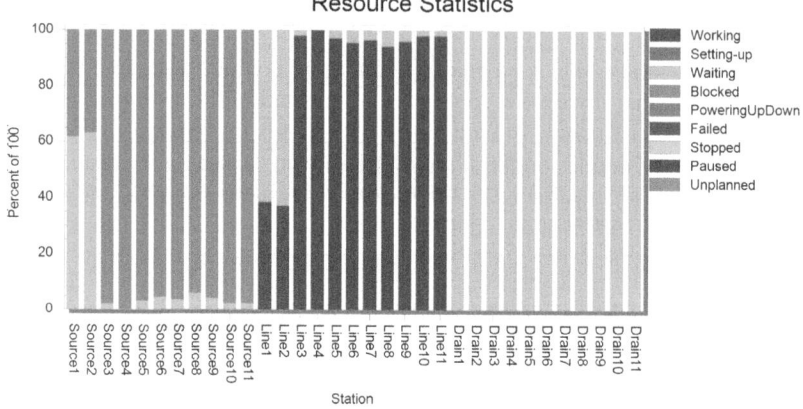

Fig. 7. Overall resource statistics of Schedule B.

of the schedule (Table 3(b)) states that 'Line1' and 'Line2' take up 61.70% and 63.11% of the overall simulated time. This can be deduced from the table where 'Line1' and 'Line2' have significantly short makespan of 8:58:51.7 and 8:39:0.0 respectively, which ultimately add up to their idle time of not processing any products until the finish time of the model at 23:26:52.8.

Table 3. Machine utilization analysis with (a) working time for Schedule B and (b) waiting time for Schedule B.

(a) Working time			(b) Waiting time		
Object	Portion	Sum	Object	Portion	Sum
Line1	38.30%	8:58:51.7	Line1	61.70%	14:28:01.5
Line2	36.89%	8:39:00.2	Line2	63.11%	14:47:52.9
Line3	98.01%	22:58:51.9	Line3	1.99%	28:01.2
Line4	100.00%	23:26:52.8	Line4	0.00%	0.3801
Line5	97.06%	22:45:32.4	Line5	2.94%	41:20.8
Line6	95.50%	22:23:34.8	Line6	4.50%	1:03:18.4
Line7	96.49%	22:37:29.5	Line7	3.51%	49:23.6
Line8	94.13%	22:04:19.0	Line8	5.87%	1:22:34.1
Line9	95.94%	22:29:42.0	Line9	4.06%	57:11.1
Line10	97.94%	22:57:54.0	Line10	2.06%	28:59.2
Line11	97.96%	22:58:09.4	Line11	2.04%	28:43.7

Table 4 compiles the throughput analysis from the Tecnomatix statistical report by merging based on the type of products and the count of completed units being accumulated at the 'Drain'. The condensed analysis represents the group of product types at each 'Drain' with their individual counts. All the products listed in the table indicate that the products assigned to the machine line have been successfully made. The overall throughput is also aggregated by hour and day in the table. It is discovered that 'Drain7' gathers four different orders which comprise D_26, D_28, H_23, and H_24 with 1,120 units, 546 units, 10,500 units and 18,000 units respectively. The highest throughput of product H_24 in 'Drain7' shows that H_24 is handled in a greater amount out of all.

Table 4. Summarized throughput analysis for 'Drain' on Schedule B.

Object	Product Types (product units)	Total Throughput	Throughput per Hour	Throughput per Day
Drain1	A_3 (450), B_4 (2748)	3198	136.38	3273.27
Drain2	A_1 (7125), A_2 (660)	7785	332.00	7968.23
Drain3	F_18 (15442), F_34 (4092)	19534	833.07	19993.78
Drain4	D_19 (13937), D_20 (11856)	25793	1100.00	26400.10
Drain5	B_5 (3116), E_13 (1588), I_29 (148), I_30 (314), I_31 (290)	5456	232.68	5584.42
Drain6	F_14 (11998), F_32 (2100), F_33 (448), F_35 (4488)	19034	811.75	19482.01
Drain7	D_26 (1120), D_28 (546), H_23 (10500), H_24 (18000)	30166	1286.50	30876.03
Drain8	C_25 (2025), C_6 (900), C_7 (1680), C_8 (6570), C_9 (4275)	15450	658.90	15813.65
Drain9	D_11 (8246), D_21 (9839), D_27 (1036)	19121	815.46	19571.05
Drain10	D_10 (6460), D_12 (3902), D_22 (14900)	25262	1077.35	25856.60
Drain11	G_15 (13000), G_16 (13500), G_17 (4508)	31008	1322.41	31737.84

A comparative analysis of the two scenarios reveals the obtained schedules have a notably high percentage of blocked time on their 'Source'. Schedule B, which caters to multi-objective optimization problem obtains to be more balanced in terms of line utilization with better distribution in waiting times and blocked times. For example, despite the low utilization of 'Line1' and 'Line2' from Schedule B, it exhibits a balanced completion time (8:58:51.7 and 8:39:0.0) as compared to the 'Line1', 'Line2', and 'Line5' from Schedule A which obtain a multiple range of completion time (16:23:52.8, 10:51:0.0, and 13:08:31.2). This is due to the scheduling problem of Schedule B, which is tailored to optimize two scheduling objectives (makespan and production line conversion frequency) which contributes to a better resource allocation and a more balanced distribution of operation. This shows that the multi-objective schedule effectively reduces idle times and balances the workload across the lines. Moreover, the throughput analysis discloses interesting insights on both schedules. Schedule A achieves slightly higher total throughput (201,807 units) than Schedule B (196,807 units). There is a 2.54% difference in total production throughput between the two schedules, each with distinct scenarios. However, while the built model of Schedule A tends to maximize throughput per product part, the simulated model of Schedule B attains a better balance across production lines. The simulated model in Schedule B performs proportionately well in terms of throughput per hour and per day with several drains ('Drain3', 'Drain7', and 'Drain11') closely matching the performance of Schedule A.

These findings highlight the need to improve the performance of the schedule based on different scenarios. This can be accomplished, for example, by installing buffer stations in the model system to reduce bottlenecks. Another option to achieve this is to maximize the throughput on less critical production lines to balance the working time. Hence, various experiments are performed with the described model. They are designed for a smoother production workflow and eliminate the build-up of unprocessed products at stations. The modifications of the model in Fig. 8 can be clearly evaluated by comparing it to Fig. 3 as buffer stations are added into the model. Figure 9 and Fig. 10 show the outcomes of capacity utilization for overall operations in the experiment where the buffers are utilized before the production lines. The blocked time of the 'Source' from both scenarios is resolved as 'Buffer' temporarily holds parts from the 'Source' before processing them at the 'Line'. Therefore, the flow of the entire production is improved and idle time is reduced. As a result of the modified experiment, buffer stations can be used to balance the entire system.

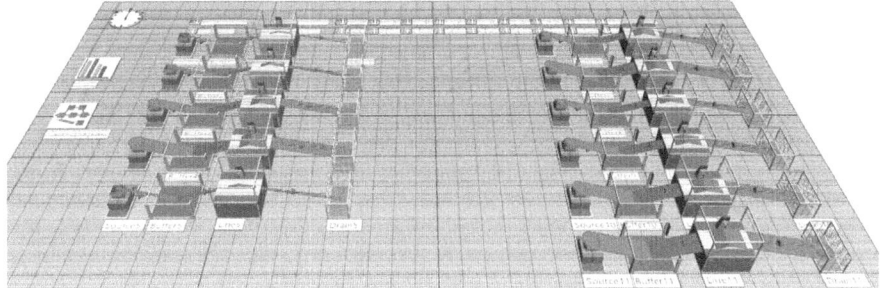

Fig. 8. Production model layout with 'Buffer' in 3D landscape view.

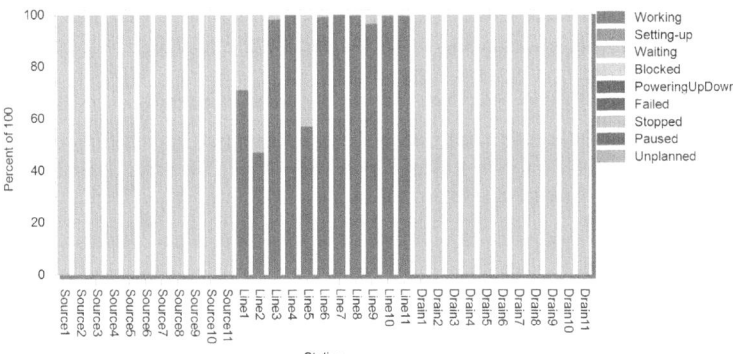

Fig. 9. Resource statistics of Schedule A after adding buffers.

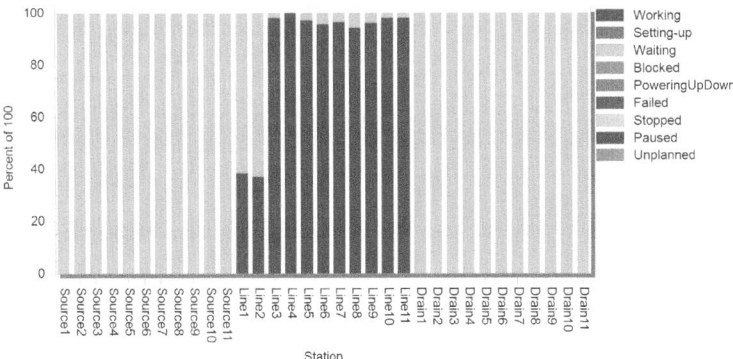

Fig. 10. Resource statistics of Schedule B after adding buffers.

5 Conclusion

The application of simulation tool has proven to be an invaluable model system for production scheduling in the manufacturing industry. The research demonstrates that the discrete-event simulation tool with Tecnomatix Plant Simulation is a potential game changer for production planners to analyze, visualize, and experiment with their planned schedules. Planners can observe how each schedule performs across a range of scenarios. This research revolves around two different analyses of schedules where the first schedule purely focuses on minimizing makespan while the second schedule concurrently levels two scheduling objectives with makespan and production line conversion frequency. By simulating the production schedule, planners can grasp the overview and further insights of the entire production flow. Also, they can further investigate how a schedule with different scenarios might create bottlenecks at certain production lines or how a schedule that emphasizes resource efficiency might leave other lines underutilized on the simulation. Moreover, Tecnomatix Plant Simulation enables the navigation of potential issues that are not apparent in static scheduling methods and suggests strategies to enhance manufacturing performance. The embedded statistical report and analytical tools within Tecnomatix Plant Simulation help production planners to make decisions in a more impactful manner by ensuring that they choose the schedule that not only meets key objectives but also minimizes potential risks.

Acknowledgment. The authors gratefully acknowledge the support of the USM-Industry Matching Grant awarded by the Division of Research and Innovation, Universiti Sains Malaysia, with Grant No: R502-KR-ARU003-0008070026-K134.

References

1. Alomair, Y., Ahmad, I., Alghamdi, A.: A review of evaluation methods and techniques for simulation packages. Procedia Comput. Sci. **62**, 249–256 (2015)
2. Bangsow, S.: Tecnomatix Plant Simulation: Modeling and Programming by Means of Examples. Springer, Cham (2020)
3. Blaga, F., Stanăşel, I., Hule, V., Pop, A.: Balancing the manufacturing lines through modelling and simulation using tecnomatix plant simulation. In: MATEC Web of Conferences, vol. 112, p. 06012 (2017)
4. Chung, C.A.: Simulation Modeling Handbook: A Practical Approach. CRC Press, Boca Raton (2003)
5. Kikolski, M.: Identification of production bottlenecks with the use of plant simulation software. Eng. Manag. Prod. Serv. **8**(4), 103–112 (2016)
6. Kikolski, M.: Study of production scenarios with the use of simulation models. Procedia Eng. **182**, 321–328 (2017)
7. Kliment, M., Popovič, R., Janek, J.: Analysis of the production process in the selected company and proposal a possible model optimization through PLM software module tecnomatix plant simulation. Procedia Eng. **96**, 221–226 (2014)
8. Kopec, J., Lachvajderová, L., Kliment, M., Trebuňa, P.: Simulation processes in companies using PLM and tecnomatix plant simulation software. Acta Simulatio Simulatio—Int. Sci. J. About Simul. **7**, 13–18 (2021)
9. Maidstone, R.: Discrete event simulation, system dynamics and agent based simulation: discussion and comparison. System **1**(6), 1–6 (2012)
10. Marasova, D., Saderova, J., Ambrisko, L.: Simulation of the use of the material handling equipment in the operation process. Open Eng. **10**(1), 216–223 (2020)
11. Mourtzis, D.: Simulation in the design and operation of manufacturing systems: State of the art and new trends. Int. J. Prod. Res. **58**(7), 1927–1949 (2020)

12. Musil, M., Laskovský, V., Fialek, P.: Analysis of logistic processes using the software tecnomatix plant simulation. In: 13th International Conference on Industrial Logistics, ICIL 2016-Conference Proceedings, pp. 195–200 (2016)
13. Siderska, J.: Application of tecnomatix plant simulation for modeling production and logistics processes. Bus. Manag. Educ. **14**(1), 64–73 (2016)
14. Siderska, J., Perkowska, K.: Application of computer simulation in improving the process of screws production. Transp. Econ. Logist. **66**, 89–96 (2017)
15. Siemens Digital Industries Software: Tecnomatix: Manufacturing operations software. https://plm.sw.siemens.com/en-US/tecnomatix/. Accessed 30 Nov 2024

Modelling of Electrical Capacitance Tomography Sensor Array for Air-Wax Detection

Shahrulnizahani Mohammad Din[2] , Pei Ling Leow[1,2]($\boxtimes$) ,
Jaysuman Pusppanathan[1,2] , Ruzairi Abdul Rahim[1,2,3] , Xian Feng Hor[1] ,
Wen Pin Gooi[4] , Yasmin Abdul Wahab[5] , and Suzanna Ridzuan Aw[6]

[1] Faculty of Electrical Engineering, Universiti Teknologi Malaysia, 81310 Johor Bahru, Johor, Malaysia
leowpl@utm.my

[2] PROTOM-I Research Group, Universiti Teknologi Malaysia, 81310 Johor Bahru, Johor, Malaysia

[3] Faculty of Electrical and Electronic Engineering, Universiti Tun Hussein Onn Malaysia, 86400 Parit Raja, Johor, Malaysia

[4] School of Mechanical and Aerospace Engineering, University of Southampton Malaysia, 79100 Iskandar Puteri, Johor, Malaysia

[5] Advanced Metrology Sensor Research Group (AMES), Faculty of Electrical and Electronics Engineering Technology, Universiti Malaysia Pahang Al-Sultan Abdullah, 26600 Pekan, Pahang, Malaysia

[6] Electronic and Instrumentation Programme, Faculty of Engineering Technology (Electrical and Automation), University College TATI, Jln Panchor, Telok Kalong, 24000 Kemaman, Terengganu, Malaysia

Abstract. Electrical Capacitance Tomography is a fast, cost-effective, non-invasive and non-intrusive technique to analyze a multiphase flow in a closed pipe region. The ECT detects the dielectric material constant of two or more materials or mediums within the pipe. One of the most important factors of an ECT system is the parameters of the sensor array. This paper focuses on the design features and specifications of the sensor array used in experiments involving multiphase flow with air and waxy crude oil. This article discusses determining the number of electrodes and the sensor's axial length using the sensitivity uniformity coefficient and sensitivity average. For a pipe with an inner diameter of 100 mm and copper electrodes, the 12 mm and 80 mm electrodes deliver the best results, as they offer the most effective sensitivity distribution within the pipe. The 12-electrode ECT system is tested with several types of multiphase flow. It produces an image error between 0.06 and 1.0 and a coefficient correlation between 0.76 and 0.96, indicating a strong correlation between image reconstruction and actual sample size.

Keywords: Electrical Capacitance Tomography · number of electrodes · axial length · sensitivity distribution

1 Introduction

Electrical capacitance tomography (ECT) is a non-invasive measurement technique that has gained interest in industries for applications such as monitoring gas separation processes, pneumatic conveying systems, and measuring multiphase flows in oil pipelines and fluidized beds [1]. In the process industry, ECT visualizes the movement of the sample in the pneumatic conveying system [2–4], the proportion of the medium composition [5–7] and the liquid accumulation in the pipeline [8]. In addition, the ECT imaging capability is used to identify the state of matter in a fluidized bed, where the solid substance could behave like a liquid under the right conditions [9] or to monitor the drying process [10]. In the pipeline industry, ECT is used for maintenance and cleaning by monitoring coating thickness [9, 11] to produce cleaner fuels [12].

The design and development of the ECT system include the geometric design and arrangement of the electrodes, the selection of electrode materials, data acquisition and the noise filter circuit [11]. A 12-electrode ECT sensor is mainly used for monitoring the physical properties of crude oil [6], monitoring multiphase flow in the pipeline [5, 6] and analyzing the formation of gelled crude oil as a function of temperature [7]. A 6-electrode ECT sensor was utilized to visualize stratified flow with different concentrations [13], while an 8-electrode ECT system was applied to visualize gas bubble formation in crude oil due to thermal shrinkage [13] and to monitor the variables of thick fluid layers of oil continuous ring flows [14].

When deciding on the number of electrodes for an ECT sensor, Yang [15] recommended using 8 or 12 electrodes in a single-plane design. This is because increasing the number of electrodes makes the ECT hardware more challenging to fabricate. Although increasing the number of electrodes improves image resolution [16], it also increases data acquisition time [17]. Peng et al. [18] stated that an ECT sensor with 12 electrodes works well for most applications because the image quality hardly improves when the number of electrodes is more than 12. However, this does not prevent researchers from considering industrial applications of ECT with more than 12 electrodes. The 16-electrode ECT sensor was adopted to monitor oil pipelines and potential leaks [19], as well as to monitor gas and liquid mixing processes in a tank [20]. Since the width of the electrode varies depending on the pipe size and number of electrodes, the axial electrode length is the geometry that can be analyzed to obtain an optimal measurement. A short electrode size restricts the measurement to achieve an optimum high signal-to-noise ratio (SNR) [21]. Peng et al. [22] indicated that the optimal axial electrode length in the ECT system is equal to the diameter of the pipe diameter. The sensitivity distributions were distorted when the electrode was smaller than the pipe width. Li et al. [23] recommended that the ideal ratio of electrode length to pipe diameter is 0.75, as it enhances the sensitive area, minimizes the sensitivity to noise, and increases the axial resolution. Process Tomography Ltd [24] suggested that the minimum axial electrode length should be 3.5 cm for 8-electrode and 5 cm for 12-electrode ECT systems. It is recommended that the combined length of the electrode and screens do not exceed the diameter of the pipe [24]. In addition, the axial electrode length is at least 0.5 of the pipe diameter [25]. Therefore, the proposed axial electrode length for a pipe with an inner diameter of 100 mm is between 50 mm and 100 mm, and this range is used as the reference value for the study to determine the optimal electrode's axial length in this research.

Taking into account the inhomogeneous properties of the sensitivity field, a mathematical model was developed to analyze the sensitivity distribution. The model uses the average sensitivity and standard deviation as indicators to assess the uniformity of the sensitivity field [26]. Brandisky et.al. [27] introduced a mathematical modelling method using the sensitivity coefficient uniformity (S_u) and the sensitivity average (S_a) for sensor optimization. The S_u is defined as the ratio of the standard deviation of the sentitivity map to the mean. A lower S_u value signifies a more uniform sensitivity distribution. Meanwhile, S_a represents the sensor's sensitivity towards small changes in the pipe area and a higher value of S_a is preferred to indicate optimal sensitivity [28].

These optimisation methods are used to identify the optimal transmitter input voltage ECT system [29]. The methods also apply to determine the electrode configuration in ECT sensors [30] as well as Electrical Impedance Tomography (EIT) [28].

This research focuses on optimizing the number of electrodes and the length of the sensor for an ECT system designed to detect air and waxy crude oil in a closed pipe. Utilizing a single-plane ECT system, the research examines multiphase flow in an acrylic pipe with an inner diameter of 100 mm. The pipe contains waxy crude oil and air, with relative permittivities of 2.30 and 1, respectively. In the first phase of analysis, the ideal number of electrodes 6, 8, 12 or 16 is determined by assessing its sensitivity. The next analysis is to determine the optimal axial lengths for the electrodes, ranging from 50 mm to 100 mm.

2 Electrode Design Modelling

The correlation between the capacitance and the permittivity distribution is derived as follows:

$$C = \frac{Q}{V} = -\frac{1}{V} \int \varepsilon(x, y) \nabla \varphi(x, y) \cdot d\Gamma \tag{1}$$

where C represents capacitance, Q is the charge per unit length, V is the voltage difference between the two electrodes, and Γ refers to the electrode surface. The terms $\varepsilon(x, y)$ and $\phi(x, y)$ denote the permittivity distribution and electric potential, respectively. The normalized Linear Back Projection (LBP) algorithm, which scales values between 0 and 1, is defined as follows.

$$g(x, y) = \frac{\sum_{i=1}^{m} \lambda_{(i,j)} S_{(i,j)}(x, y)}{\sum_{i=1}^{m} S_{(i,j)}(x, y)} \tag{2}$$

where $\hat{g}$ is the grey scale level at the pixel $\hat{g}(x, y)$, $\lambda_{i,j}$ represents the normalized capacitance reading of electrode pairs i and j. $S_{(i,j)}(x.y)$ is the sensitivity map between the electrode pairs. The normalized capacitance measurement is determined from:

$$\lambda_{i,j} = \frac{C_{m(i,j)} - C_{(i,j)low}}{C_{(i,j)high} - C_{(i,j)low}} \tag{3}$$

where $C_{m(i,j)}$ is the measured capacitance value, $C_{(i,j)low}$ is the capacitance of the electrode-pair i and j when the measuring chamber is filled with the material with the lowest permittivity. $C_{(i,j)high}$ is the capacitance of the electrode-pair i and j when the sensing chamber is filled with a material with the highest permittivity. The images generated from the experiments are evaluated using the image error [31] which can be determined from the following equation:

$$\text{Image error} = \frac{\|\hat{g} - g\|}{\|g\|} \tag{4}$$

where g is the true permittivity distribution and $\hat{g}$ denotes the reconstructed permittivity distribution. The image error measures the deviation between the actual and the reconstructed image. In tomography imaging, a typical image error value is usually below 0.1, though values as high as 0.3 are commonly observed for LBP algorithms. [32]. The correlation coefficient (CC) calculates the linear relationship between two variables [31]; in this study the permittivity distributions of the true and reconstructed images. The CC can be calculated as:

$$CC = \frac{\sum_{i=1}^{N}(g_i - \underline{g})(\hat{g}_i - \underline{\hat{g}})}{\sqrt{\sum_{i=1}^{N}(g_i - \underline{g})^2 \sum_{i=1}^{N}(\hat{g} - \underline{\hat{g}})^2}} \tag{5}$$

where the $\underline{g}$ and $\underline{\hat{g}}$ is the average of g and $\hat{g}$ respectively. The value of the correlation coefficient (CC) ranges from -1 to 1 [33]. The correlation coefficient interpretation is as follows: 1 means perfect correlation, 0.7–0.9 suggests strong or excellent correlation. The value of 0.4–0.6 is moderate, 0.1–0.3 indicates a weak correlation, and 0 implies no correlation [34].

2.1 Sensitivity Map and Image Reconstruction

A sensitivity map is the correlation sensitivity between electrodes, generated using electric field data from the forward model using COMSOL Multiphysics. This data is then processed in MATLAB to create a full set of sensitivity maps. The number of sensitivity maps corresponds to the number of electrodes used. The sensitivity map can be stated as:

$$S_{(i,j)}(x,y) = -\frac{1}{V_i V_j} \int E_i(x,y) \times E_j(x,y) \times dx \times dy \tag{6}$$

where $S_{(i,j)}$ is the sensitivity map between the electrode pair of i and j, E_i and E_j is the electric field measured between electrode pair i and j when the voltage V_i and V_j are applied to electrodes i and j, respectively. The sensitivity coefficient uniformity (S_u) can be derived from the following equation:

$$S_u = \frac{S_{i,j}^{dev}}{S_{i,j}^{avg}} \tag{7}$$

$$S_{i,j}^{avg} = \frac{1}{n}\sum_{p=1}^{n} S_{i,j}(p), S_{i,j} > 0$$

$$S_{i,j}^{dev} = \sqrt{\frac{1}{n-1}\sum_{p=1}^{n}\left[S_{i,j}(p) - S_{i,j}^{avg}\right]^2}, S_{i,j} > 0$$

The average sensitivity (S_a) is determined by the vector of positive sensitivities within the sensitivity distribution. It reflects the sensor's ability to detect small variations in the pipe area. The S_a is be derived as:

$$S_a = \frac{1}{l}\sum_{k}^{K}\sum_{n=1}^{N} S_{kn}, S_{kn} > 0 \qquad (8)$$

where k is the six typical sensitivity distributions and $k = 1, 2, \cdots, 6$. l is the total number of positive sensitivities in respective distributions.

3 Result and Discussion

3.1 Number of Electrodes

As recommended in [22], the axial length of the sensor should be equal to the diameter of the pipe. This is because the sensitivity distribution is distorted when the electrode length is smaller than the pipe diameter. In this experiment, the axial length is therefore set to 100 mm. This test focuses on determining the optimal number of electrodes for the ECT system. The numbers of electrodes tested are 6, 8, 12 and 16, as shown in Fig. 1.

From the model, the electric potential is applied to each electrode, and the resulting electric field data is obtained through numerical computations. The data extraction method is consistent across all ECT models, where the extraction plane is at the centre of the pipe, as indicated in Fig. 2. Figure 3 shows the sensitivity maps generated from the electric field data. The sensitivity map establishes the correlation between electrodes based on the number of electrodes.

All possible electrode connections are shown in Fig. 3. For 6-electrode ECT systems, only 3 sensitivity map correlations are required to represent all projections, and compared to the 16-electrode ECT system, 8 sensitivity map correlations are required to represent all projections. More sensitivity maps improve the image quality of the tomogram, but more electrodes can weaken the electric field. This is because smaller electrode widths reduce the resolution of the ECT system at the centre of the pipe [8]. The sensitivity map is utilized to determine the sensitivity coefficient uniformity (S_u) using Eq. (7). The resulting value is presented in Fig. 4.

The S_u value reflects the uniformity of the sensitivity map relative to its mean. Therefore, a lower S_u is preferred, as it indicates that the sensitivity is evenly distributed across the entire pipe cross-section. The 6-electrode ECT system exhibits s the highest S_u value of 2.309, while the 12-electrode ECT system provides the lowest S_u value of 1.768. Based on the results, the 12-electrode ECT system provides a more uniform sensitivity distribution compared to the 6, 8, and 16-electrode systems. Therefore, the 12-electrode configuration is selected for use in this research.

Modelling of Electrical Capacitance Tomography Sensor Array 273

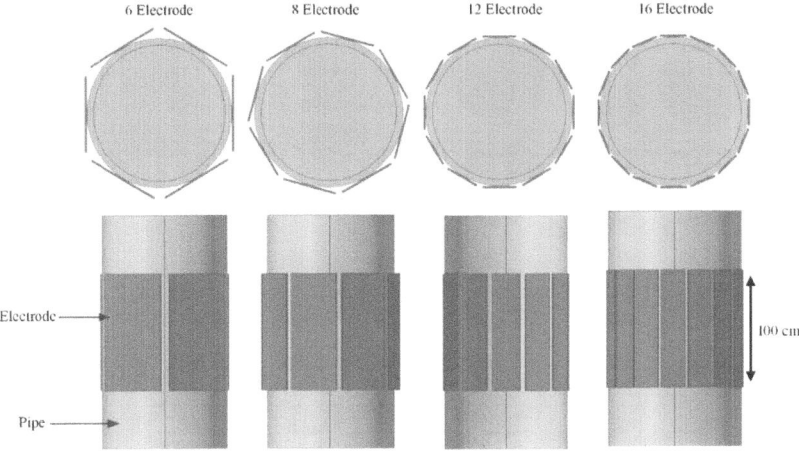

Fig. 1. ECT electrode array with (a) 6-electrode, (b) 8-electrode (c) 12-electrode and (d) 16-electrode models

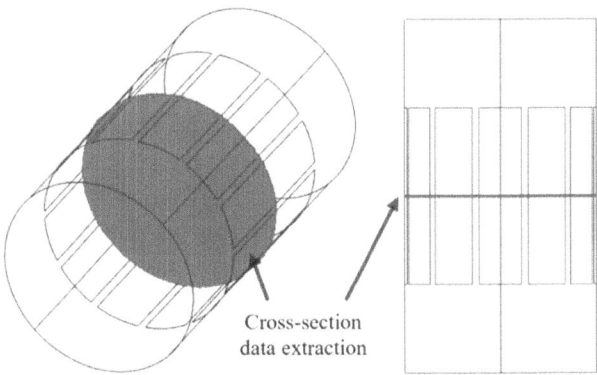

Fig. 2. The data extraction cross-section

3.2 The Sensor's Axial Length

Since the number of electrodes is fixed at 12 electrodes, the next dimension to examine is the axial length. The electrode is made up of several components, including the radial and axial screens. As a result, the width is set to 25 mm, considering the size of the screens between the electrodes. Figure 5 shows the tested lengths, ranging between 50 cm to 100 cm.

Figure 5 shows 6 simulation models for different axial lengths of 50 mm, 60 mm, 70 mm, 80 mm, 90 mm, and 100 mm. The same procedure is performed where the electric field data is extracted and sensitivity maps are generated. Then, sensitivity map data is exported for analysis of the sensitivity coefficient uniformity (S_u) and sensitivity average (S_a). The standard deviation of the sensitivity matrix is calculated for each electrode pair: E1-2, E1-3, E1-4, E1-5, E1-6 and E1-7. These combinations of sensitivity maps

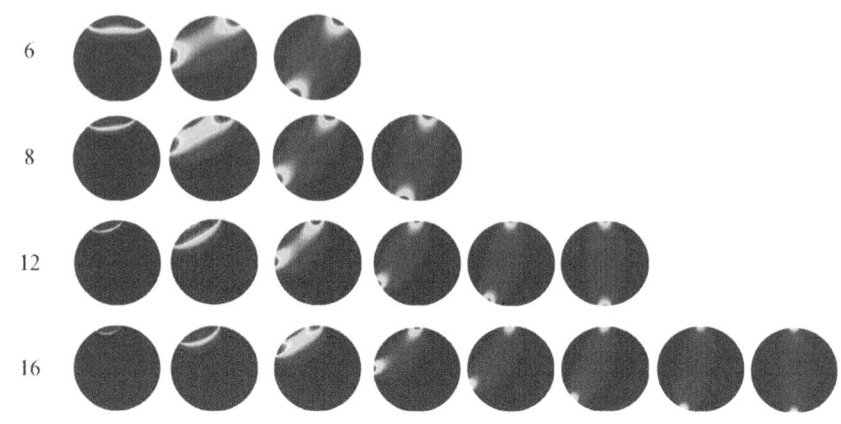

Fig. 3. The sensitivity map for different numbers of an electrode array

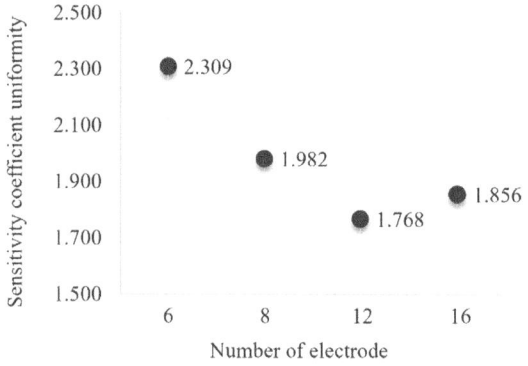

Fig. 4. The sensitivity coefficient uniformity for different electrode numbers

represent all the electrode pairs between the transmitter and receivers. The calculated sensitivity coefficient uniformity (S_u) is shown in Fig. 6. Since S_u reflects the distribution of sensitivity values relative to their mean, a lower S_u indicates a more uniform sensitivity distribution across the pipe region. [29].

E1-2 is the correlation analysis value between electrode 1 and electrode 2, and E1-3 is the correlation analysis value between electrode 1 and electrode 3. The initial evaluation of the axial length selection focuses on electrode pairs E1-2 and E1-7, which represent adjacent and opposing pairs, as shown in Fig. 6(a) and 6(f). For E1-2, the electrode with an axial length of 80 mm shows the lowest value with a value of 0.975. For the E1-7 pair, the 80 mm electrode recorded the initial saturation value with a value of 0.069. The length of 90 mm and 100 mm indicates that the value is in the range between 1.062 and 1.080. This shows a high value for both 50 mm and 60 mm electrode lengths compared to an axial length of 80 mm. For this initial evaluation, an axial electrode length of 80 mm is good potential for the choice of ECT system in this research. The average sensitivity

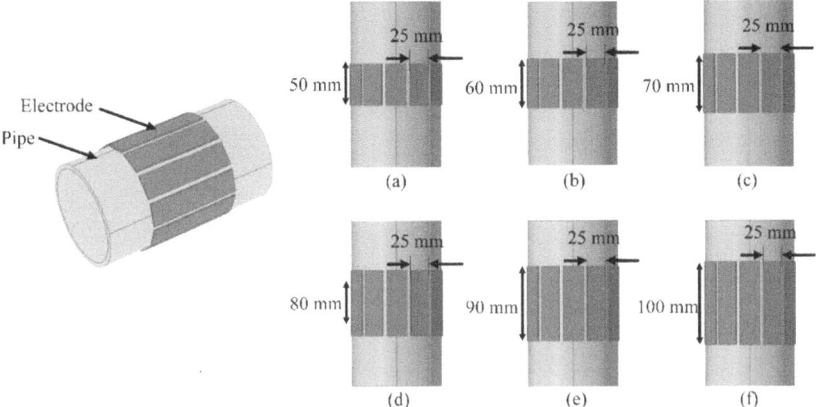

Fig. 5. The electrode model for different electrode axial lengths (a) 50 mm (b) 60 mm (c) 70 mm (d) 80 mm (e) 90 mm (f) 100 mm

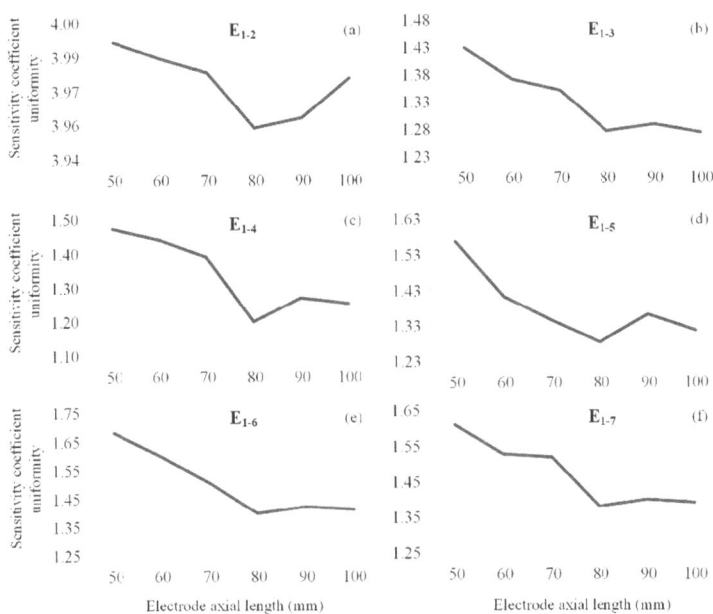

Fig. 6. The sensitivity coefficient uniformity (S_u) of different electrode axial length for different electrode axial lengths for (a) E1-2 (b) E1-3 (c) E1-4 (d) E1-5 (e) E1-6 (f) E1-7 electrode pair

(S_a) reflects how responsive the sensor is to slight changes in the area of the pipe. S_a calculates the average sensitivity value in the detection and is defined by the vector of positive sensitivities within its region.

Figure 7 shows that an axial length of 50 mm has the lowest average sensitivity (S_a) value with a value of 0.00295. From an axial length of 80 mm, the measurement reaches its initial saturation value of 0.00334. The average sensitivity (S_a) signifies the average

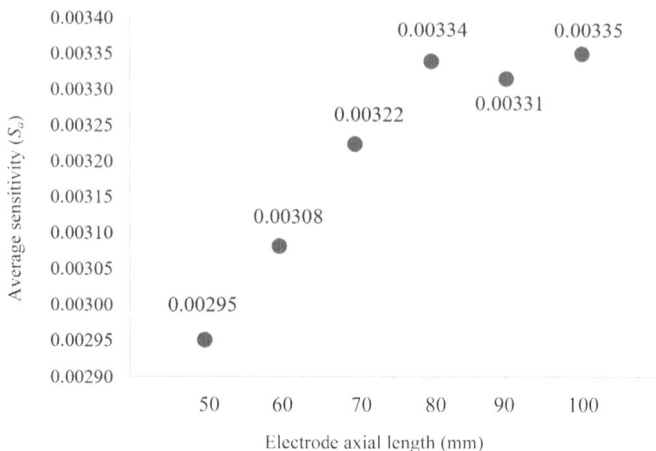

Fig. 7. The sensitivity average (S_a) for different axial electrode lengths

sensitivity value within the detection range, while the sensitivity coefficient uniformity (S_u) assesses the overall variations in sensitivities across the range. As a result, for selecting the axial electrode length, it is better to have a higher average (S_a) and a lower sensitivity coefficient uniformity (S_u). Based on both analyses, the electrode axial length of 80 mm was selected to be fabricated to complete the ECT system.

3.3 Sensor Fabrication

Figure 8 shows the design and sensor fabrication using a copper-based FR4 PCB with a detection area of 25 mm × 80 mm. The total electrode length of the electrode is 110 mm with a width of 29 mm. The axial screens are attached to the end of the electrode and the radial shields are along the length of the sensor. A total of 12 electrodes are fabricated, each integrated with a sensor measuring circuit. The complete ECT system is shown in Fig. 9.

The complete ECT sensors are integrated into the ECT system, which includes a DAQ (Data Acquisition) unit with a computer for data processing and image display. The DAQ features an Arduino Mega 2560 microcontroller and a function generator that provides a 20 V peak-to-peak 300 kHz sine wave signal to the system. The collected data is then transferred to the computer for processing and image reconstruction using Data Streamer and MATLAB.

3.4 Image Reconstruction

Figure 10 shows the image reconstruction of waxy crude oil and air in different flow types including core and annular flow, and stratified flow in static conditions. The image reconstruction is obtained from the normalized capacitance measurement (Eq. 3) and LBP algorithm (Eq. 2).

In Fig. 10, the blue colour shows the air area in the pipe and the red colour shows the waxy crude oil sample. The fabricated sensor and circuit successfully reconstruct

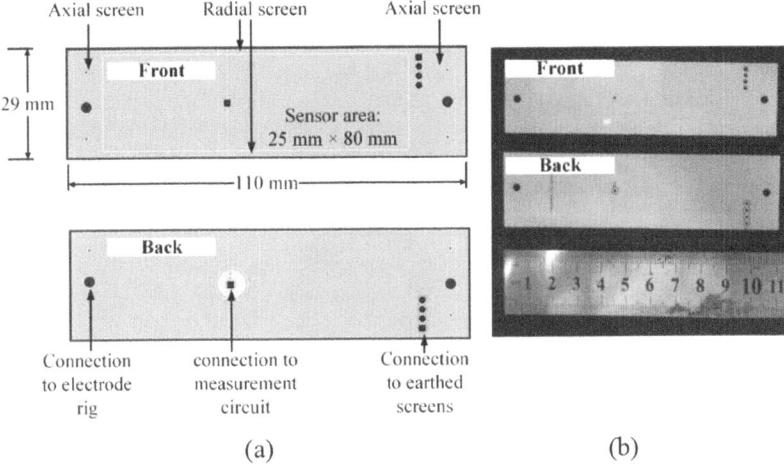

Fig. 8. FR4 double-sided PCB electrode (a) design dimension and (b) the fabricated electrode

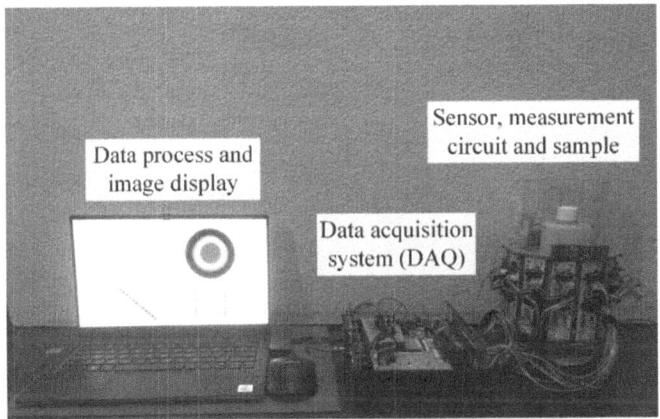

Fig. 9. The ECT system

the multiphase flow. The image error is between 0.06 and 0.10, indicating that the image reconstruction range is consistent with the actual sample size. The CC value ranges from 0.76 to 0.96, indicating that image reconstructions have a strong correlation with the actual sample size and position.

Flow	Experiment setup	Image reconstruction	Image error	correlation coefficient
Core flow - sample near wall	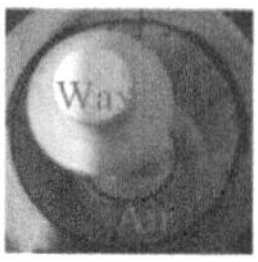		0.08	0.81
Core flow		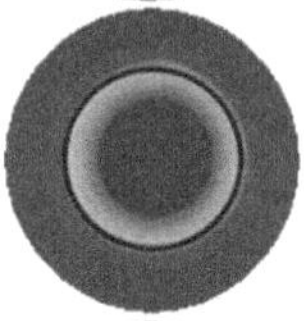	0.06	0.96
Annular flow			0.09	0.76
Startified flow			0.10	0.79

Fig. 10. Image reconstruction of waxy crude oil and air. (Color figure online)

4 Conclusion

The number of sensors and the electrode's axial length are determined using the sensitivity coefficient uniformity (S_u) and sensitivity average (S_a). From the analysis, based on the sensitivity analysis, the 12-electrode and 80 mm axial length is selected. The 12 electrodes are fabricated using double-sided FR4 PCB. The ECT system is equipped with an AC-based measurement circuit and a linear back-projection algorithm. The image reconstruction results demonstrate that the sensor is capable of detecting various types of multiphase flows, with a high level of correlation between the reconstructed images and the actual sample.

Acknowledgement. The authors express their gratitude to Universiti Teknologi Malaysia and the Ministry of Higher Education Malaysia for their generous financial support through the Research University Grants (Votes Q.J130000.2451.09G14, Q.J130000.2451.04G93, Q.J130000.2551.20H93) and the FRGS project with project code FRGS/1/2020/TK0/UTM/02/47). Additionally, the authors would also like to thank Petroliam Nasional Berhad for supplying the research sample.

References

1. Yang, W.: Key issues in designing capacitance tomography sensors. IEEE Sens. 497–505 (2006). https://doi.org/10.1109/ICSENS.2007.355515
2. Samsun Zaini, N.A.H., et al.: A preliminary result on electrical capacitance tomography for gas-solid flow in pneumatic conveyor system. Elektrika **20**(2), 73–78 (2021)
3. Grudzień, K., Chaniecki, Z., Niedostatkiewicz, M., Romanowski, A., Sankowski, D.: Measurement of dynamic pulsations in bulk solid during silo discharging using ECT method. Elektryka **114**, 35–40 (2009)
4. Abrar, U., Shi, L., Jaffri, N.R., Li, Q., Sindhu, H.R., Omar, M.W.: Solids velocity measurement using electric capacitance sensor assemblies. IOP Conf. Ser. Mater. Sci. Eng. **715**, 1–9 (2020). https://doi.org/10.1088/1757-899X/715/1/012097
5. Al-Hosani, E., Zhang, M., Soleimani, M.: A limited region electrical capacitance tomography for detection of wax deposits in pipelines. IEEE Sens. J. **15**(11), 6089–6099 (2015). https://doi.org/10.1109/JSEN.2015.2453361
6. Mei, I.L.S., Ismail, I., Shafquet, A., Abdullah, B.: Real-time monitoring and measurement of wax deposition in pipelines via non-invasive electrical capacitance tomography. Meas. Sci. Technol. **27**, 1–11 (2016). https://doi.org/10.1088/0957-0233/27/2/025403
7. Shafquet, A., Ismail, I., Jaafar, A.: Thermal effect on void formation of waxy crude oil using electrical capacitance tomography. J. Teknol. (Sci. Eng.) **63**(1), 105–113 (2013). https://doi.org/10.11113/jt.v63.1509
8. Okoro, E.E., Rachael, J.E., Sanni, S.E., Emetere, M.E.: Liquid holdup measurement in crude oil transportation using capacitance sensors and electrical capacitance tomography: concept review. IOP Conf. Ser. Earth Environ. Sci. **655**(1), 1–14 (2021). https://doi.org/10.1088/1755-1315/655/1/012037
9. Che, H.Q., Ye, J.M., Tu, Q.Y., Yang, W.Q., Wang, H.G.: Investigation of coating process in wurster fluidised bed using electrical capacitance tomography. Chem. Eng. Res. Des. **132**, 1180–1192 (2018). https://doi.org/10.1016/j.cherd.2018.02.015
10. Voss, A.: Imaging Moisture Flows in Cement-Based Materials Using Electrical Capacitance Tomography. The University of Eastern Finland (2020)
11. Abd Rahman, N.A., et al.: A review: tomography systems in medical and industrial processes. J. Teknol. **73**(6), 1–11 (2015). https://doi.org/10.11113/jt.v73.4398
12. Li, Z., Chen, Y., Yang, Y., Liu, C., Lucquiaud, M., Jia, J.: Flow regime transition in counter-current packed column monitored by ECT. Chem. Eng. J. **420**, 1–7 (2021). https://doi.org/10.1016/j.cej.2021.129841
13. Shafquet, A., Ismail, I., Japper-Jaafar, A., Sulaiman, S.A., Chala, G.T.: Estimation of gas void formation in statically cooled waxy crude oil using online capacitance measurement. Int. J. Multiph. Flow **75**, 257–266 (2015). https://doi.org/10.1016/j.ijmultiphaseflow.2015.06.005
14. Li, Y., et al.: Gas/oil/water flow measurement by electrical capacitance tomography. In: IEEE International Conference on Imaging Systems and Techniques, Proceedings, pp. 83–88 (2012). https://doi.org/10.1109/IST.2012.6295481
15. Yang, W.: Design of electrical capacitance tomography sensors. Meas. Sci. Technol. **21**(4), 42991–42015 (2010). https://doi.org/10.1088/0957-0233/21/4/042001
16. Amizan, N., et al.: A review on electrical capacitance tomography sensor development. J. Teknol. **3**(73), 35–41 (2015)
17. Goh, C.L., et al.: Hardware development of electrical capacitance tomography (ECT) system with capacitance sensor for liquid measurements. J. Teknol. **73**(6), 13–22 (2015)
18. Peng, L., Ye, J., Lu, G., Yang, W., Sensor, A.E.C.T.: Evaluation of effect of number of electrodes in ECT Sensors on image quality. IEEE Sens. J. **12**(5), 1554–1565 (2012). https://doi.org/10.1109/jsen.2011.2174438

19. Mohamad, E.J., Rahim, R.A.: Multiphase flow reconstruction in oil pipelines by portable capacitance tomography. In: Proceedings of IEEE Sensors, pp. 273–278 (2010). https://doi.org/10.1109/ICSENS.2010.5689865
20. Tantiyani, N., Othman, A., Danial, M., Ronizan, N.: Simulation study on determination of gas-liquid profile in an aeration tank for mixing process by using an electrical capacitance tomography system. J. Tomogr. Syst. Sensors Appl. **4**(1), 70–79 (2021)
21. Kryszyn, J., Wanta, D., Smolik, W.T.: Evaluation of the electrical capacitance tomography system for measurement using 3D sensor. J. Inform. Autom. Pomiary w Gospod. i Ochr. Środowiska **9**(4), 52–59 (2019). https://doi.org/10.35784/iapgos.205
22. Peng, L., Mou, C., Yao, D., Zhang, B., Xiao, D.: Determination of the optimal axial length of the electrode in an electrical capacitance tomography sensor. Flow Meas. Instrum. **16**, 169–175 (2005). https://doi.org/10.1016/j.flowmeasinst.2005.02.015
23. Li, Y., Holland, D.J.: Optimizing the geometry of three-dimensional electrical capacitance tomography sensors. IEEE Sens. J. **15**(3), 1567–1574 (2015). https://doi.org/10.1109/JSEN.2014.2363901
24. Process Tomography Ltd, Electrical Capacitance Tomography System Type TFLR5000 Operating Manual Volume 1 - Fundamentals of ECT, no. 1 (2009)
25. Wang, C., Pang, R., Cao, Q., Ye, J.: Optimization of the structure of electrode array of voltage-driven electrical resistance tomography. In: IEEE Instrumentation and Measurement Technology Conference, pp. 1–6 (2021). https://doi.org/10.1109/I2MTC50364.2021.9459895
26. Wang, H.X., Zhang, L.F., Zhu, X.M.: Optimum design of array electrode for ECT system. J. Tianjin Univ. **36**(3), 307–310 (2003)
27. Brandisky, K., Sankowski, D., Banasiak, R., Dolapchiev, I.: ECT sensor optimization based on RSM and GA. COMPEL Int. J. Comput. Math. Electr. Electron. Eng. **31**(3), 858–869 (2012). https://doi.org/10.1108/03321641211209753
28. Jiang, Y., He, X., Wang, B., Huang, Z., Soleimani, M.: On the performance of a capacitively coupled electrical impedance tomography sensor with different configurations. Sensors **20**(20), 1–18 (2020). https://doi.org/10.3390/s20205787
29. Zhao, Q., Li, J., Liu, S., Liu, G., Liu, J.: The sensitivity optimization guided imaging method for electrical capacitance tomography. IEEE Trans. Instrum. Meas. **70**, 1–15 (2021). https://doi.org/10.1109/TIM.2021.3106131
30. Array, S., Design, O., Image, R.: Optimum design of an internal 8-electrode electrical capacitance tomography sensor array. Adv. Mater. Res. **508**, 84–87 (2012). https://doi.org/10.4028/www.scientific.net/AMR.508.84
31. Ye, J., Wang, H., Yang, W.: Image reconstruction for electrical capacitance tomography based on sparse representation. IEEE Trans. Instrum. Meas. **64**(1), 89–102 (2015). https://doi.org/10.1109/tim.2014.2329738
32. Stephenson, D.R., Mann, R., York, T.A.: The sensitivity of reconstructed images and process engineering metrics to key choices in practical electrical impedance tomography. Meas. Sci. Technol. **19**, 1–15 (2008). https://doi.org/10.1088/0957-0233/19/9/094013
33. Inamdar, D., Leblanc, G., Soffer, R.J., Kalacska, M.: The correlation coefficient as a simple tool for the localization of errors in spectroscopic imaging data. Remote Sens. **10**(2), 1–26 (2018). https://doi.org/10.3390/rs10020231
34. Ichijo, N., et al.: Resolution enhancement of electrical resistance tomography by iterative back projection method. J. Vis. **19**(2), 183–192 (2016). https://doi.org/10.1007/s12650-015-0308-8

Modeling and Simulation of an ECT Sensor for Non-invasive Agarwood Inspection

Muhammad Aiqil Sarudin[1], Yasmin Abdul Wahab[1(✉)],
Nurhafizah Abu Talip Yusof[1,2], Mohd Mawardi Saari[1], Suzanna Ridzuan Aw[3],
Mohd Shafie Bakar[1], Nurul Wahidah Arshad[1], and Sia Yee Yu[4]

[1] Faculty of Electrical and Electronics Engineering Technology, Universiti Malaysia Pahang Al-Sultan Abdullah, 26600 Pekan, Pahang, Malaysia
yasmin@umpsa.edu.my
[2] Centre for Research in Advanced Fluid and Processes (Fluid Centre), Universiti Malaysia Pahang Al-Sultan Abdullah, Lebuhraya Tun Razak, 26300 Gambang, Kuantan, Pahang, Malaysia
[3] Faculty of Electrical and Automation Engineering Technology, Terengganu Advance Technical Institute University College (TATiUC), Jalan Panchor, Telok Kalong, 24000 Kemaman, Terengganu, Malaysia
[4] LOGO Solution Sdn. Bhd., Suite 0525, Level 5, Wisma SP Setia, Jalan Indah 15, Bukit Indah, 79100 Iskandar Puteri, Johor, Malaysia

Abstract. Electrical Capacitance Tomography (ECT) is a promising non-invasive imaging technique with potential applications in agriculture, particularly for assessing tree health. This study explores the modeling and simulation of an ECT sensor tailored for detecting agarwood, a highly valued aromatic wood. The objective is to develop a finite element model using COMSOL Multiphysics, featuring an array of eight electrodes arranged in a two-dimensional configuration. The sensor's performance is evaluated through simulations based on basic and complex geometric shapes, aiming to detect capacitance variations indicative of agarwood presence. Results are presented graphically, demonstrating the sensor's capability to differentiate between agarwood and non-agarwood regions. The graph shows that sample A closely follows the baseline (no agarwood), indicating minimal distortion, while Ssample C exhibits slightly more variation, suggesting a higher presence or uneven distribution of agarwood. The sensor was also able to identify the complex agarwood shapes. These findings highlight the potential of ECT as a diagnostic tool for non-destructive testing in forestry, offering a foundation for future advancements in the non-invasive inspection of valuable tree species. Further refinement of the sensor design is recommended to enhance sensitivity and accuracy, with future work exploring different shapes and configurations, including tomographic imaging, to improve detection capabilities.

Keywords: ECT · FEM · Agarwood

1 Introduction

Electrical Capacitance Tomography (ECT) is an advanced non-invasive imaging technique that has shown considerable promise in agriculture, particularly for assessing tree health. The technique reconstructs the permittivity distribution within an object by measuring the capacitance between multiple electrodes placed around its perimeter [1, 2]. ECT has attracted significant attention due to its broad potential applications, particularly in agriculture. In agricultural environments, ECT facilitates real-time, non-destructive imaging, enabling the internal health of plants and trees to be assessed without inflicting any harm [3, 4]. This capability is crucial for disease detection, resource optimization, and continuous plant health monitoring, making ECT a valuable tool in modern farming practices. By integrating ECT, farming operations can become more efficient and sustainable, potentially leading to improved crop yield and quality.

Agarwood, a highly prized aromatic wood, presents unique challenges for assessment and detection due to its uneven distribution within the tree [5]. Traditional methods for detecting agarwood often involve invasive techniques that can damage the tree and diminish its value [6, 7]. These methods include chemical analysis and physical inspection, both of which are time-consuming and destructive. As the demand for agarwood continues to grow, there is an increasing need for non-destructive methods, as in Ref. [3, 8–10] that can accurately detect its presence without compromising the tree's structural integrity.

This study aims to explore the use of ECT as a non-destructive method for agarwood detection. By developing and simulating an ECT sensor using a finite element method, this research seeks to establish a foundation for non-invasive agarwood inspection. The goal is to create a sensor capable of precisely detecting the presence of agarwood, thereby preserving the tree's integrity and value. This approach not only addresses the limitations of current methods but also offers a sustainable alternative for agarwood management.

The preliminary findings of this research focus on the design and performance evaluation of the ECT sensor. By evaluating different agarwood samples, the study demonstrates the sensor's ability to detect variations in permittivity associated with agarwood presence. These findings lay the groundwork for future advancements in ECT technology, with the potential to enhance non-destructive testing methods in forestry and agriculture. The insights gained from this study could lead to more effective and sustainable practices in the management of valuable tree species such as agarwood. To enable such non-invasive detection, a solid understanding of the electrostatic principles behind ECT is essential, particularly the relationship between permittivity variations and capacitance measurements. The following section outlines the theoretical framework for ECT and its application in agarwood detection.

2 Principles of ECT in Agarwood Detection

ECT is a non-invasive imaging technique that utilizes electrostatic fields to detect spatial variations in permittivity within an object, thereby reconstructing its internal structure. The fundamental principles behind ECT are rooted in the relationship between the permittivity distribution and the measured capacitance. This relationship, as outlined in

Refs. [11, 12], is derived from Maxwell's equations, where the dielectric flux density D is expressed through Gauss's Law as described in Eq. (1).

$$\nabla \cdot D = \rho_v \qquad (1)$$

Here, ρ_v represents the volume charge density and ∇ is the divergence operator. In an ECT system, typically, only one electrode is energized at a time, with the remaining electrodes functioning as receivers. Given the assumption of no net charge at the electrode surfaces, the total electric flux across all electrode surfaces is zero, which implies that the volume charge density is also zero. This simplification enables a more straightforward analysis of the electrostatic field within the system. The electric flux density, D, and the electric field intensity, E, are related through Eq. (2). Where ε denotes the spatial permittivity of the medium, and the electric field intensity E can further be described in terms of the electric potential φ expressed in Eq. (3).

$$D = \varepsilon E \qquad (2)$$

$$E = -\nabla \varphi \qquad (3)$$

With ∇ represents the gradient operator. Substituting Eq. (3) into Eq. (2), we derive the expression for the electric flux density in Eq. (4).

$$D = -\varepsilon \nabla \varphi \qquad (4)$$

In the ECT sensor system, the spatial permittivity distribution ε(x,y) varies within the object being examined, while φ(x,y) represents the electric potential distribution. These parameters are vital for modeling the permittivity variations that correspond to the presence of agarwood or other internal features within the tree. By combining Eq. (4) with Gauss's Law (Eq. (1)), we arrive at Poisson's equation, which governs the electrostatic field in the ECT system as described in Eq. (5).

$$\nabla \cdot (\varepsilon(x, y) \cdot \nabla \varphi(x, y)) = 0 \qquad (5)$$

In a two-dimensional scenario, this equation relates the spatial permittivity distribution ε(x,y) to the electric potential φ(x,y). The applied voltage at the electrodes typically sets the boundary conditions for this system. Specifically, for the first electrode, $\varphi = V_C$, and for all other electrodes, $\varphi = 0$ is assumed. This mathematical framework is essential for the finite element modeling of the ECT sensor. By simulating this model, we can predict capacitance measurements based on the internal permittivity distribution. The resulting data is crucial for detecting and visualizing the presence of materials such as agarwood, which exhibit distinct permittivity characteristics compared to surrounding tissue. Understanding this electrostatic field and its relation to permittivity variations is key to developing a sensor capable of accurately detecting agarwood without damaging the tree.

3 Methodology

COMSOL Multiphysics software was utilized to develop the geometry for the ECT model and create and define the various elements and parameters involved. This software provides a versatile platform for simulating multiphysics and single-physics phenomena and is widely used across engineering, manufacturing, and scientific research. It integrates powerful simulation capabilities, data management tools, and a user-friendly interface to build customized applications for specific research needs [13].

The first step in developing the numerical model involved designing a circular representation of the tree trunk (wood) with a specified diameter using COMSOL. Parameters such as the diameter of the agarwood, electrode dimensions, and the relative permittivity values for various materials were based on a comprehensive literature review [13]. The electrode geometry was created using circular arc segments from COMSOL's library, ensuring an accurate representation of the electrodes' shape and placement around the object. Correctly setting up these parameters before initiating simulations is crucial for ensuring reliable results. The primary geometry parameters used in this study are summarized in Table 1.

Table 1. Parameters of geometry.

List of parameters	
Items	Parameters
Number of electrodes	8
Material of electrode	Gold
Electrode stretch angel, θ (90% of angle)	$\frac{360°}{8} = 45°$ and $\frac{90°}{100} \times 45° = 40.5°$
Thickness of electrode	2.5 mm
Diameter of wood (pine)	600 mm
Diameter of each agarwood (sample double phantom)	181 mm
Relative Permittivity, ε_r:	
Gold	1.143
Wood (pine)	3.17944
Air	1
Agarwood (sample A)	2.84180
Agarwood (sample B)	1.60002
Agarwood (sample C)	1.71432
Supply voltage	20Vdc

After defining the geometry and input parameters, the model was visualized within COMSOL, with two key figures displaying the overall electrode arrangement and detailed electrode design, as shown in Figs. 1 and 2.

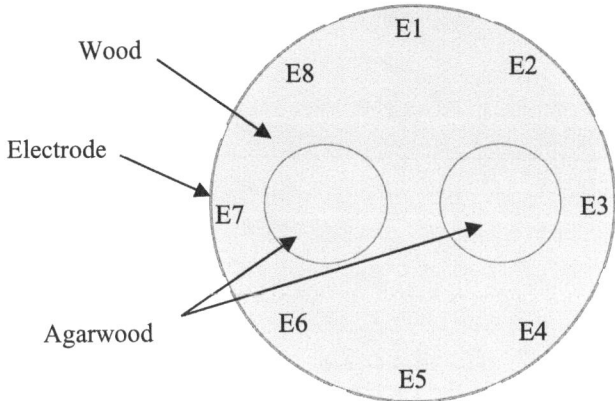

Fig. 1. The 2-Dimensional geometry of the 8 electrodes.

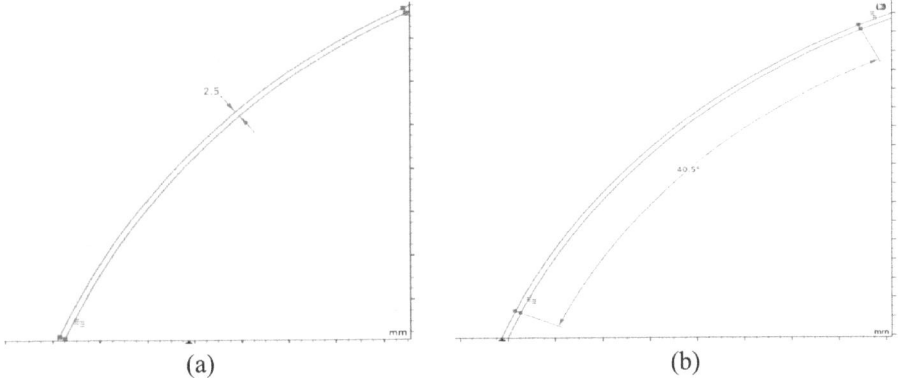

Fig. 2. Closed-up single electrode: (a) 2.5 mm thickness and (b) 40.5° stretch angle.

For this simulation, a DC voltage was used exclusively to streamline the modeling process, emphasizing the electrostatic behavior of the ECT system. A 20 V DC source was selected due to its capability to produce stable and measurable electric fields while avoiding challenges associated with low voltage fluctuations. The setup involved defining boundary conditions for the electrodes: Electrode 1 was designated as the excitation channel, responsible for distributing the electrical potential, while the outer electrodes were grounded to a potential of 0 V (V = 0 V) (Fig. 5). These configurations ensured a clear distinction between the excitation and grounding conditions, simplifying the interpretation of results. Figures 3, 4, and 5 illustrate the labeling of channels (electrode 1 through electrode 8), the application of electrical potential at the excitation channel, and the grounding of the outer electrodes, respectively.

After defining the boundary conditions and subdomains, a fine mesh was generated to discretize the ECT model for numerical analysis. Figure 6 displays the two-dimensional finite element mesh applied to the system, focusing on the eight external electrodes. The mesh was optimized for accuracy near the electrodes, where permittivity variations

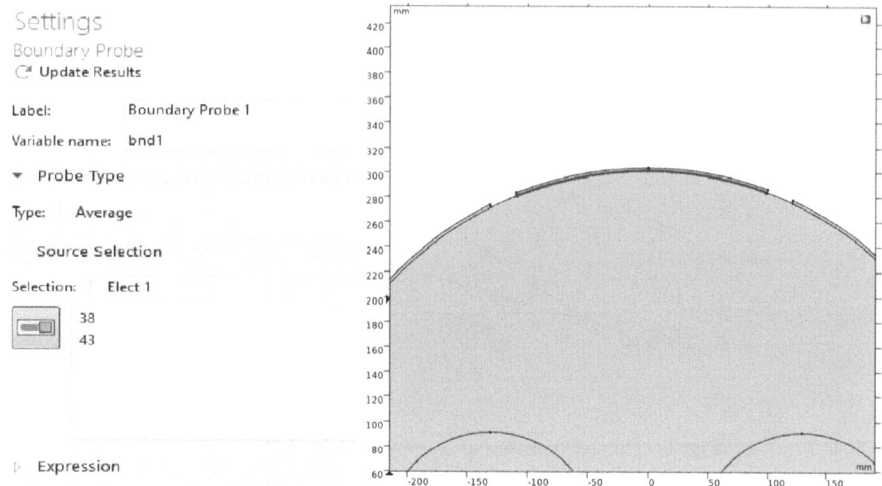

Fig. 3. Example of setting up the boundary probe for channel 1.

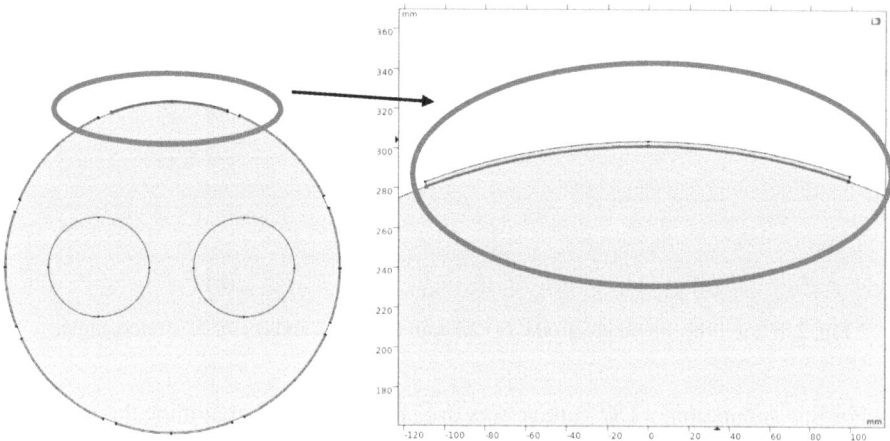

Fig. 4. Selection of the electrical potential at excitation channel.

are critical. Simulations were conducted for two scenarios: without agarwood and with agarwood present. These results provided surface distributions, streamlines, and contour maps, enabling a detailed analysis of the internal permittivity variations. Additionally, receiver data were captured through probes, ensuring reliable imaging of the modeled agarwood structure.

The electrostatic solver in COMSOL was employed to calculate the potential distribution within the model. This solver, optimized for solving Laplace and Poisson equations, ensures high accuracy in modeling the electrostatic field. The resulting electric potential distribution between the electrode pairs was analyzed using contour and streamline

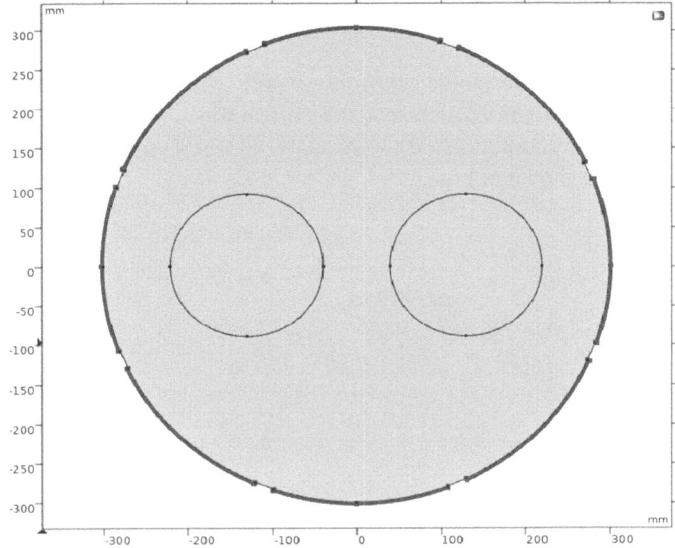

Fig. 5. Set the ground for each outer electrode.

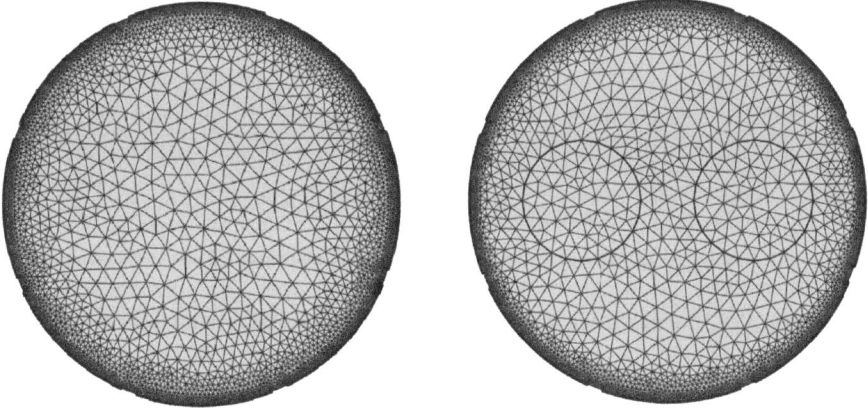

Fig. 6. Fine mesh with no agarwood (left) and with agarwood (right).

plots. These visualizations highlight variations in the field and the flow of electric potential across the domain, enabling a detailed examination of the sensor's performance. Streamline plots illustrated the direction of the electric field, revealing how the presence of agarwood influences the potential flow. Contour plots depicted the magnitude of electric potential variations, providing a clear view of permittivity changes within the sensor. These analyses are crucial for assessing the sensor's capability to detect agarwood and its distribution within the wood, reinforcing the potential of ECT for non-destructive detection applications.

4 Results and Discussion

In the initial simulation, the model with only wood (without agarwood) was tested. The electrical field distribution, represented by streamlines (black), and the electrical potential, represented by contours (red), were analyzed based on the relative permittivity of the wood alone, as shown in Fig. 7.

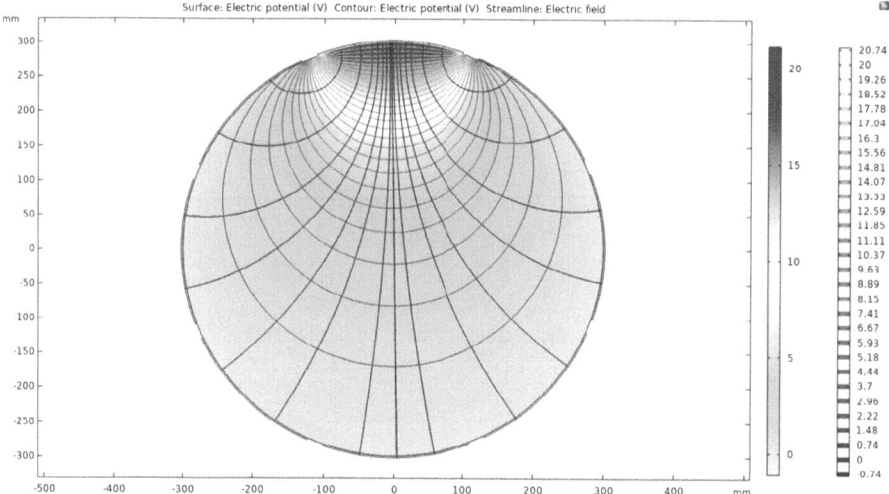

Fig. 7. Electrical potential distribution when channel 1 is set as an excitation signal. (Color figure online)

The electrical potential values obtained from this simulation were plotted in a line graph, as seen in Fig. 8. The graph illustrates the electrical potential (in volts) measured across eight receiver electrodes in the absence of agarwood. The potential begins at approximately 0.4 V at electrode 1, decreases to a minimum of around 0.05 V between electrodes 3 and 5, and then rises again to approximately 0.4 V at electrode 8. This symmetrical U-shaped curve indicates a uniform distribution of electrical potential when no agarwood is present.

Following the simulation with wood, three different agarwood samples were tested using the same procedure. Each sample exhibited a distinct electrical relative permittivity value. The results are summarized in Table 2, which shows the streamline (green) and contour (black) plots for each agarwood sample. The electrical potential values for all samples were compared to the baseline results (wood without agarwood), as shown in Fig. 9. From Table 2, it can be observed that agarwood sample A caused minimal distortion compared to the other samples. Samples B and C showed little difference in distortion, although sample C exhibited slightly more distortion than sample B. These observations align with the physical phenomenon where variations in relative permittivity led to changes in the streamline and contour plots. The images in Table 2 further confirm that sample C caused the most distortion, followed by sample B, with sample A showing the least.

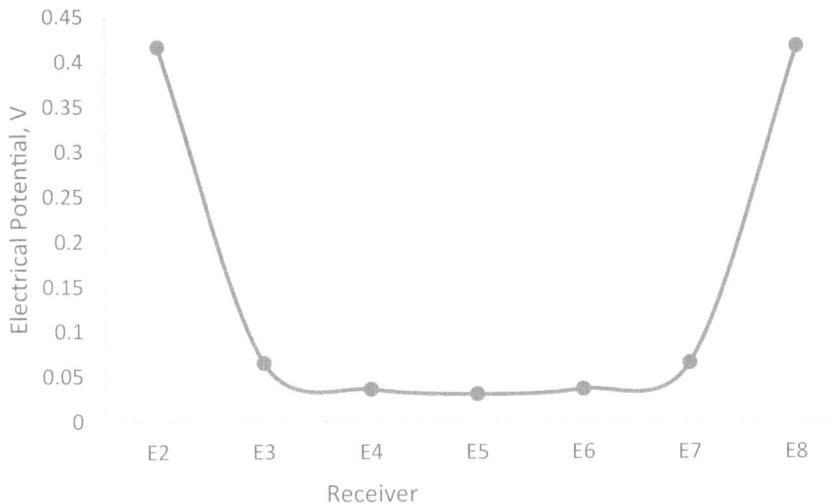

Fig. 8. Graph plot for receivers when channel 1 is set as excitation signal.

The graph in Fig. 9 displays the electrical potential across eight receiver electrodes for the different scenarios: no agarwood and three agarwood samples (A, B, and C). The baseline (no agarwood) shows a U-shaped curve, starting and ending at approximately 0.4 V, with a dip near 0.05 V between electrodes 3 and 5. Sample A follows this baseline closely, indicating minimal distortion and a permittivity similar to the baseline. Sample B also aligns with the baseline, showing negligible deviation, suggesting a similar distribution of agarwood as sample A. However, sample C shows slightly more distortion, particularly between electrodes 3 and 5, which suggests greater variation in permittivity and a higher presence or uneven distribution of agarwood. These results indicate that the ECT sensor can detect small permittivity differences, but further refinement may be needed to enhance its sensitivity for more distinct differentiation. Overall, ECT demonstrates potential as a non-invasive tool for agarwood detection, with opportunities for optimization to improve accuracy and sensitivity.

Table 2. Result for 3 samples of agarwood (samples A, B, and C).

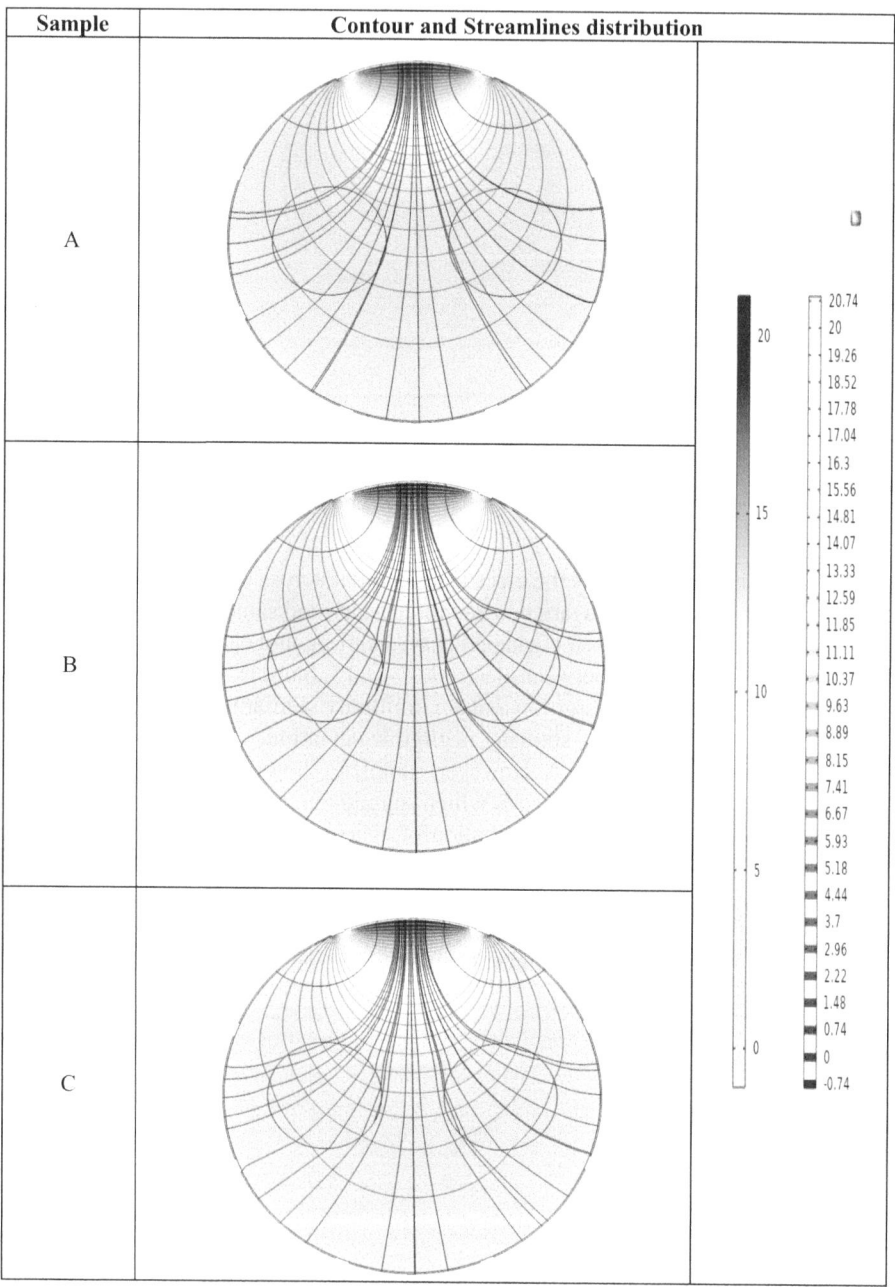

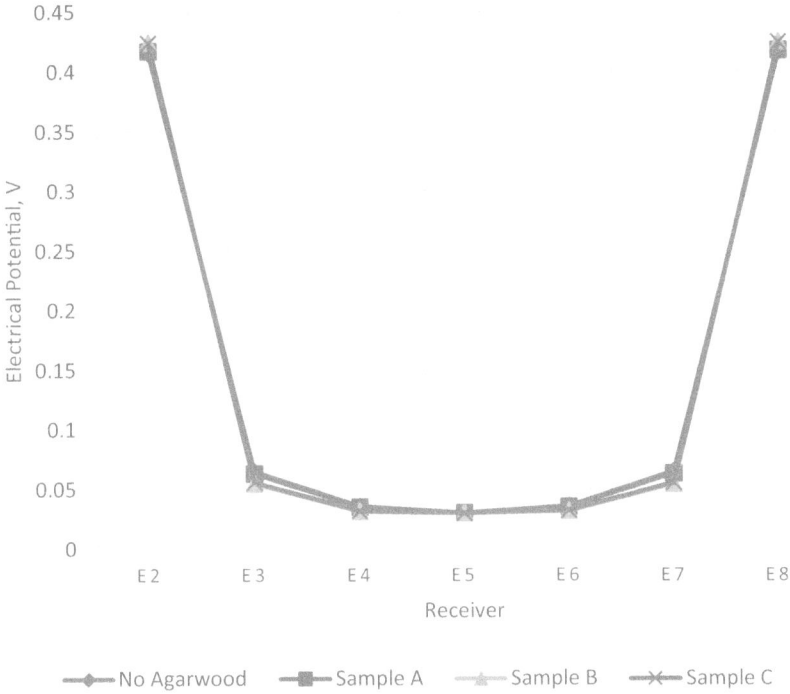

Fig. 9. Graph plot for receivers when channel 1 set as excitation signal with different samples of double agarwood.

To investigate the impact of agarwood shape further, additional simulations were conducted using the same permittivity as sample C (which showed the most distortion) but with more complex agarwood geometries (as outlined in Table 3). The shapes were selected based on prior literature, as described in Ref. [14]. Table 3 shows how the presence of agarwood affected the electric field (streamlines) and electrical potential (contours) for each sample. With Channel 1 acting as the transmitter and the remaining channels as receivers, Fig. 10 displays the initial sensor readings. The sensor's ability to distinguish between different geometries was demonstrated by variations in the received potential across the samples. Compared to samples D and E, sample F consistently generated the highest received potential (E2-E8), suggesting a higher permittivity or a larger volume of agarwood. Additionally, the area between channels E3 and E5 exhibited the most distortion for all samples, indicating the proximity of agarwood in that region. These results highlight the sensor's sensitivity to the distribution and form of agarwood.

Table 3. Complex shapes of agarwood (Sample D, E, and F).

	Sample D	Sample E	Sample F
Geometry			
Streamline and Contour			

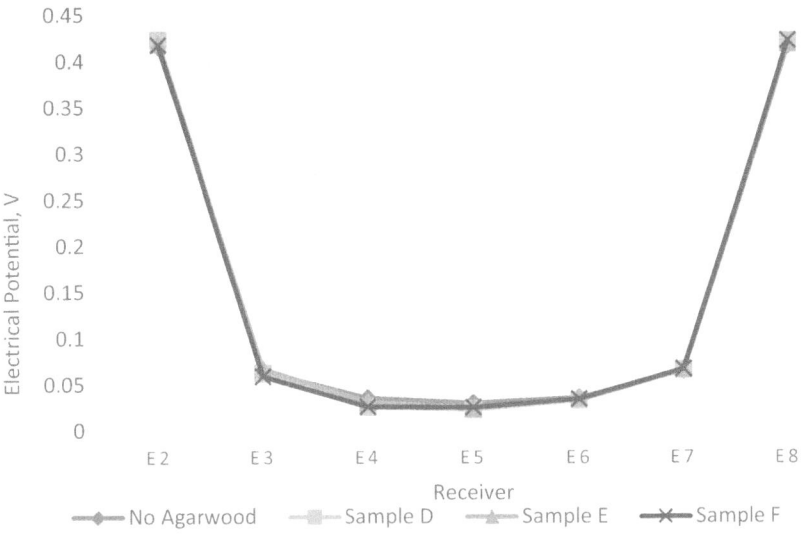

Fig. 10. Received electrical potential for complex agarwood shapes.

5 Conclusion

This study explored the use of Electrical Capacitance Tomography (ECT) for non-invasive agarwood detection through the modeling and simulation of an ECT sensor using a finite element approach. The results demonstrated that the sensor successfully measured electrical potential across various agarwood samples. Minimal distortion was observed in samples A and B, which closely resembled the baseline (no agarwood), while sample C exhibited slightly more variation, indicating a higher presence or uneven distribution of agarwood. The simulation model also showed the ability to detect more complex agarwood shapes. These findings highlight the potential of ECT as a promising tool for assessing agarwood presence, offering a non-destructive alternative to traditional methods. However, further refinement of the sensor design is recommended to improve sensitivity and accuracy. Future work should focus on testing with a wider variety of agarwood shapes and configurations, incorporating tomographic imaging techniques to enhance detection capabilities. This would provide deeper insights into agarwood distribution and improve the overall effectiveness of the method. The research presented in this study lays the foundation for future developments in non-invasive agarwood inspection and contributes to advancing sustainable and efficient forestry practices.

Acknowledgement. The authors would like to thank the Universiti Malaysia Pahang Al-Sultan Abdullah under Tabung Persidangan Dalam Negara (TPDN) for providing financial support.

References

1. Mousazadeh, H., Tarabi, N., Taghizadeh-Tameh, J.: A fusion algorithm for mass flow rate measurement based on neural network and electrical capacitance tomography. Meas. J. Int. Meas. Confed. **231**, 114573 (2024). https://doi.org/10.1016/j.measurement.2024.114573
2. Lei, J., Liu, Q.: Multitasking optimization for the imaging problem in electrical capacitance tomography. Expert Syst. Appl. **257**, 125105 (2024). https://doi.org/10.1016/j.eswa.2024.125105
3. Gut, Z., Lisowski, A., Klonowski, J., Świętochowski, A.: Potential of the electrical capacitance tomography method to monitor the mass flow rate of chopped maize in a field forage harvester. Sustainability **11**, 1–14 (2019)
4. Yu, L., Zhang, M., Yang, D., et al.: Grain moisture sensing using electrical capacitance tomography. IEEE Sens. J. **24**, 2038–2048 (2024). https://doi.org/10.1109/JSEN.2023.3335366
5. Karlinasari, L., Nandika, D.: Acoustic-based technology for agarwood detection in aquilaria trees. In: Agarwood Science Behind the Fragrance, pp. 137–148. Springer, Singapore (2016)
6. Ngadiran, S., Baba, S., Nor, N.F.A., et al.: The induction techniques of resinous agarwood formation: a review. Bioresource Technol. Rep. **21**, 101337 (2023). https://doi.org/10.1016/j.biteb.2023.101337
7. Shivanand, P., Arbie, N.F., Krishnamoorthy, S., Ahmad, N.: Agarwood—the fragrant molecules of a wounded tree. Molecules **27**, 1–23 (2022). https://doi.org/10.3390/molecules27113386
8. Karlinasari, L., Indahsuary, N., Kusumo, H.T., et al.: Sonic and ultrasonic waves in agarwood trees (aquilaria microcarpa) inoculated with fusarium solani. J. Trop. For. Sci. **27**, 351–356 (2015)
9. Xie, Y., Li, L., Chen, Y., et al.: Rapid authentication of agarwood by using liquid extraction surface analysis mass spectrometry (LESA-MS). Phytochem. Anal. **31**, 801–808 (2020)
10. Zhang, N., Xue, S., Song, J., et al.: Effects of various artificial agarwood-induction techniques on the metabolome of Aquilaria sinensis. BMC Plant Biol. **21**, 1–13 (2021). https://doi.org/10.1186/s12870-021-03378-8
11. Zeeshan, Z., Zuccarelli, C.E., Acero, D.O., et al.: Enhancing resolution of electrical capacitive sensors for multiphase flows by fine-stepped electronic scanning of synthetic electrodes. IEEE Trans. Instrum. Meas. **68**, 462–473 (2019). https://doi.org/10.1109/TIM.2018.2847918
12. Ji, Z., Liu, J., Tian, H., Zhang, W.: ECT sensor simulation and fuzzy optimization design based on multi index orthogonal experiment. IEEE Access **8**, 190039–190048 (2020). https://doi.org/10.1109/access.2020.3029839
13. Ishak, N., Lee, C.K., Mohd Muji, S.Z.: A simulation magnetic induction tomography (MIT) for agarwood using COMSOL multiphysics. Int. J. Eng. Adv. Technol. **10**, 67–71 (2021). https://doi.org/10.35940/ijeat.c2174.0210321
14. Fazalul Rahiman, M.H., Jack, S.P.: Microwave tomography for agarwood detection. Universiti Malaysia Perlis, Malaysia (2021)

Hardware Accelerator Simulation for Colour Correction Algorithm

Shu Ting Loh[1], Mohd Shahrizal Rusli[2]([✉]), Muhammad Nadzir Marsono[3], Ab Al-Hadi Ab Rahman[3], Norlina Paraman[3], Shahidatul Sadiah[3], Michael Loong Peng Tan[3], Izam Kamisian[3], Jaysuman Pusppanathan[3], and Nur Diyana Kamarudin[4]

[1] Intel Microelectronics (M) Sdn Bhd, Bayan Lepas, Penang, Malaysia
[2] Faculty of Artificial Intelligence, Universiti Teknologi Malaysia, Kuala Lumpur, Malaysia
shahrizalr@utm.my
[3] Faculty of Electrical Engineering, Universiti Teknologi Malaysia, Johor Bahru, Johor, Malaysia
[4] Faculty of Science and Defense Technology, The National Defence University of Malaysia, Kuala Lumpur, Malaysia

Abstract. An algorithm for color correction is essential for processing colour data. In colour correction algorithms, the range of statistical techniques continues to expand to achieve higher accuracy and reproducibility across various applications. One of the methods that is most frequently applied in practice is polynomial color correction. However, timing performance is significantly impacted by the computationally demanding and time-consuming aspect of implementing a sophisticated colour correction technique in Hardware Description Language, particularly for critical diagnostic applications. In order to enhance the temporal performance of repetitive operations in the polynomial algorithm while preserving accuracy with the least degree of deterioration, this study suggests a hardware accelerator. By employing a combination of pipelining, array partitioning, and loop unrolling techniques, a speedup of 22.05 times has been achieved, albeit at the expense of hardware resources. In conclusion, the various approach configurations proposed in this project offer multiple design alternatives, balancing hardware cost and performance to meet diverse usage and performance objectives.

Keywords: Colour correction · hardware accelerator · high level synthesis · polynomial algorithm

1 Introduction

Over the past few decades, stable digital computers and sophisticated image processing techniques have made digital images very versatile with many benefits, including processing flexibility, reliable transmission, ease of reproduction, storage and retrieval, and compatibility with digital networks [1]. In a variety of domains, such as computer games, virtual reality applications, biological diagnosis, and visual quality evaluation, color images have become increasingly popular [2]. However, picture quality [4] can be affected by nonlinear contrast and brightness fluctuations caused by factors like lighting, image acquisition devices, and image angles [3]. To achieve optimal color images, researchers have developed numerous image processing approaches.

This research focuses on Polynomial Color Correction (PCC), a widely used approach. While Linear Color Correction (LCC) can approximate color values to the sRGB standard through linear adjustments in scene radiance or exposure, it can lead to substantial mapping inaccuracies for some surfaces. PCC addresses this issue by incorporating straightforward additions to the linear approximation [5]. However, the compute-intensive nature of such methods can result in longer execution times, particularly with the growing demand for low power consumption and high timing performance in processing massive datasets. Furthermore, implementing complex algorithms using Hardware Description Language (HDL) can reduce flexibility and extend design timelines.

Recent advances in hardware accelerators have demonstrated significant improvements in computing performance for image processing tasks. For instance, Bruno *et al.* [22] developed a hardware accelerator that reduced latency considerably while maintaining high accuracy for real-time video color correction in underwater robots. Similarly, Khalid *et al.* [24] proposed an efficient FPGA-based architecture for real-time video color correction, focusing on resource management and pipeline optimization to achieve high throughput with minimal hardware overhead. Additionally, Kumar *et al.* [23] highlighted the versatility of hardware accelerators in resource-constrained environments by designing a hardware-efficient color correction algorithm optimized for low-power applications in wearable technology.

High-Level Synthesis (HLS) tools have revolutionized the development process, enabling more streamlined workflows. Tsiktsiris *et al.* [6] demonstrated the effectiveness of HLS tools in design exploration by developing and evaluating an image processing accelerator. Furthermore, techniques like polynomial approximation acceleration on GPUs, as explored by Janosek and Nemec [16], have inspired similar enhancements in FPGA-based systems to further improve computational efficiency for large datasets.

Building on these advancements, this work aims to utilize Vivado HLS to design a hardware accelerator for the PCC algorithm. The proposed approach employs optimization techniques such as array partitioning, pipelining, and loop unrolling to significantly enhance timing performance while maintaining low power consumption. This study contributes to the growing field of hardware-accelerated color correction methods by addressing trade-offs between computational performance and resource utilization.

2 Related Works

This research has examined a number of color correcting algorithms. Gray World [8], Multi-Scale Gray World [9], Retinex [10], Multi-Scale Retinex with Color Restoration (MSRCR) [11], and MSCRCR with Autolevels [12] are a few of the well-known methods. By converting the camera's device-dependent RGB color representation to a device-independent XYZ color representation, polynomial regression models have also been investigated [1]. The fuzzy 3D filter, which was first developed in 2012, aids in decreasing noise while maintaining the image's fine features, edges, and chromaticity [15].

Colour correction algorithms have been implemented more often in recent years using hardware accelerators, which offer energy conservation and speed improvements. They work especially well for portable and real-time applications. For example, Bruno et al. [22] created a hardware accelerator for real-time video colour correcting for underwater robotics. Similarly, a hardware-efficient colour correcting technique for low-power wearable technology was described by Kumar et al. [23].

Field-Programmable Gate Arrays (FPGAs) have garnered a lot of attention recently for integration into image processing workflows due to their ability to efficiently execute parallel operations, which has led to noticeable improvements in processing speed and energy efficiency. An obvious illustration of this tendency is the speed at which Convolutional Neural Networks (CNNs) are being developed for image classification tasks. For instance, Ali et al. demonstrated a method that reduces design time and enables rapid adaptation to evolving CNN designs by using high-level specifications to develop efficient FPGA-based CNN accelerators [25]. The development of theoretical models for FPGA-based hardware accelerators has also facilitated a deeper comprehension of their computational capabilities. In a paradigm that combines standard CPUs and FPGAs, Hora et al. demonstrated algorithms that perform better than their word-RAM equivalents in tasks like dynamic programming and sorting [26]. This theoretical underpinning highlights how FPGAs can improve computational efficiency in a range of applications.

In the field of colour image enhancement, hardware accelerators have been developed to fulfil real-time processing requirements. Khalid et al. [24] proposed an efficient FPGA-based architecture for real-time video colour correction. Janosek and Nemec, in 2012, introduced the implementation of polynomial approximation using the Graphics Processing Unit (GPU) parallel computing architecture provided by NVIDIA's Compute Unified Device Architecture (CUDA). This approach significantly enhanced computational power for handling large datasets. Further optimization of memory bandwidth was achieved by investigating the allocation of pageable memory, page-locked memory, and memory mapped into CUDA's address space [16].

Compiler directives, or pragmas, available in HLS tools, were utilized to optimize performance through techniques such as pipelining and loop unrolling, thereby improving the system's memory bandwidth [6]. Optimization methods like array partitioning, pipelining, and loop unrolling play a vital role in hardware accelerator design. Pipelining allows for temporal overlap of processes, while loop unrolling operates by dividing the set of operations within a for-loop into multiple distinct operations, enhancing the efficiency of the system. By dividing arrays into smaller arrays, array partitioning can expand the band-width of local memory [17]. Directives for array partitioning, pipelining, and loop unrolling are included in the Vivado HLS tool [6, 17, 18]. Although there are trade-offs to take into account, hardware acceleration approaches have generally been seen to boost speed. Loop unrolling, pipelining, and array partitioning can raise critical path latency and lower the maximum operating frequency, and they can also increase the cost of hardware resources [13, 17, 18].

These developments' convergence points to a bright future for FPGA-based hardware accelerators in image processing. High-level synthesis tools and theoretical models can be used by academics to create effective accelerators that meet the increasing needs of real-time applications. Building on previous advancements, this work investigates the application of polynomial colour correction algorithms on FPGAs, employing modern design techniques to attain peak performance.

3 Methodology

The second order PCC method is implemented in both hardware and software, with hardware outcomes being assessed and software parameters being recorded as a reference. The following illustrates a basic LCC or PCC transformation. Let each pixel's RGB values be represented by vector p, and let q represent the tri-stimulus values that correspond to those values. Below is an example of a basic LCC or PCC transformation:

$$q = Mp \qquad (1)$$

The least-squares mapping, denoted as matrix M, is commonly determined using the least-squares regression method for both Linear Color Correction (LCC) and Polynomial Color Correction (PCC). This process involves color target-based characterization, where a defined number of known color samples serve as the reference target. In this approach, a set of N known XYZ values for a reflectance target is represented by Q, while the corresponding camera responses are expressed as a $3 \times N$ matrix R. The least-squares mapping M, which maps R to Q, is derived using the Moore-Penrose inverse method [23], as follows:

$$M = QRT(RRT) - 1 \qquad (2)$$

In Polynomial Color Correction (PCC), the vector p can be extended by incorporating higher-order terms such as $r2$, $g2$, rg, rb a constant term like 1. Here, r, g, and b represent the RGB color channels. The vector p denotes the responses from N sensors. For a typical scenario where $N = 3$, the set of K-th degree polynomial terms in N variables is defined as vectors, denoted by $p_{K,N}$, as shown below:

i. $p1,3 = (r, g, b)^T$
ii. $p2,3 = (r, g, b, r2, g2, b2, rg, gb, rb)^T$
iii. $p3,3 = (r, g, b, r2, g2, b2, rg, gb, rb, r3, g3, b3, rg2, gb2, rb2, gr2, bg2, br2, rgb)^T$

A linear regression is indicated by a vector p with degree 1. The RGB values of a pixel are now recorded by the polynomial expansion in the representation of 9 and 19, respectively, for polynomial orders of 2 and 3. For order 2 and order 3, the equivalent matrices M have formats of 3×9 and 3×19, respectively. We test the transformation matrix M of sizes $3 \times N$.

3.1 Software Algorithm

One type of predictive modeling that looks into the relationship between an independent and dependent variable is regression analysis. While multiple linear regression is utilized as the number of independent variables increases, polynomial regression fits a polynomial line with a minimum error or cost function. In Fig. 1, the software algorithm is displayed. The input image, which has 362 columns and 237 rows, is read and saved in matrix I.

Matrix M, which has nine rows and three columns, contains the polynomial coefficients, whereas the input pixels are integers that range from 0 to 255. This project's matrix multiplication algorithm is denoted by

$$I\_out = S\_I * M \tag{3}$$

The output matrix, I_out, is obtained by multiplying the input matrix, S_I, by the matrix of polynomial coefficients, M. After values are entered into the final output matrix, time is spent examining the time-consuming portion. The overall execution time, output pixel values, and image are documented.

3.2 Proposed Hardware Algorithm

Both computationally demanding and repetitive processing operations can be accelerated by hardware techniques. Prior to accessing the matrix multiplication logic, the *colour_correction_polynomial()* function pre-defines polynomial coefficients and other constant integers. Hardware design is coded to perform *I1_out*, *I2_out*, and *I3_out* multiply-and-add operations in a single row (r) and column (c) iteration. *I_out* is the output matrix with N rows and 3 columns that represents the colour-corrected pixels of the R, G, and B channels. According to a modification, the coefficients are now stated as long long and multiplied by 10^{15}, which is used as an integer during the multiplication procedure. The hardware used for rounding off is calculated by dividing 10^{15} by 2. Upon completion of the output array, they are released from the result struct of type un-signed short.

Loop Unrolling. Loop unrolling is used to split the for-loop's single group of activities into multiple separate operations. The color correcting algorithm has two types of loops: ROW and COL. There are 237 iterations in the ROW loop and 362 iterations in the COL loop. Loop unrolling can be used with the first level operations from the signal flow graph (SFG) (Fig. 2) since they are data independent. Because of the HLS synthesis tool's limited resources, a specific unrolling factor is chosen to unroll the loops rather than fully unrolling them (as stated in Table 1).

Pipelining. Pipelining aims to reduce the overall execution time and encourage concurrent execution of subtasks. The COL loop is split up into multiple pipeline stages for concurrent execution, and pipelining is applied using the Vivado HLS tool. Data dependencies between the pipeline stages are added, and the stages must be balanced to allow each to be completed in the same amount of time. As shown in Table 2, the pipeline approach is used in conjunction with the loop unrolling method, and the tool inserts pipeline registers while taking stage balancing into account.

Array Partitioning. Array partitioning can be used to increase local memory bandwidth and, when combined with loop pipelining and loop unrolling, improve system performance. A single port memory resource is de-faulted in each array in Vivado HLS, which is not the best memory architecture for algorithms that prioritize speed. If the array's first dimension reflects its row, array partitioning can be used to divide the array's

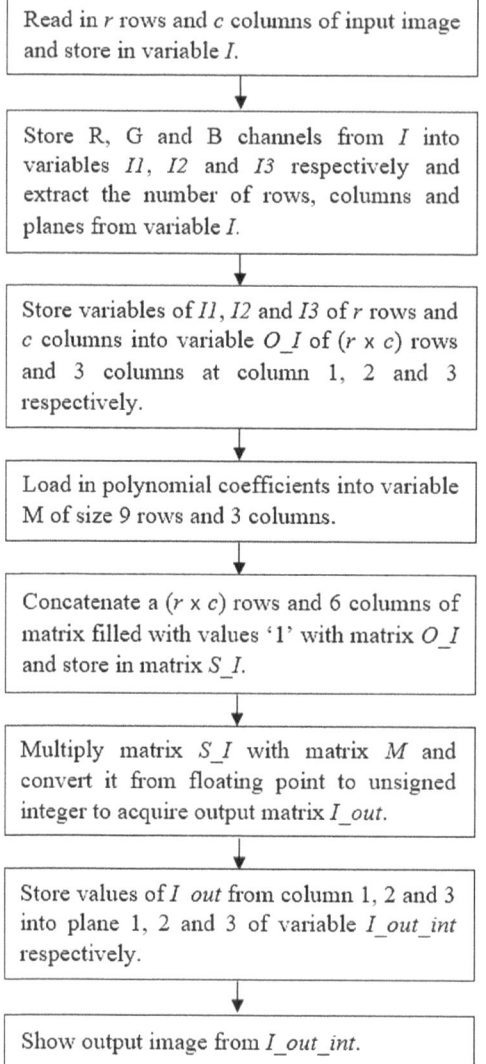

Fig. 1. Software flowchart of colour correction algorithm.

Table 1. Configurations for unrolling various loops with various factors

Optimization Approach	Unrolling Factor
Unroll COL	4
Unroll ROW	3
Unroll ROW and COL	3 + 4

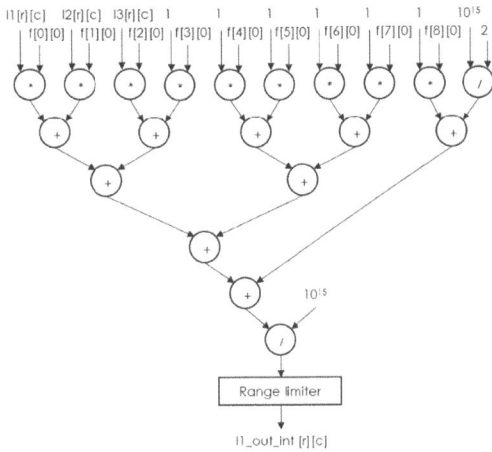

Fig. 2. Signal flow graph for an *I_out_int[i][n]* mathematical operation

Table 2. Pipelining setups using various unrolling techniques

Optimization Approach	Unrolling Factor	Pipelining
Original Loop	-	Yes
Unroll COL	4	Yes
Unroll ROW and COL	3 + 4	Yes

second dimension by its column, increasing the number of read and write ports. Using the same partitioning technique, all three input arrays (I1, I2, and I3) are used. Array partitioning is also used on the output ports to look into potential improvements from memory writing.

4 Evaluation Setup

The software algorithm and hardware implementation performance evaluations comprise the project performance evaluation. A number of matrix multiplication works using various hardware accelerating approaches are seen in [18] and [27], but the outcomes of this project are not compared with those of those works because of the differences in the matrices' sizes and elements. The performance evaluation in this work consists of hardware implementation performance evaluation and software algorithm performance evaluation. The software algorithm is evaluated using an input image of a color-distorted Munsell X-Rite Color Checker that is 237 high by 362 wide. The hardware implementation performance is evaluated using three text files that were extracted from the identical input image using MATLAB. The output parameters that are tracked for further evaluation include cycle duration, maximum operating frequency (Fmax), clock cycle latency, and resource consumption of flip-flops and look up table. The overall execution time

5 Result and Discussion

5.1 Software Algorithm Results and Evaluation

Figure 3 displays the color-distorted input image and the original Munsell X-Rite color checker. The output image after color correction deviates from cool colors. Table 3 lists the machine specs of the machine that was utilized.

State 7 involves converting floating points to integers and placing the I_out array's elements in the I_out_int array. In state six, the matrix is multiplied and the result is converted from double-precision floating point to an unsigned integer. By calling elements from three arrays and '1' straight into the computation rather than storing them in a separate large array, time spent on states three and five can be reduced (Table 4). Table 4 shows that, in comparison to the other five stages, state 3 uses the most processing time, consuming 0.519471 s. The sequence of operations is depicted in Fig. 3.

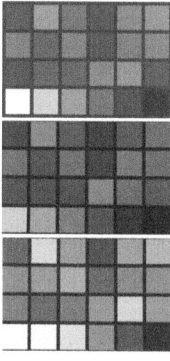

Fig. 3. Original Munsell X-Rite Color Checker (top) with color distortion [19] Munsell X-Rite Color Checker [14] in the middle, with the color bottom adjusted: The Munsell X-Rite Color Checker

5.2 Hardware Simulation Results and Evaluation

Performance measurements, resource utilization estimation, and serial processing or rolling loop are all included in the hardware implementation performance comparison. Table 5 lists and discusses the outcomes of serial processing and the three unroll settings for loop unrolling. The finest two loop unrolling performances are combined with the pipelining approach for pipelining, and the outcomes are examined. The top two pipelining performances are combined with various array partitioning method configurations for array partitioning, and the outcomes are assessed in Table 6.

Table 3. Specification of machine used for software simulation

System Specification	Unit
Processor Frequency	2.6 GHz
Period	0.385 ns
RAM	8 GB
Operating System	64-bit

Table 4. The duration of each software algorithm operation state

States	Cumulative time (s)	Individual time (s)
1	0.075436	0.075436
2	0.077135	0.001699
3	0.596606	0.519471
4	0.597521	0.000915
5	0.608005	0.010484
6	0.619361	0.011356
7	0.846802	0.227441
8	2.023082	1.176280

Parallel Processing Results. Our concept takes advantage of the FPGA's special architectural advantage of configurable logic blocks by using Verilog HDL to construct bespoke processing elements for parallel processing. Table 5 illustrates how closely the loop unrolling strategy performs when compared to the unroll COL by 4 and ROW by 3 approaches. Nevertheless, the unroll ROW by 3 approach performs better since it performs more actions in parallel inside a single COL iteration. 18 multipliers per 6 input components and 24 multipliers per 2 input elements are the maximum memory ports per output that the HLS tool allows, with a restriction of 2 ports per clock cycle. The unrolling approach's parallelism is exemplified by the subsequent COL operations being carried out only after the first COL is finished. This trade-off results in more resources being utilized.

Pipelining Results. Table 5 lists the performance metrics for several pipeline configurations with loop unrolling in comparison to serial processing. The pipeline directive is implemented in the COL loop. Loop unrolling is accelerated by 11.4 times and the overall execution time is reduced by nearly 50% when pipeline is added. However, there are the same number of adders and multipliers for both unrolling COL by 4 and unrolling COL by 4 and ROW by 3.

Array Partitioning Results. Table 6 lists the array splitting implementations' resource usage and performance data. With a speedup of over ten times, all the methods demonstrate an improvement in timing. In contrast to non-array-partitioned approaches, the method that simply partitions the input array *Ix* exhibits a modest drop in speedup. Moreover, splitting the output arrays of *I1_out_int*, *I2_out_int*, and *I3_out_int* by 4 has greatly reduced latency. The design that has the largest speedup, 22.05 times, is the array partitioning of input and output arrays by 4, in addition to unrolling ROW by 3 and COL by 4.

Table 5. Comparison of performance estimation between baseline serial processing and various pipelining topologies with loop unrolling.

Optimization Approach		Baseline	Pipeline	Unroll COL by 4 + Pipeline	Unroll ROW by 3 COL by 4 + Pipeline
F_{max} (MHz)		120.45	120.45	120.45	120.45
Latency (clock cycles)		515239	85801	45031	44873
Total Execution Time (ms)		4.278	0.712	0.374	0.373
Resource Utilization	Flip-flops	1063	1180	12502	12502
	LUTs	1521	1593	18049	18049
Speedup		-	6.01X	11.44X	11.47X
Adder		9	9	36	108
Multiplier		12	12	48	144

5.3 C/RTL Co-simulation Results Simulation Results and Evaluation

The validation of the hardware accelerator is done by using a testbench and input text files in C language. Co-simulation between C and RTL is used to produce the output waveform. The values in the waveform are compared to a sample of output pixels from the program code. 16-bit variables called *agg_result_I2_out_int_d0[15:0]*, *agg_result_I3_out_int_d0[15:0]* and *agg_result_I1_out_int_d0[15:0]* represent the output arrays from hardware design. The numbers are same when comparing the first 12 output pixels between the hardware and software designs. To confirm the hardware accelerator's operation, a complete comparison of all three output array values is made.

Table 6. Evaluation of pipelining performance for various setups using array partitioning and loop unrolling techniques.

Optimization Approach		Unroll COL by 4 + Pipeline + Partition I_x by 4	Unroll COL by 4 + Pipeline + Partition I_x by 4 $I_{x\_out\_int}$ by 4	Unroll ROW by 3 COL by 4 + Pipeline + Partition I_x by 4	Unroll ROW by 3 COL by 4 + Pipeline + Partition I_x by 4 $I_{x\_out\_int}$ by 4
F_{max} (MHz)		120.45	120.45	120.45	120.45
Latency (clock cycles)		45268	23464	45110	23306
Total Execution Time (ms)		0.376	0.195	0.375	0.194
Resource Utilization	Flip-flops	4511	4314	12769	12253
	LUTs	6133	5773	18592	17872
Speedup		11.38X	21.94X	11.41X	22.05X
Adder		36	36	108	108
Multiplier		48	48	144	144

6 Conclusion

In conclusion, this project has successfully developed a hardware accelerator for the second-order PCC algorithm. The optimization methods of pipelining, loop unrolling and array partitioning have greatly improved the algorithm's temporal performance. When used, they unroll COL by 4 and 3 of ROW, and divide the input and output arrays by 4. The speedups have ranged from two to 22.05 times. The computational core of the program consists of memory accessing activities and time-consuming repetitive procedures, which are the focus of these optimizations. High-Level Synthesis tools, like Vivado, have reduced design time and increased design space exploration, providing effective solutions while striking a balance between robustness and ease of use. The hardware accelerator's performance, as measured by resource usage, maximum operating frequency, latency, and total execution time, indicates its market value. Furthermore, it has achieved a 100% hit rate in accuracy validation against software baselines. To further improve the existing work, a number of possible changes or enhancements could be made. First, additional code optimization, such the Strassen implementation described in [18], can be carried out. The reason is because throughout the creation, it was discovered that certain library functions and coding styles would restrict the tool's ability to optimize efficiency. Therefore, there is a chance that by refining the fundamental algorithm, this project could be improved even further.

Acknowledgement. The authors would like to thank UTM for the funding of this project, with the Flagship CoE/RG research grant number Q.J130000.5023.10G06, Q.J130000.5023.10G08 and Q.J130000.5023.10G09.

References

1. Hong, G., Luo, M.R., Rhodes, P.A.: A study of digital camera colorimetric characterization based on polynomial modeling. Color. Res. Appl. **26**(1), 76–84 (2000)
2. Zemcik, P.: Hardware acceleration of graphics and imaging algorithms using FPGAs. In: Proceedings of the 18th Spring Conference on Computer Graphics - SCCG 2002 (2002)
3. Chen, R., Xie, J.-W., Li, C.-H.: Research on color correction algorithm for mobile-end tongue images. In: 2017 International Symposium on Intelligent Signal Processing and Communication Systems (ISPACS) (2017)
4. Obukhova, N., Motyko, A., Pozdeev, A.: Modern methods and algorithms in digital processing of endoscopic images. In: 2017 21st Conference of Open Innovations Association (FRUCT) (2017)
5. Wang, X., Zhang, D.: An optimized tongue image color correction scheme. IEEE Trans. Inf. Technol. Biomed. **14**(6), 1355–1364 (2010)
6. Tsiktsiris, D., Ziouzios, D., Dasygenis, M.: A high-level synthesis implementation and evaluation of an image processing accelerator. Technologies **7**(1), 4 (2018)
7. Li, S.-A., Chen, C.-Y., Chen, C.-H.: Design of a shift-and-add based hardware accelerator for color space conversion. J. Real-Time Image Proc. **10**(2), 193–206 (2013)
8. Buchsbaum, G.: A spatial processor model for object colour perception. J. Franklin Inst. **310**(1), 1–26 (1980)
9. Kyung, W.-J., Kim, D.-C., Ha, H.-G., Ha, Y.-H.: Color enhancement for faded images based on multi-scale gray world algorithm. In: 2012 IEEE 16th International Symposium on Consumer Electronics (2012)
10. Land, E.H.: Recent advances in retinex theory and some implications for cortical computations: color vision and the natural image. Proc. Natl. Acad. Sci. **80**(16), 5163–5169 (1983)
11. Jobson, D., Rahman, Z., Woodell, G.: A multiscale retinex for bridging the gap between color images and the human observation of scenes. IEEE Trans. Image Process. **6**(7), 965–976 (1997)
12. Jiang, B., Woodell, G.A., Jobson, D.J.: Novel multi-scale retinex with color restoration on graphics processing unit. J. Real-Time Image Proc. **10**(2), 239–253 (2014). https://doi.org/10.1007/s11554-014-0399-9
13. Yamakabe, R., Monno, Y., Tanaka, M., Okutomi, M.: Tunable color correction between linear and polynomial models for noisy images. In: 2017 IEEE International Conference on Image Processing (ICIP) (2017)
14. Kamarudin, N.D., et al.: Performance comparison of colour correction and colour grading algorithm for medical imaging applications. Int. J. Eng. Technol. **7**, 353–356 (2018)
15. Ponomaryov, V., Montenegro, H., Rosales, A., Duchen, G.: Fuzzy 3D filter for color video sequences contaminated by impulsive noise. J. Real-Time Image Proc. **10**(2), 313–328 (2012)
16. Janosek, L., Nemec, M.: Fast polynomial approximation acceleration on the GPU. In: The Sixth International Conference on Digital Society (ICDS), pp. 69–72 (2012)
17. Hani, M.K.: Design of Digital Systems II: RTL System Verilog and High-Level Synthesis, Skudai, Johor (2019)
18. Ryan, M.V.: FPGA Hardware Accelerators - Case Study on Design Methodologies and Trade-Offs. M. S. thesis, Department of Electrical Engineering Rochester Institute of Technology, Rochester, New York (2013)
19. Using colorcheck. http://www.imatest.com/docs/colorcheck/
20. Zhang, C., Li, P., Sun, G., Guan, Y., Xiao, B., Cong, J.: Optimizing FPGA-based accelerator design for deep convolutional neural networks. In: Proceedings of the 2015 ACM/SIGDA International Symposium on Field-Programmable Gate Arrays, pp. 161–170 (2015)

21. Cong, J., Liu, B., Neuendorffer, S., Noguera, J., Vissers, K., Zhang, Z.: High-level synthesis for FPGAs: from prototyping to deployment. IEEE Trans. Comput. Aided Des. Integr. Circuits Syst. **30**(4), 473–491 (2011)
22. Bruno, F., et al.: Real-time video color correction for underwater robots. IEEE Robot. Autom. Lett. **3**(4), 3372–3379 (2018)
23. Kumar, A., et al.: Hardware-efficient color correction algorithm for wearable applications. In: 2017 27th International Symposium on Power and Timing Modeling, Optimization and Simulation (PATMOS), Thessaloniki, pp. 1–8 (2017). https://doi.org/10.1109/PATMOS.2017.8106968
24. Khalid, M., Ahmad, H.F., Arshad, K.: Efficient FPGA-based architecture for color correction in real-time video processing. In: 2019 IEEE 9th Annual Computing and Communication Workshop and Conference (CCWC), Las Vegas, NV, USA, pp. 0821–0826 (2019). https://doi.org/10.1109/CCWC.2019.8666601
25. Ali, N., Philippe, J.-M., Tain, B., Coussy, P.: Generating efficient FPGA-based CNN accelerators from high-level descriptions. J. Signal Process. Syst. **94**(9), 945–960 (2022)
26. Hora, M., Končický, V., Tětek, J.: Theoretical model of computation and algorithms for FPGA-based hardware accelerators. In: Theory and Applications of Models of Computation, pp. 295–312. Springer (2019)
27. Chen, Z.: Hardware accelerator of matrix multiplication on FPGAs. Thesis, Department of Information Technology, Uppsala Universitet, Uppsala, Sweden (2018)

Analysis of the Impact of Watermarking Technique in Neural Network Models to Predict Lung Diseases

Tuan Nguyen-Thanh[1,2](✉) and Kiet Vo-Tuan[1,2]

[1] Ho Chi Minh City University of Technology (HCMUT), Ho Chi Minh City, Vietnam
nttuan@hcmut.edu.vn
[2] Vietnam National University Ho Chi Minh City (VNU-HCM), Ho Chi Minh City, Vietnam

Abstract. With current developments in machine learning, deep learning, and artificial intelligence (AI), significant advances have been achieved across various fields, especially in the healthcare sector. Besides, watermarking techniques are also widely applied to store and secure data, allowing data to be seamlessly embedded into images that are invisible to the human eye, facilitating the transmission of secure messages. However, there is still no analysis and evaluation of the impact of watermarking techniques on the prediction results of models. In this study, the Least Significant Bit-based image watermarking technique is extensively analyzed for 1-bit, 2-bit, and 3-bit embedded planes with 4 different neural network models in predicting lung pathology, including Vanilla, VGG16, Densenet-121, and Capsule. The performance indexes are considered before and after embedding to conclude the impact of embedding techniques on the predictive ability of these models. First, the impact of embedding at different bit levels on the prediction ability of the original models is investigated. To better understand the impact of embedding on the supervised learning process, these models are retrained with 1-bit and 2-bit embedded training data; then the suitable thresholds are determined for each model to help improve the models' ability to classify and predict diseases.

Keywords: image watermarking · neural network models · lung diseases

1 Introduction

Due to climate change, increasing pollution, and lifestyles, various respiratory diseases have emerged, especially during the recent COVID-19 pandemic, which has seen a dramatic rise in the number of respiratory patients. The increasing number of deaths related to pulmonary diseases presents a significant challenge in meeting the diagnostic demands. Currently, with the development of artificial intelligence, there have been numerous applications and models deployed for disease prediction. Neural net-work models have become a valuable tool in predicting and diagnosing diseases in the field of healthcare. Breakthroughs in basic research and applied research on deep learning in healthcare are continuously announced and put into practice [1–6]. In short, deep learning models can assist doctors in the entire clinical examination and treatment process based on medical images. Specifically, they excel at handling large volumes of medical image data, such as lung X-rays, and can aid healthcare professionals in making rapid and accurate diagnoses, even before obvious clinical symptoms appear. This means they can detect early preclinical signs of diseases and predict the risk of illness for individuals, enabling doctors and healthcare personnel to make timely and effective decisions. The use of deep learning models and artificial intelligence in disease diagnosis has significantly saved time and effort for doctors and hospitals, especially during disease outbreaks.

In an increasingly digitized world, watermarking techniques are becoming integral to data security and information storage. This technique allows data to be covertly embedded into images, rendering it imperceptible to the naked eye and facilitating the transmission of confidential messages without easy detection. Applying watermarking techniques in securing and managing data in the healthcare field, particularly when dealing with critical medical images, can ensure data integrity while safeguarding patient privacy [7–11].

In the current context of digitalization, information embedding techniques applied in these technical jobs are also used to store information and secure information. This information embedding technique refers to inserting data, information, or signals into an object or environment. This can be done in a variety of fields, including digital, media, commerce, and many others. This allows data to be inserted into images that are undetectable to the naked eye, which facilitates the transmission of secret messages without being easily perceived.

By combining watermarking techniques in neural network models, this paper conducts an analysis and evaluation of the impact of the amount of embedded information on the predictive capabilities of neural network models in lung disease diagnosis. This necessitates the embedding of multi-bit information into an entire dataset of images, followed by an analysis and comparison of the prediction results of these models before and after information embedding. The ultimate goal of this research is to provide valuable insights into the significance of information embedding techniques and their interaction with neural network models in the domain of lung disease diagnosis. This is a new method for evaluating the effectiveness of information embedding. Instead of relying solely on changes in image quality, evaluating the capabilities of neural network models helps us recognize the impact of embedded information on data classification, prediction, or processing. This method provides further insight into the influence of embedded information. By doing this, we have the opportunity to evaluate the true importance of embedded information in maintaining data integrity and the feasibility of prediction. This method opens up a new approach to better understanding the impact of information embedding techniques and their interaction with neural network systems.

2 Literature Review

2.1 Neural Network Models

CNN is a fundamental and widely used model in artificial intelligence, especially in tasks related to computer vision. The structure of a common CNN network, as shown in Fig. 1, includes convolutional layers, pooling layers, and fully connected layers. With these layers, CNN can easily learn features from input data, such as images, and build a deep learning model to classify different cases.

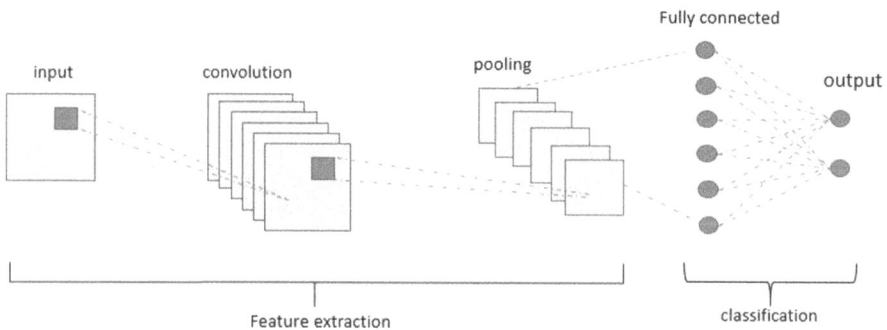

Fig. 1. CNN architecture.

Absolutely, with CNNs, we can easily create deep learning models to perform lung disease predictions. CNNs are highly effective for image classification tasks like diagnosing lung diseases from medical images, as they can automatically learn and extract relevant features from the images, making them a popular choice in medical image analysis. These models can significantly contribute to medical diagnosis and healthcare. Through the analysis of specific characteristics, four CNN models are selected to be employed for lung disease prediction in this study. These models comprise Vanilla, Capsule, Densenet-121, and VGG16.

Vanilla CNN is the term for the most basic type of CNN, without complex components or special architecture. As shown in Fig. 2, it is a simpler model than common CNN with some structural layers removed to facilitate faster learning and prevent overfitting, a phenomenon where a model learns too much from the training data, resulting in accurate predictions only on the training data but failing to generalize well to real-world data. Due to its simple architecture, Vanilla CNN is often used for basic image classification problems or as a foundation for more complex CNN models such as ResNet, Inception, or VGG. Although it is not very effective in complex problems, Vanilla CNN is still an important concept and is the foundation for many deep learning research [12].

In CNN networks, the pooling function (especially the max pooling function) is often used mainly to down-sample (reduce the size) of the feature map, creating a new feature map that generalizes the characteristics of the input. This makes it possible to increase the field-of-view of neurons in higher-level layers, helping them learn higher-level features. In addition, using the pooling function also helps the CNN network identify an object in

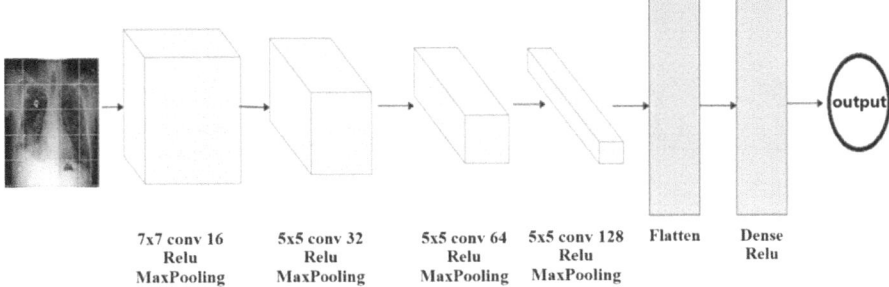

Fig. 2. Vanilla CNN architecture.

different locations by using the good weight obtained at one location with other locations in the input. However, the way the pooling function works also creates its weakness: the pooling function does not keep information about the relative positions between objects.

Thus, to solve the above problem of CNN networks, Hinton et al. [13] proposed using Capsule, a specialized neural network architecture, instead of max pooling. Capsule is defined as a group of neurons whose vector is used to represent the parameters of a certain entity such as an object or part of that object. It works according to the inverse graphics mechanism. In Capsule Net, capsules are groups of neurons organized in pairs, with each capsule assigned the task of identifying specific features in an image. Capsules store information about the orientation and relative relationships between these features.

The structure of the Capsule network is presented in Fig. 3. The encoder part consists of three main layers, including Conv+ReLU, PrimaryCaps, and DigitCaps. Conv+ReLU is responsible for detecting features in the input image. Primarycaps creates capsules with information about the extracted features. DigitCaps is the final layer tasked with predicting the class of the object in the image. The decoder part has the same structure; its purpose is to reconstruct the input image from the information obtained from DigitCaps. It consists of 3 fully connected layers, with the first two layers using ReLU activation functions and the final layer using a Sigmoid activation function. During training, the activity vectors of the capsules that belong to different numbers in the input image will be zero-masked, so that the decoder only recreates the image of the number in the input. This training helps the Capsule network to recreate a new image that is similar to the original image.

With its ability to learn multiple features along with the position and size relationships of objects, the Capsule network can predict various lung images from different patients at various angles and shapes [14, 15].

Densenet-121 is an advanced network derived from the basic CNN architecture [16]. Densenet-121 has a structure as shown in Fig. 4, which extracts features from all previous layers, not just from the nearest previous connection. These characteristics help Densenet-121 reduce the vanishing gradient problem and make better use of features. The structure of Densenet-121 allows it to learn more comprehensively and not miss any extracted features from any layer. In the medical field, where high precision is required, Densenet-121 is suitable for building deep learning models for lung disease prediction [17].

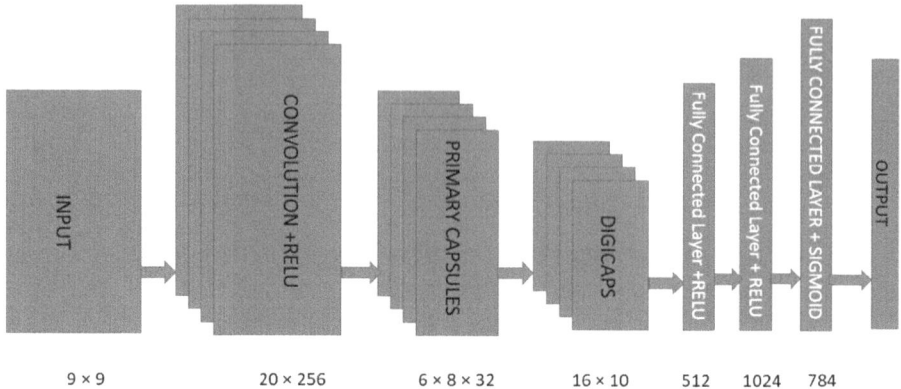

Fig. 3. Capsule network architecture.

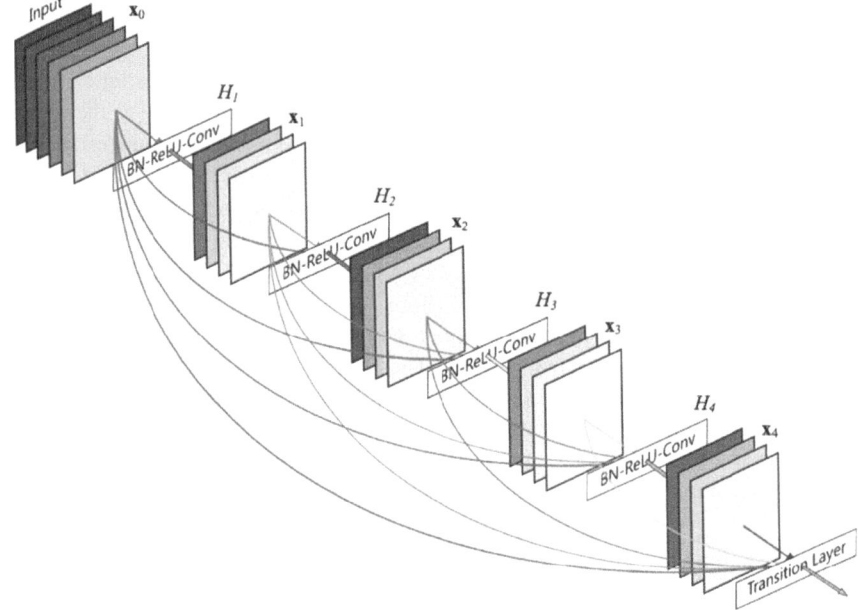

Fig. 4. Features extraction in Densenet-121.

VGG16 (Visual Geometric Group-based neural network) has a deep architecture with a total of 16 layers, as shown in Fig. 5. Among these, there are 13 convolutional layers and 3 fully connected layers (or dense layers) [18].

All the convolutional layers in VGG16 use a 3x3 kernel size. This results in convolutional layers with a dynamic combination of features in the image. Max-pooling layers are employed after each convolutional layer to reduce the size of feature maps and simplify the model's complexity. Finally, following the convolutional layers, VGG16 has three fully connected layers combined with a softmax function to perform the ultimate

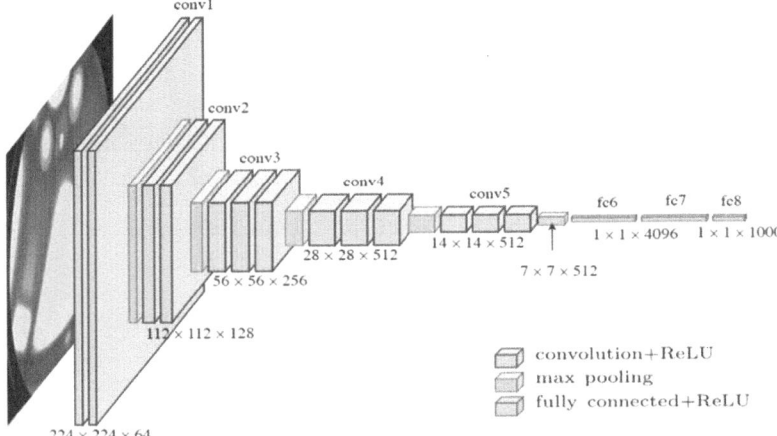

Fig. 5. VGG16 architecture.

classification of the image into different classes. These fully connected layers are often connected to one or more dropout layers to prevent overfitting.

VGG16 has been evaluated to have the ability to learn deeper than conventional CNN architectures thanks to its substantially increased depth. Some diseases require a model to be deep enough to detect and predict the disease accurately. Therefore, VGG16 is also suitable for building deep-learning models for lung disease prediction [19].

2.2 LSB Watermarking Technique

In the early 1990s, watermarking technology began to receive attention and developed rapidly in many areas such as copyright protection, copy control, data authentication, fingerprinting, indexing, broadcast monitoring, and data hiding. The first studies of image watermarking were carried out directly in the spatial domain based on the grayscale value adjustment methods. The LSB (Least Significant Bit) replacement technique was proposed by Schyndel et al. [20] by embedding the watermark as a binary random string into the remaining LSB of the image after 7-bit gray-level histogram compression and using bit sequence comparators to detect the watermark. Other authors embed binary information directly in LSB planes of images. Because LSBs carry little information and have negligible impact on image quality, this technique achieves high embedded information capacity and good imperceptible embedded image [21, 22].

In image processing, each pixel in an image contains information about its color or brightness represented as numbers. These numbers are stored as bits. The least significant bit is the bit with the smallest value and plays the least significant role in the value of the number.

To perform embedding, these pieces of information must be converted into the specified ASCII code. Afterward, from ASCII code, it is converted to decimal, and finally, the last value is in binary form.

Subsequently, we will access the pixel values and convert these values into binary format. The 8-bit pixels will have a value range of [0; 255], equivalent to binary values in the range [00000000; 11111111]. Therefore, we need 8 binary bits to represent pixel values. In the final step, we will take the binary values of the embedded information, meaning that one character of information will have 8 binary values of 0 or 1. From these values of 0 or 1, we will embed them into the least significant bit (LSB) of the pixel, equivalent to 1 character of information requiring 8 pixels to execute as shown in Fig. 6.

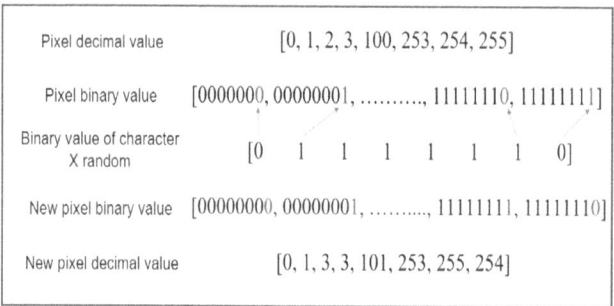

Fig. 6. Embedding of character X.

As per the above process, the final value of the pixel will be replaced with either 0 or 1 depending on the embedded information. Therefore, at each pixel value after embedding the information, this value changes by only 1 bit. This does not significantly affect the image quality or the diagnostic capability.

3 Materials and Methods

In this paper, we use two X-ray lung image datasets: ChestXray-NIHCC from NIH [23] and Pneumonia ChestXray from Kaggle [24]. The ChestXray-NIHCC dataset contains over 100,000 images of 14 different lung diseases as well as images of normal lungs. The Pneumonia ChestXray dataset includes more than 5,000 images of pneumonia cases and normal cases. The data splits for training are detailed in Table 1.

Table 1. Data distribution.

Data	Pneumonia chestXray	ChestXray NIHCC
Disease	4273	60412
Normal	1583	51708
Train	5232	624
Test	90000	22120

The research focuses on using CNN models (Vanilla, Capsule, Densenet-121, VGG16) to predict lung diseases and embedding text information into images to analyze

Analysis of the Impact of Watermarking Technique 315

the impact before and after embedding with LSB method. The research focuses on the impact of embedding 1-bit, 2-bit, and 3-bit planes into lung X-ray images in the prediction process. First, the impact of embedding at different bit levels on prediction ability of the original models is investigated. To better understand the impact of embedding on the supervised learning process, these models are retrained with 1-bit and 2-bit embedded training data; then the suitable thresholds are determined for each model to help improve the models' ability to classify and predict diseases.

Since each model may require different input sizes, we performed image resizing before embedding the information. This is done to assess the actual impact of the embedding technique on predictions without being affected by the resizing function. We analyze the impact of watermarking technique in images with lung diseases as shown in Fig. 7 and without lung diseases as shown in Fig. 8 on neural network models for lung disease prediction.

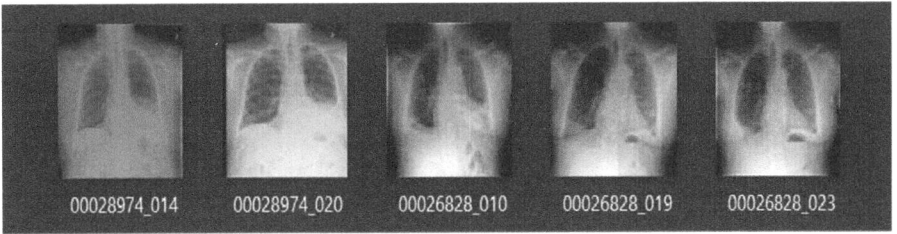

Fig. 7. Disease images.

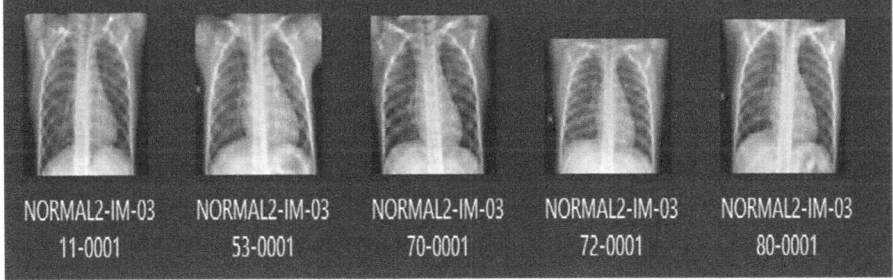

Fig. 8. Normal images.

We embed these images with random text files to provide the most comprehensive results. For each case, we collect the performance indexes before and after embedding, obtained from the predictions of the models.

4 Result and Discussion

4.1 Original Models

In this case, we use the original models to make predictions with the test images in Sect. 2. First, the performance indexes of the original models after training are presented in Table 2.

Table 2. Performance of the original models after training.

Model	Accuracy	Precision	Recall	Fscore	Training time (min)
VaCNN	0.689	0.697	0.611	0.678	13.51
VGG16	0.707	0.685	0.684	0.685	66.97
Densenet-121	0.825	0.879	0.836	0.869	44.35
Capsule	0.627	0.613	0.473	0.569	131.54

The Vanilla CNN model has the fastest training time. It has relatively good accuracy and F-score; however, recall is still not high, showing that there is still potential for improvement. The VGG16 model is quite accurate, but the precision and recall metrics do not reach the same high level. This could suggest that the model may be having difficulty identifying some specific classes. The Densenet-121 model has good accuracy and the precision and recall indexes are also high. This shows that the model has good ability to combine information from multiple layers of the network to make predictions. The Capsule Net model has the worst performance than other models, implying that the model is not effective in capturing all the data that actually belongs to the class to be predicted.

In next step, the confidence scores are collected for test images with the disease and test images without the disease to evaluate the impact of the embedding technique. After embedding 10 different text files, we obtain the maximum and minimum confidence scores of images with the disease and images without the disease as shown in Table 3.

Table 3. Max and min confidence scores in the embedded cases of the original models.

Model	Max_normal_1	Min_disease_1	Max_normal_2	Min_disease_2	Max_normal_3	Min_disease_3
Vanilla	0.5849	0.7074	0.5888	0.7065	0.5942	0.7038
VGG16	0.5832	0.7160	0.5966	0.7129	0.6274	0.6933
Densenet-121	0.5100	0.5864	0.5460	0.5776	0.5935	0.5455
Capsule	0.5627	0.4991	0.5632	0.4977	0.5641	0.4957

Based on the results in Table 3, we reselected the confidence threshold to improve the classification performance of the models and mitigate the impact of the embedding technique. To select suitable threshold T for use in the models in order to ensure prediction reliability after embedding, we need to satisfy the following condition:

$$C_0 < T < C_1 \qquad (1)$$

where C_0 is maximum confidence score in case of normal and C_1 is minimum confidence score in case of disease.

Choosing the appropriate threshold will depend on the difference Δ between C_1 and C_0, determined according to formula (2). If the difference Δ is a positive value and is larger, it is easier to choose an appropriate threshold in order to ensure prediction reliability after embedding.

$$\Delta = C_1 - C_0 \qquad (2)$$

Table 4 shows the difference for threshold selection of embedded cases with the original model. It can be seen that the Vanilla CNN model is least affected by information embedding, followed by VGG16 and Densenet-121. Moreover, 1-bit and 2-bit embeddings do not significantly affect the reliability, as is the case with 3-bit embeddings. However, there is an issue with the Capsule Net model as it cannot select a suitable threshold in order to ensure prediction reliability after embedding because the maximum predicted value for normal is greater than the minimum predicted value for disease. Therefore, to mitigate the impact of the embedding technique on this model, further improvements in model quality are needed.

Table 4. The Δ for threshold selection of embedded cases with the original model.

Model	1-bit	2-bit	3-bit
Vanilla	0.1225	0.1177	0.1096
VGG16	0.1328	0.1163	0.0659
Densenet-121	0.0764	0.0316	−0.0480
Capsule	−0.0636	−0.0655	−0.0684

4.2 Retrained Models with 1-Bit Embedded Training Data

In this case, the models are retrained using the dataset that has been 1-bit embedded to analyze the impact of the embedding technique on the supervised learning of these models and how it affects the disease prediction results. After retraining the models and collecting prediction confidence data from images with the disease and images without the disease, embedding with 10 different text files, we obtain the data as shown in Table 5.

From the results in Table 5, we analyze the appropriate thresholds for each embedding case for the models by determining the difference Δ as shown in Table 6.

In the case of retraining the models from a 1-bit embedded image, the Vanilla CNN model has the ability to perform well in all three embedding cases of 1-bit, 2-bit, and

Table 5. Max and min confidence in the embedded cases of the models which have 1-bit embedded training data.

Model	Max_normal_1	Min_disease_1	Max_normal_2	Min_disease_2	Max_normal_3	Min_disease_3
Vanilla	0.5185	0.6383	0.5238	0.6378	0.5411	0.6268
VGG16	0.6553	0.7461	0.6716	0.7518	0.7022	0.7405
Densenet-121	0.7212	0.7531	0.7477	0.7194	0.7887	0.6971
Capsule	0.5690	0.4930	0.5696	0.4917	0.5706	0.4900

Table 6. The Δ for threshold selection of embedded cases with 1-bit embedded training data.

Model	1-bit	2-bit	3-bit
Vanilla	0.1198	0.114	0.0857
VGG16	0.0908	0.0802	0.0383
Densenet-121	0.0319	−0.0283	−0.0916
Capsule	−0.076	−0.0779	−0.0806

3-bit. The VGG16 model also has good performance in 1-bit and 2-bit embedding cases and is lower in 3-bit embedding cases. The Densenet-121 model only works relatively well in the 1-bit embedding case and is gradually negatively affected in the 2-bit and 3-bit embedding cases. The Capsule model alone does not work well in all 3 cases.

4.3 Retrained Models with 2-Bit Embedded Training Data

In this case, the models are retrained using the dataset that has been 2-bit embedded to analyze the impact of the embedding technique on the supervised learning of these models and how it affects the disease prediction results. After retraining the models and collecting prediction confidence data from images with the disease and images without the disease, embedding with 10 different text files, we obtain the data as shown in Table 7.

Table 7. Max and min confidence in the embedded cases of the models which have 2-bit embedded training dat.

Model	Max_normal_1	Min_disease_1	Max_normal_2	Min_disease_2	Max_normal_3	Min_disease_3
Vanilla	0.5477	0.6669	0.5476	0.6660	0.5485	0.6659
VGG16	0.5752	0.7106	0.5905	0.7107	0.6408	0.6703
Densenet-121	0.7043	0.6138	0.7303	0.5731	0.7701	0.4941
Capsule	0.5632	0.5196	0.5638	0.5182	0.5647	0.5159

From the results in Table 7, we analyze the appropriate thresholds for each embedding case for the models by determining the difference Δ as shown in Table 8.

Table 8. The Δ for threshold selection of embedded cases with 2-bit embedded training data.

Model	1-bit	2-bit	3-bit
Vanilla	0.1192	0.1184	0.1174
VGG16	0.1354	0.1202	0.0295
Densenet-121	−0.0905	−0.1572	−0.2760
Capsule	−0.0436	−0.0456	−0.0488

In the case of retraining the models from a 2-bit embedded image, the Vanilla CNN model still has the ability to perform well in all three embedding cases of 1-bit, 2-bit, and 3-bit. The VGG16 model also has good performance in 1-bit and 2-bit embedding cases and is lower in 3-bit embedding cases. Meanwhile, both Densenet-121 and Capsule Net models do not perform well in all three embedding cases of 1-bit, 2-bit, and 3-bit. In particular, the Densenet-121 model is significantly affected by increasing the embedded bit plane.

5 Conclusion

This paper has successfully analyzed and evaluated the technique of image watermarking in neural network models (Vanilla, VGG16, Densenet-121, and Capsule) for diagnosing lung diseases. By embedding 1-bit, 2-bit, and 3-bit planes in both disease and non-disease cases in the original models, we can observe that the influence of embedding a 3-bit plane is the most significant and greater than that of embedding 1-bit and 2-bit planes. Therefore, it can be concluded that embedding higher-order bits from 3 bits onward can significantly affect image quality and, consequently, alter the prediction scores of the models. We also embed 1-bit and 2-bit planes into the training image data and retrain the models. From the experimental results, it is still possible to select the threshold to minimize the impact of data embedding after retraining the Vanilla and VGG16 models with 1-bit and 2-bit embedded training data. However, retraining the model with 1-bit and 2-bit embedded training data makes it more challenging to choose thresholds with Densenet-121. In many cases, the threshold selection criteria are still not met with Capsule model. Therefore, it's necessary to improve the model's quality before conducting further investigations involving embedding more information capacity.

Acknowledgment. This research is funded by Vietnam National University Ho Chi Minh City (VNU-HCM) under grant number C2022-20-12. We would like to thank Ho Chi Minh City University of Technology (HCMUT), Vietnam National University Ho Chi Minh City (VNU-HCM) for the support of time and facilities for this study.

References

1. Al-qaness, M.A.A., Zhu, J., AL-Alimi, D., et al.: Chest X-ray images for lung disease detection using deep learning techniques: a comprehensive survey. Arch. Comput. Methods Eng. **31**, 3267–3301 (2024)

2. Vinta, S.R., Lakshmi, B., Safali, M.A., Kumar, G.S.C.: Segmentation and classification of interstitial lung diseases based on hybrid deep learning network model. IEEE Access **12**, 50444–50458 (2024)
3. Badawy, M., Ramadan, N., Hefny, H.A.: Healthcare predictive analytics using machine learning and deep learning techniques: a survey. J. Electr. Syst. Inf. Technol. **40**(10) (2023)
4. Arumugam, K., Naved, M., Priyanka, P.S., Orlando, L.C., Antonio, H.O., Tatiana, G.Y.: Multiple disease prediction using machine learning algorithms. Mater. Today Proc. **80**(3), 3682–3685 (2023)
5. Nia, N.G., Kaplanoglu, E., Nasab, A.E.: Evaluation of artificial intelligence techniques in disease diagnosis and prediction. Discov. Artif. Intell. **3**(5) (2023)
6. Kermany, D.S., et al.: Identifying medical diagnoses and treatable diseases by image-based deep learning. Cell **172**(5), 1122–1131 (2018)
7. Gull, S., Parah, S.A.: Advances in medical image watermarking: a state of the art review. Multimedia Tools Appl. **83**, 1407–1447 (2024)
8. Ayuba, S., Zainon, W.M.N.W.: Medical image watermarking: a survey on applications, approach and performance requirement compliance. Int. J. Multimedia Inf. Retr. **12**(33) (2023)
9. Ahmadi, S.B.B., Zhang, G., Wei, S., Boukela, L.: An intelligent and blind image watermarking scheme based on hybrid SVD transforms using human visual system characteristics. Vis. Comput. **37**(2), 385–409 (2020). https://doi.org/10.1007/s00371-020-01808-6
10. Liu, J., Ma, J., Li, J., Huang, M., Sadiq, N., Ai, Y.: Robust watermarking algorithm for medical volume data in internet of medical things. IEEE Access **8**, 93939–93961 (2020)
11. Liu, J., et al.: A novel robust watermarking algorithm for encrypted medical image based on DTCWT-DCT and chaotic map. Comput. Mater. Continua **61**(2), 889–910 (2019)
12. Bharati, S., Podder, P., Mondal, M.R.: Hybrid deep learning for detecting lung diseases from X-ray images. Inform. Med. Unlocked **20** (2020)
13. Hinton, G.E., Krizhevsky, A., Wang, S.D.: Transforming auto-encoders. In: ICANN 2011, Part I, LNCS, vol. 6791, pp. 44–51. Springer, Heidelberg (2011)
14. Mensah, K.P., Adebayo, F.A., Ayidzoe, A.M., Baagyire, Y.E.: Capsule networks – a survey. J. King Saud Univ. Comput. Inf. Sci. **34**(1), 1295–1310 (2022)
15. Yadav, S., Dhage, S.: TE-CapsNet: time efficient capsule network for automatic disease classification from medical images. Multimedia Tools Appl. **83**, 49389–49418 (2024)
16. Huang, G., Liu, Z., Van Der Maaten, L., Weinberger, K.Q.: Densely connected convolutional networks. In: Proceedings of the IEEE Conference on Computer Vision and Pattern Recognition, pp. 4700–4708 (2017)
17. Chutia, U., Tewari, A.S., Singh, J.P., et al.: Classification of lung diseases using an attention-based modified densenet model. J. Imaging Inform. Med. **37**, 1625–1641 (2024)
18. Simonyan, K., Zisserman, A.: Very deep convolutional networks for large-scale image recognition. In: 3rd International Conference on Learning Representations, pp. 1–14 (2015)
19. Gehani, R., Victor, A.: X-ray based lung disease classification using fine-tuned VGG16 model. In: 2nd International Conference on Intelligent Data Communication Technologies and Internet of Things (IDCIoT), pp. 533–538 (2024)
20. Schyndel, R.G.V., Tirkel, A.Z., Osborne, C.F.: A digital watermark. In: Proceedings of IEEE International Conference on Image Processing, pp. 86–90 (1994)
21. Benyoucef, A., Hamadouche, M.: RONI-based medical image watermarking using DWT and LSB algorithms. In: Lejdel, B., Clementini, E., Alarabi, L. (eds.) Artificial Intelligence and Its Applications, AIAP 2021. LNNS, vol. 413. Springer, Cham (2021)
22. Nazari, M., Mehrabian, M.: A novel chaotic IWT-LSB blind watermarking approach with flexible capacity for secure transmission of authenticated medical images. Multimedia Tools Appl. **80**, 10615–10655 (2021)

23. "CXR8 ChestXray-NIHCC" dataset. https://nihcc.app.box.com/v/ChestXray-NIHCC
24. "Chest X-Ray Images (Pneumonia)" dataset. https://www.kaggle.com/datasets/paultimothymooney/chest-xray-pneumonia

Adaptive Genetic Algorithm Based LQR for Optimal Control of Nonlinear Double Pendulum Gantry Crane

Mohamed O. Elhabib[1], Herman Wahid[1(✉)], Zaharuddin Mohamed[1], and Hussein Shutari[2]

[1] Control and Mechatronics Engineering Division, Faculty of Electrical Engineering, Universiti Teknologi Malaysia, 81310 Johor Bahru, Malaysia
herman@utm.my
[2] School of Electrical and Electronic Engineering, Universiti Sains Malaysia, 14300 Nibong Tebal, Penang, Malaysia

Abstract. Gantry cranes are widely employed to handle heavy loads in construction projects and critical industries such as petrochemical and nuclear power stations. Achieving precise trolley positioning and minimizing sway oscillations are crucial for ensuring safety and operational efficiency. This paper presents an optimal control strategy based on a Linear Quadratic Regulator (LQR), where controller performance is highly sensitive to the selection of its weighting matrices, Q and R. Determining these matrices is often a challenging and time-consuming task. To address this issue, an Adaptive Genetic Algorithm (AGA) is applied to automatically compute the LQR parameters for a nonlinear double pendulum crane model implemented in MATLAB. Simulation results reveal that the proposed AGA-based LQR controller outperforms a conventional LQR tuned with a standard Genetic Algorithm (GA), delivering robust performance across a range of payload masses.

Keywords: Adaptive genetic algorithm · LQR · Double pendulum gantry crane · Optimal control

1 Introduction

Gantry cranes have become indispensable in contemporary industrial operations, serving as critical equipment in sectors such as construction, maritime logistics, and port management. They excel at transporting heavy and hazardous materials with both efficiency and precision [1]. Typically, a crane consists of two key components: a hoisting mechanism-incorporating the cable and hook-and a supporting structure, which can range from a trolley-girder configuration to a trolley-jib or boom arrangement. By coordinating these elements, the crane can maneuver suspended payloads throughout its operational area, enabling the system to navigate around obstacles and position loads with remarkable accuracy [2]. Despite these advantages, cranes present considerable challenges related to their inherent nonlinear and underactuated dynamics. A single control input, generally the force applied to the trolley, must regulate multiple outputs, including the trolley's displacement and the angles of the hook and the payload, respectively. As the trolley is moved, the payload tends to oscillate, which can destabilize the overall system and increase the likelihood of accidents [3]. Hence, ensuring precise trolley positioning and minimizing load sway are paramount for enhancing safety and operational efficiency. The persistent frequency of crane-related workplace incidents underscores the importance of developing robust, effective control strategies.

Over the past few decades, various control methodologies have been explored to address these issues. Singhose [4] introduced input shaping to mitigate residual and transient oscillations in planar gantry cranes with hoisting mechanisms, markedly reducing unwanted vibrations. Huang et al. [5] proposed a command smoothing approach tailored to double pendulum bridge cranes, achieving stable and robust sway suppression. Similarly, Chen et al. [6] combined sliding mode control and input shaping in a hybrid robust control framework that provided precise trolley positioning and minimized payload sway, while also reducing chattering. Sun et al. [7] applied adaptive control techniques to underactuated cranes, demonstrating accurate cart positioning and effective attenuation of residual swings even under parametric uncertainties.

Among the various optimal control methods, the Linear Quadratic Regulator (LQR) has shown particular promise for crane control applications. Corriga [8] employed a gain-scheduling LQR strategy to handle varying cable lengths, successfully damping initial disturbances, though the approach faced certain limitations in steady-state accuracy and response speed. Adeli [9] utilized a Takagi-Sugeno fuzzy model in conjunction with LQR for position regulation and swing reduction, achieving superior results compared to conventional fuzzy controllers. Zawawi [10] conducted comparative analyses of LQR, Proportional-Derivative (PD), and Direct Feedback Stabilization (DFS) methods, highlighting LQR's advantages in limiting overshoot and reducing settling time under varying load conditions. However, a key drawback of LQR lies in determining the weighting matrices Q and R, a process that often relies on trial-and-error and can be both time-consuming and suboptimal. Although soft computing techniques have been introduced to refine matrix selection, consistently identifying the best configuration remains a complex challenge. This paper addresses these issues by applying an Adaptive Genetic Algorithm (AGA) to optimize the LQR weighting matrices. Genetic algorithms, rooted in evolutionary principles, effectively search complex optimization spaces, while adaptive probabilities of crossover and mutation further enhance their ability to balance exploration and exploitation. Leveraging the strengths of AGAs allows this study to overcome long-standing tuning difficulties in LQR-based designs. The proposed approach aims to achieve more precise trolley positioning and efficient sway suppression in a gantry crane system. Simulation results validate the superiority of the proposed controller compared to conventional techniques.

2 Mathematical Modeling of Double Pendulum Crane System

Figure 1 shows a double pendulum gantry crane system, in which horizontal motion is induced by an external force F acting on the trolley. The system consists of three main components: the trolley, the hook (the first pendulum), and the payload (the second pendulum). Its configuration is described by three generalized coordinates: the horizontal position of the trolley x (in meters), the hook angle θ_1 (in radians) relative to the vertical, and the payload angle θ_2 (in radians) relative to the vertical.

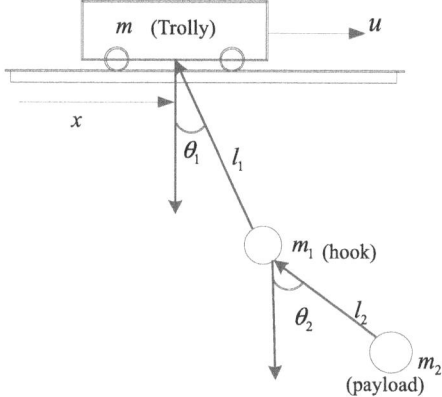

Fig. 1. Gantry Crane System with Double Pendulum

In Fig. 1, m (in kg) represents the mass of the trolley, while m_1 (in kg) and m_2 (in kg) denote the masses of the hook and payload, respectively. The lengths L_1 (in m) and L_2 (in m) are the distances from the trolley to the hook and from the hook to the payload, respectively. The following modeling assumptions are made:

- Nonlinear friction forces are neglected.
- The hook and payload are treated as point masses.
- The connecting rods or cables are massless and rigid.
- The motion is confined to a vertical plane.

To apply the Lagrangian approach is employed to the model, let L The Lagrangian is defined as the difference between the kinetic energy K and the potential energy P:

$$L = K - P \tag{1}$$

The equations of motion are obtained from Lagrange's equations:

$$\frac{d}{dt}\left(\frac{\partial L}{\partial \dot{q}_i}\right) - \frac{\partial L}{\partial q_i} = Q_i \tag{2}$$

where q_i represents the generalized coordinates (x, θ_1, θ_2) and Q_i denotes the generalized forces corresponding to each q_i.

The kinetic energy K of the system is the sum of the kinetic energies of the trolley, the hook mass, and the payload mass:

$$K = \frac{1}{2}M\dot{x}^2 + \frac{1}{2}m_1 v_1^2 + \frac{1}{2}m_2 v_2^2 \qquad (3)$$

Here, $\dot{x}$ is the horizontal velocity of the trolley, v_1 is the velocity of the hook mass, and v_2 is the velocity of the payload mass.

Because the trolley moves only in a horizontal plane, its potential energy remains constant. Thus, the total system potential energy depends solely on the hook and payload:

$$P = m_1 g L_1 (1 - \cos\theta_1) + m_2 g \left[L_1 (1 - \cos\theta_1) + L_2 (1 - \cos\theta_2)\right] \qquad (4)$$

Substituting the expressions for kinetic and potential energy, the Lagrangian L becomes:

$$\begin{aligned}&\frac{1}{2}M\dot{x}^2 + \frac{1}{2}m_1 v_1^2 + \frac{1}{2}m_2 v_2^2 - m_1 g L_1 (1 - \cos\theta_1) \\ &- m_2 g \left[L_1 (1 - \cos\theta_1) - L_2 (1 - \cos\theta_2)\right]\end{aligned} \qquad (5)$$

Applying Lagrange's equation with respect to x:

$$\begin{aligned}(M + m_1 + m_2)\ddot{x} + (m_1 + m_2) L_1 \ddot{\theta}_1 \cos\theta_1 - (m_1 + m_2) L_1 \dot{\theta}_1^2 \sin\theta_1 \\ + m_2 L_2 \ddot{\theta}_2 \cos\theta_2 - m_2 L_2 \dot{\theta}_2^2 \sin\theta_2 = F\end{aligned} \qquad (6)$$

With respect to θ_1:

$$\begin{aligned}(m_1 + m_2) L_1 \ddot{x} \cos\theta_1 + (m_1 + m_2) L_1^2 \ddot{\theta}_1 + m_2 L_1 L_2 \ddot{\theta}_2 \cos(\theta_1 - \theta_2) \\ + m_2 L_1 L_2 \dot{\theta}_2^2 \sin(\theta_1 - \theta_2) + (m_1 + m_2) g L_1 \sin\theta_1 = 0\end{aligned} \qquad (7)$$

With respect to θ_2:

$$\begin{aligned}m_2 L_2 \ddot{x} \cos\theta_2 + m_2 L_2^2 \ddot{\theta}_2 + m_2 L_1 L_2 \ddot{\theta}_1 \cos(\theta_1 - \theta_2) - m_2 L_1 L_2 \dot{\theta}_1^2 \\ \sin(\theta_1 - \theta_2) + m_2 g L_2 \sin\theta_2 = 0\end{aligned} \qquad (8)$$

Equations (6), (7), and (8) capture the nonlinear dynamic behavior of the double pendulum gantry crane system.

3 LQR Controller Design

An optimal control strategy is crucial for achieving precise trolley positioning and effectively suppressing swings in a double pendulum gantry crane system. In this study, we employ a Linear Quadratic Regulator (LQR) due to its proven capability in handling multi-variable dynamics and delivering optimal control actions.

The LQR approach is applicable to linear time-invariant (LTI) systems described in state-space form:

$$\dot{\mathbf{x}} = \mathbf{A}\mathbf{x} + \mathbf{B}u$$

Here, $\mathbf{x} \in \mathbb{R}^n$ is the state vector, $u \in \mathbb{R}^m$ is the control input, $\mathbf{A}$ is the system matrix, and $\mathbf{B}$ is the input matrix. The objective is to determine a state feedback control law that minimizes the following quadratic cost function [11]:

$$J = \int_0^\infty \left(\mathbf{x}^T Q \mathbf{x} + u^T R u\right) dt \tag{9}$$

where $Q \geq 0$ is the state weighting matrix and $R > 0$ is the control weighting scalar. Appropriate selection of Q and R ensures an optimal balance between system performance and control effort.

$\mathbf{K}$ defined as The matrix of optimal gain and can be determined by solving the continuous-time Algebraic Riccati Equation (ARE):

$$\mathbf{A}^T \mathbf{P} + \mathbf{P}\mathbf{A} - \mathbf{P}\mathbf{B}\mathbf{R}^{-1}\mathbf{B}^T\mathbf{P} + \mathbf{Q} = 0$$

Here, $\mathbf{P}$ is the positive definite solution of the ARE. Once $\mathbf{P}$ is determined, the gain matrix $\mathbf{K}$ is given by:

$$\mathbf{K} = \mathbf{R}^{-1}\mathbf{B}^T\mathbf{P}$$

This leads to the optimal control law:

$$u = -\mathbf{K}\mathbf{x} \tag{10}$$

Q and R matrices play a critical role in the LQR optimization process, exerting a significant influence on overall system performance. The dimensionality of Q and R is determined by the number of state variables $[x, \dot{x}, \theta_1, \dot{\theta}_1, \theta_2, \dot{\theta}_2]$ and the number of input variables $[F]$, respectively. When these weighting matrices are chosen to be diagonal, the quadratic performance index simplifies to a weighted integral of the squared states and inputs. Traditionally, selecting the elements of Q and R has relied on manual trial-and-error, which often fails to yield optimal performance. To overcome this limitation, a bio-inspired optimization method-namely, the Adaptive Genetic Algorithm (AGA)-is integrated into the LQR design framework to automatically and optimally determine the Q and R matrices.

4 Adaptive Genetic Algorithm Optimization

Genetic Algorithms (GAs) are a class of optimization techniques inspired by the principles of natural evolution. They operate by generating a population of candidate solutions that iteratively evolve toward improved performance through the processes of selection, crossover (recombination), and mutation [12,13]. In the context of control system tuning, GAs have proven effective in identifying parameter values that enhance system performance [14]. Each candidate solution is evaluated using a fitness function that measures how well it meets the desired criteria and select the solutions with higher fitness values the next generations, ensuring that advantageous characteristics are retained and refined over time. However, traditional GAs maintain fixed crossover (p_c) and mutation (p_m) rates throughout the optimization process. While this approach can yield acceptable results, it also has notable drawbacks. For example, fixed rates may lead to premature convergence on suboptimal solutions, introduce unnecessary randomness in the search, and limit the algorithm's ability to balance exploration of new solutions against the exploitation of high-quality ones. Such flexibility is vital for efficiently navigating complex, multimodal optimization landscapes [15].

Adaptive Genetic Algorithms (AGAs) address these limitations by dynamically adjusting the probabilities of crossover and mutation based on the fitness of each solution in the population. This adaptability helps the algorithm avoid premature convergence and ensures a balanced search across the solution space. In an AGA [16], crossover and mutation rates are inversely related to the fitness of solutions, allowing high-fitness solutions to be preserved while encouraging exploration in low-fitness areas. This adaptability makes AGA particularly effective for challenging optimization tasks, such as tuning a Linear Quadratic Regulator (LQR) for complex, nonlinear systems like the double pendulum gantry crane. The following formulas are used by the AGA to adjust its crossover operator and mutation rate as well. [16]:

$$p_c = k_1 \frac{f_{\max} - f'}{f_{\max} - \overline{f}}, \quad p_c = k_3, \; f' < \overline{f}$$

$$p_m = k_2 \frac{f_{\max} - f}{f_{\max} - \overline{f}}, \quad p_m = k_4, \; f < \overline{f}$$

where: - $f_{\max}$ is the highest fitness value in the population, - f is the fitness of an individual solution, - $\overline{f}$ is the population's average fitness, - f' represents the higher fitness value among two parent solutions in crossover.

Constants k_1, k_2, k_3, and k_4 are used to set bounds on the adaptive rates of crossover and mutation, typically within $[0, 1]$. These parameters ensure that the AGA's adaptability is controlled, allowing for effective diversity without excessive disruption.

AGA Implementation Steps for LQR Tuning:

1. Encoding and Initialization: In the tuning process for an LQR controller, the parameters to be optimized (such as weighting matrices Q and R) are encoded as genes within a population of candidate solutions.
2. Fitness Evaluation: Each candidate solution is evaluated through simulation using the corresponding LQR parameters. The objective is to minimize the following cost function:

$$FitnessFunction = \int_0^{T_{sim}} t|x_d - x_a|dt + \int_0^{T_{sim}} |\theta_1|dt + \int_0^{T_{sim}} |\theta_2|dt \quad (11)$$

Here, T_{sim} is the total simulation time, while x_d and x_a are desired and actual positions of the trolly, respectively. θ_1 and θ_2 are sway angles of the hook and payload.

3. Adaptive Crossover and Mutation: AGA adjusts crossover and mutation probabilities based on fitness. For high-fitness solutions that effectively stabilize the system, p_c and p_m are reduced, preserving promising solutions. Low-fitness solutions experience higher mutation and crossover rates, which facilitates exploration.
4. Selection and Replacement: The optimization procedure prioritized solutions with high fitness for selection, ensuring that the most promising candidates shaped subsequent generations. The process ran for 100 generations in MATLAB, using a population size of 50 and the binary tournament method for selection. Table 1 presents the optimized LQR parameters obtained under these conditions. To enhance the realism of the study, the simulation employed actual parameter values from the Inteco Gantry Crane System, detailed in Table 2.

Table 1. LQR Parameters

	Parameter	Value
LQR-AGA	Q	$diag[0.51, 0.04, 0.5, 0.0013, 0.62, 0.012]$
	R	0.0049
	K	$[10.19, 4.49, -24.72, 1.101, 3.19, -1.48]$
LQR-GA	Q	$diag[0.342, 0.001, 0.0057, 0.0235, 0.0187, 0.049]$
	R	0.0061
	K	$[7.46, 2.886, -31.11, -0.041, 11.95, -2.051]$

Table 2. Gantry crane parameter values

Symbol	Numerical Value
m (cart mass)	1.155 kg
m_1 (Hook Mass)	0.2 Kg
m_2 (Payload Mass)	0.1 kg
L_1	0.4 m
L_2	0.2 m
b	72 Ns/m
g	9.8 m/s^2

5 Simulation Results and Discussion

This section details the implementation and evaluation of the proposed control scheme in the MATLAB/Simulink environment. First, a nonlinear dynamic model of the double pendulum gantry crane system is constructed from the previously derived equations of motion. This model serves as a realistic testbed for examining the system's performance under operational conditions.

In addition to selecting appropriate Q and R matrices, a key challenge in designing the LQR controller is obtaining an accurate linear approximation of the inherently nonlinear system. To address this, the MATLAB Linearization App is used to derive a state-space model around the crane's operating points. This procedure ensures the extraction of a high-fidelity linear model, which is essential for designing an effective controller. To confirm the accuracy of the linearized model, trolley position, hook angle, and payload angle responses are compared against those of the original nonlinear system. Specifically, Fig. 2 and Fig. 3 present these comparisons under a 0.6 N force applied for the first 2 s. The close agreement between the linearized and nonlinear responses confirms that the state-space model accurately captures the system's dynamics.

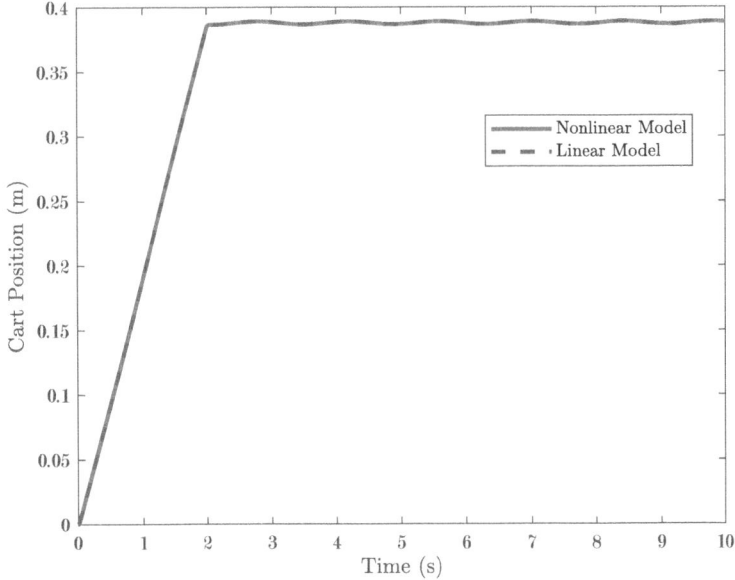

Fig. 2. Response of gantry crane without controller: trolley position

After that, LQR is designed based on the obtained linear model to accurately position the trolley to desired position and eliminate sway angles of hook and payload and the weighting matrices are determined using AGA for optimal performance. Then, the controller is then applied to the nonlinear model in Simulink as in Fig. 4.

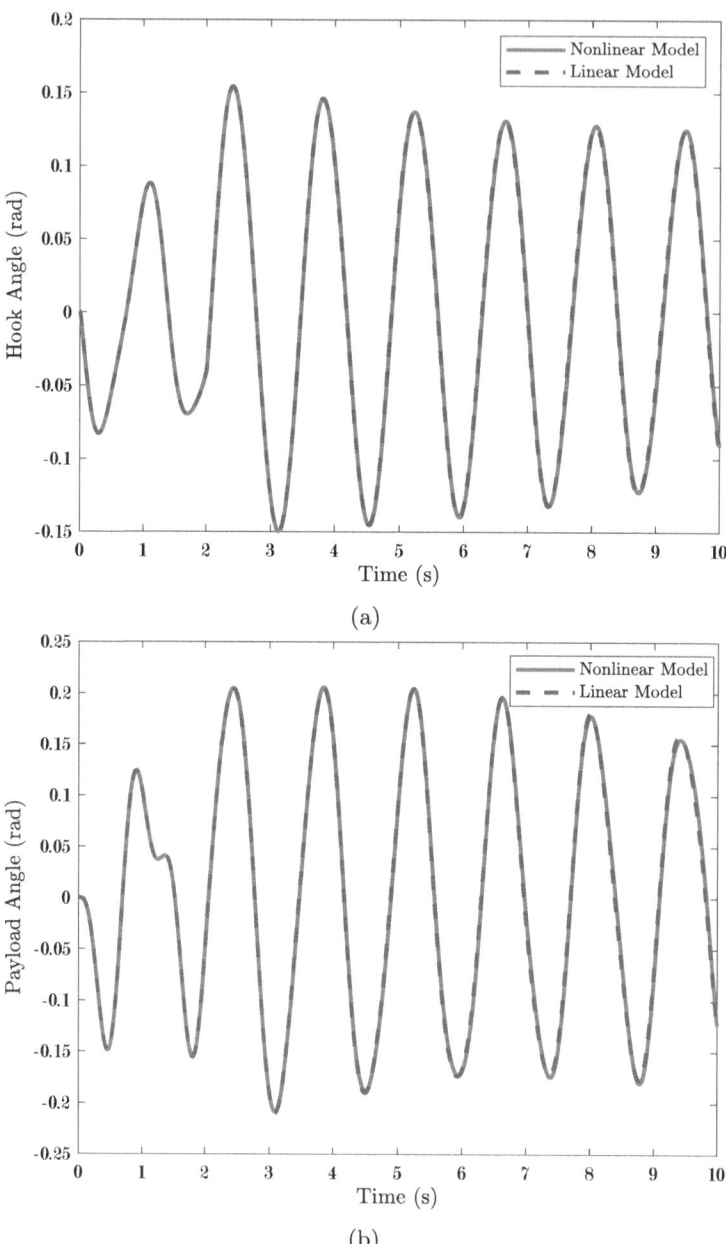

Fig. 3. Response of gantry crane without controller: hook and payload oscillations

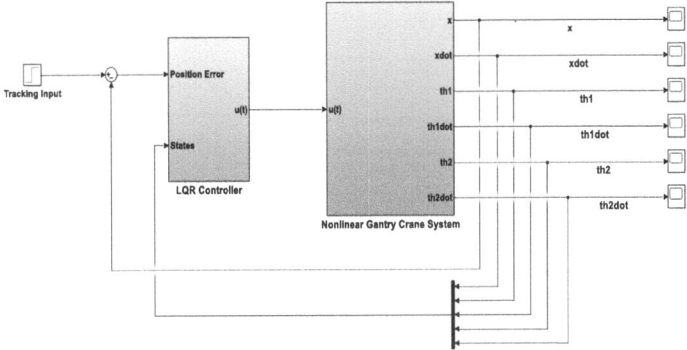

Fig. 4. Simulink diagram of LQR with nonlinear double pendulum crane

Figure 5 and 6 illustrate the system responses under the LQR controller optimized using a Genetic Algorithm (LQR-GA) and the Adaptive Genetic Algorithm-based LQR (LQR-AGA), respectively. The comparative analysis focuses on key performance metrics such as settling time, overshoot, and oscillation amplitudes of the trolley, hook, and payload.

For the LQR-AGA controller, the trolley reached steady state within approximately 1.92 s with no observable overshoot, indicating a swift and stable positioning performance. The maximum sway angle of the hook was limited to 0.065 rad, settling around 2.2 s, while the payload exhibited a peak oscillation amplitude of 0.12 rad. These results suggest that the LQR-AGA effectively minimizes oscillations, enhancing overall system stability. In contrast, the LQR-GA controller resulted in a trolley settling time of about 2.5 s, which is noticeably longer than that achieved with the LQR-AGA controller. Additionally, the maximum sway angle of the hook was higher at 0.08 rad, and the payload oscillations peaked at 0.145 rad. These larger oscillation amplitudes indicate that the LQR-GA controller is less effective in suppressing swings compared to the LQR-AGA controller.

To further evaluate the controller's effectiveness in reducing oscillations, simulations were conducted with varying payload masses. In real world applications, the hook mass (m_1) is typically constant, while the payload mass (m_2) varies depending on the load being handled. In this study, the payload mass was increased incrementally to 0.2 kg and 0.4 kg to assess the controller's performance under different loading conditions.

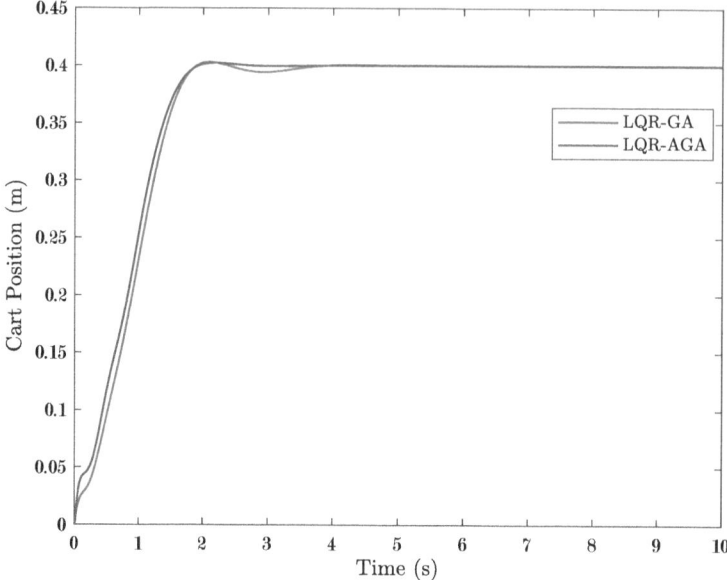

Fig. 5. LQR controller response of gantry crane: trolley position

Table 3 summarizes the system responses for these varying masses, highlighting the controller's capability to maintain stability and suppress oscillations despite changes in payload.

Table 3. Performance index of the system against various masses

	Hook Angle		Payload Angle	
	θ_{max} (rad)	Settling Time (seconds)	θ_{max} (rad)	Settling Time (seconds)
LQR-AGA (m = 2 kg)	−0.078	2.8	−0.11	2.4
LQR-GA	−0.083	2.5	−0.14	2.3
LQR-AGA (m = 4 kg)	−0.081	2.5	−0.085	2.7
LQR-GA	−0.1 rad	2.3	−0.13	2.4

As depicted in Fig. 7, the LQR-AGA controller consistently outperformed the LQR-GA controller in suppressing oscillations for both hook and payload angles across different payload masses. For a payload mass of 2 kg, the LQR-AGA achieved a maximum hook angle of −0.078 rad with a settling time of 2.8 s, and a maximum payload angle of −0.11 rad with a settling time of 2.4 s. In contrast, the LQR-GA exhibited slightly higher maximum angles and shorter settling times, indicating quicker responses but less precision in oscillation suppression.

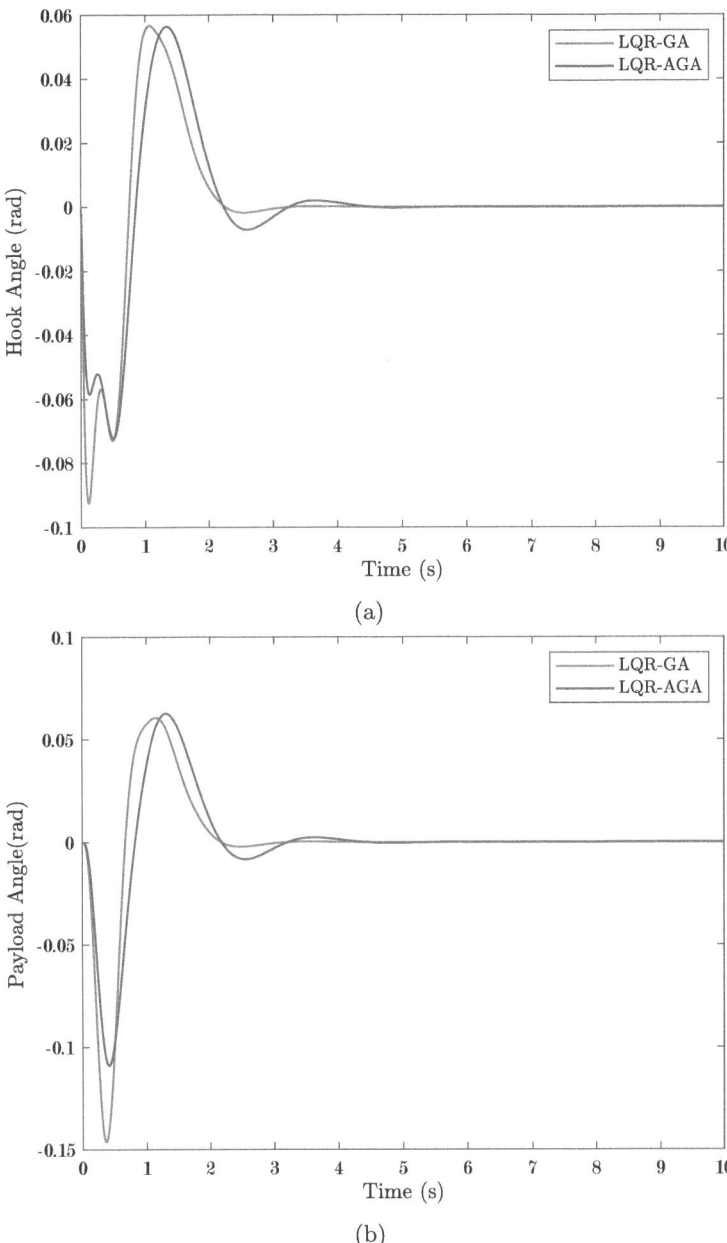

Fig. 6. LQR controller response of gantry crane: hook and payload oscillations

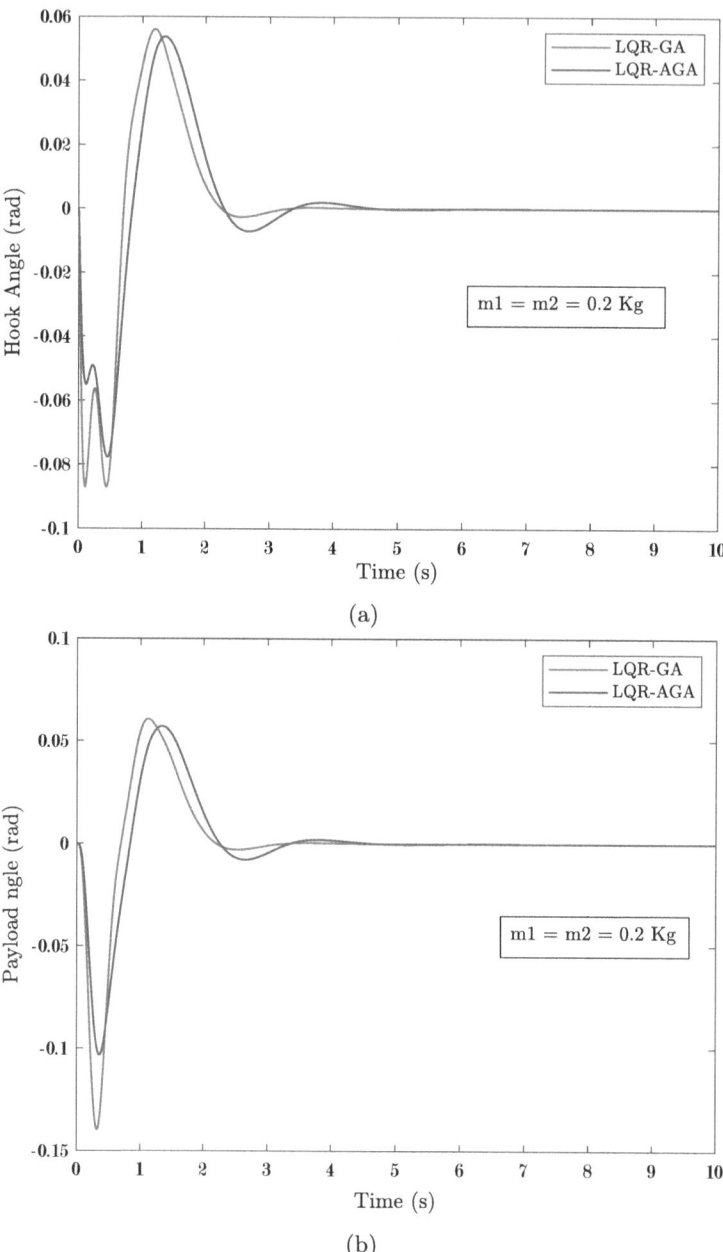

Fig. 7. LQR Step response of double pendulum crane: hook and payload oscillations ($m_2 = 0.2$ Kg)

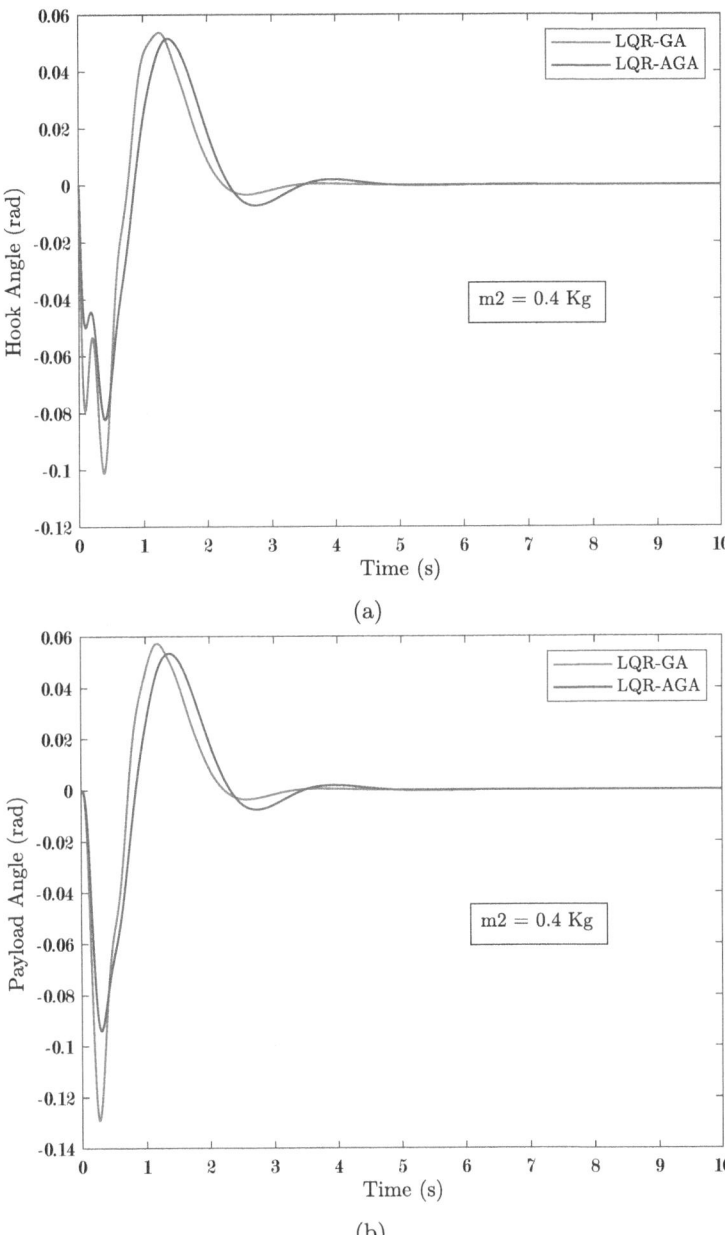

Fig. 8. LQR Step response of double pendulum crane: hook and payload oscillations ($m_2 = 0.4$ Kg)

When the payload mass was increased to 4 kg as in Fig. 8, the LQR-AGA maintained superior performance, with a maximum hook angle of -0.081 rad and a payload angle of -0.085 rad. The settling times remained relatively stable, demonstrating the controller's robustness against changes in payload mass. The LQR-GA, however, showed increased maximum angles of -0.10 rad for the hook and -0.13 rad for the payload, suggesting reduced effectiveness under heavier loads.

These results highlight the LQR-AGA controller's capability to maintain stability and effectively suppress oscillations despite variations in payload mass. The consistent performance across different loading conditions is crucial for ensuring safety and efficiency in crane operations within construction environments. The improved suppression of oscillations minimizes the risk of payload swing, thereby enhancing operational precision and reducing potential hazards.

6 Conclusion

In this work, we proposed a Linear Quadratic Regulator (LQR) controller optimized using an Adaptive Genetic Algorithm (AGA) for a nonlinear double pendulum gantry crane system. The primary objective was to maintain accurate trolley positioning while effectively suppressing sway angles of the hook and payload. We began by deriving the system's mathematical model through the Lagrangian method, followed by the design of the LQR controller based AGA. The complete framework was implemented and tested in the MATLAB/Simulink environment. Simulation results indicated that the proposed controller successfully fulfilled the desired control objectives and delivered superior performance compared to a conventional LQR controller tuned with a standard Genetic Algorithm.

Acknowledgment. This work is supported by the Universiti Teknologi Malaysia under UTMFR grant, with vote No: 22H59.

References

1. Hyla, P.: The crane control systems: a survey. In: 2012 17th International Conference on Methods & Models in Automation & Robotics (MMAR), pp. 505–509. IEEE (2012)
2. Abdel-Rahman, E.M., Nayfeh, A.H., Masoud, Z.N.: Dynamics and control of cranes: a review. J. Vib. Control **9**(7), 863–908 (2003)
3. Ramli, L., Mohamed, Z., Abdullahi, A.M., Jaafar, H.I., Lazim, I.M.: Control strategies for crane systems: a comprehensive review. Mech. Syst. Signal Process. **95**, 1–23 (2017)
4. Singhose, W., Porter, L., Kenison, M., Kriikku, E.: Effects of hoisting on the input shaping control of gantry cranes. Control. Eng. Pract. **8**(10), 1159–1165 (2000)
5. Huang, J., Liang, Z., Zang, Q.: Dynamics and swing control of double-pendulum bridge cranes with distributed-mass beams. Mech. Syst. Signal Process. **54**, 357–366 (2015)

6. Chen, Z.M., Meng, W.J., Zhao, M.H., Zhang, J.G.: Hybrid robust control for gantry crane system. Appl. Mech. Mater. **29**, 2082–2088 (2010)
7. Sun, N., Fang, Y., Chen, H.: Adaptive control of underactuated crane systems subject to bridge length limitation and parametric uncertainties. In: Proceedings of the 33rd Chinese Control Conference, pp. 3568–3573. IEEE (2014)
8. Corriga, G., Giua, A., Usai, G.: An implicit gain-scheduling controller for cranes. IEEE Trans. Control Syst. Technol. **6**(1), 15–20 (1998)
9. Adeli, M., Zarabadipour, H., Zarabadi, S.H., Shoorehdeli, M.A.: Anti-swing control for a double-pendulum-type overhead crane via parallel distributed fuzzy LQR controller combined with genetic fuzzy rule set selection. In: 2011 IEEE International Conference on Control System, Computing and Engineering, pp. 306–311. IEEE (2011)
10. Zawawi, M., Zamani, W.W., Ahmad, M., Saealal, M., Samin, R.: Feedback control schemes for gantry crane system incorporating payload. In: 2011 IEEE Symposium on Industrial Electronics and Applications, pp. 370–375. IEEE (2011)
11. Kirk, D.E.: Optimal Control Theory: An Introduction. Courier Corporation (2004)
12. Mitchell, M.: An Introduction to Genetic Algorithms. MIT Press (1998)
13. Samsuria, E., Mahmud, M., Wahab, N.A., Romdlony, M.Z., Abidin, M., Buyamin, S.: Solving an integrated job-shop-mobile robot scheduling problem in flexible manufacturing system using enhanced genetic algorithm structure with local search method. Appl. Model. Simul. **8**, 225–238 (2024)
14. Elhabib, M.O., Wahid, H., Mohamed, Z., Jaafar, H.: Optimal control of double pendulum crane using FOPID and genetic algorithm. In: Asia Simulation Conference, pp. 408–420. Springer, Cham (2023)
15. Zames, G.: Genetic algorithms in search, optimization and machine learning. Inf. Tech. J. **3**(1), 301 (1981)
16. Srinivas, M., Patnaik, L.M.: Adaptive probabilities of crossover and mutation in genetic algorithms. IEEE Trans. Syst. Man Cybern. **24**(4), 656–667 (1994)

Performance Evaluation Customer Access Network Using Difference Modulation Technique

Sivaguru Mugunthan[1], Juwairiyyah Abdul Rahman[1(✉)], Hasliza Abu Hassan[1], Norliana Muslim[2], and Mohammad Syuhaimi Ab Rahman[3]

[1] Department of Engineering, Faculty of Engineering and Life Sciences, Universiti Selangor, 45600 Bestari Jaya, Selangor, Malaysia
juwairiyyah@unisel.edu.my

[2] Department of Computer and Communication Technology, Universiti Tunku Abdul Rahman, 31900 Kampar, Perak, Malaysia

[3] Department of Electrical, Electronic and System Engineering, Faculty of Engineering and Built Environment, Universiti Kebangsaan Malaysia, Bangi, Malaysia

Abstract. Modeling Fiber To The Home (FTTH) networks involves several technical aspects and design considerations to enable effective and high-quality service delivery. This research seeks to boost the received signal quality for end-users inside a FTTH network. In order to create suitable and high-quality FTTH networks, this study will simulate FTTH networks utilizing network feasibility criteria, with an emphasis on BER and Q-factor. The signal distribution inside access networks is improved using Wavelength Division Multiplexing (WDM) and 10G-Passive Optical Network (PON) technology is used since it can deliver a fourfold improvement in downstream data transfer rates per user. Simulation result using OptiSystem 21 software, by setting transmitter wavelengths of 1490 nm and 1550 nm with a loss of 0 dB for a maximum of 2 users for total access based on the Q-factor calculation, which must be more than 6, while the BER must be less than $1 \times e-9$. System performance parameters that demonstrate Q-factor and Bit Error Rate (BER) give good quality and good performance is also shown by eye pattern in optical simulation. For 1490 nm, the highest Q-factor for 2 users is 35.8226, with a minimum BER of 2.4360×10^{-281}. For 1550 nm, the highest Q-factor is 15.3668 (in phase) and 14.8087 (in quadrature), with minimal BER values of $1.3674 \times e-53$ (in phase) and $6.4291 \times e-50$ (in quadrature). Good performance is also shown by eye pattern in optical simulation that displays amplitude signals.

Keywords: 10G-PON FTTH · BER · Q-factor · ISI · Internet VoIP · Broadcasting video

1 Introduction

Fiber to the Home (FTTH) represents a cutting-edge telecommunications architecture that leverages advanced optical fiber technology to deliver a wide array of multimedia services as well as high-speed internet connectivity directly to residential homes. As the demand for higher bandwidth and faster internet speeds continues to grow, FTTH is specifically designed to meet these evolving needs and expectations of consumers. Within the framework of FTTH deployment, several key components play a crucial role, including Optical Line Terminals (OLT), which serve as the central point of connection, Optical Network Terminals (ONT), which are installed at individual residences to facilitate access, and Optical Distribution Points (ODP), which help distribute the optical signals throughout the network. Together, these components to ensure a reliable and efficient fiber optic network that enhances the overall internet experience for users. These components help distribute signals to end user [1, 2]. FTTH represents the closest type of optical access network to end-users and serves as a primary communication method for delivering high-speed broadband services. This communication architecture offers high bandwidth with high Quality of Service (QoS) compared with previous DSL technology [3, 4].

In contrast with FTTH technology, different wired and wireless technology are analyzed to compare the advantages and disadvantages of each technologies as tabulated in Table 1 and 2. Based on the comparison, wired technology including features such as high speed, reliability, low interference and secure transmission channels and connections. Wireless technology including features such as mobility, affordability and simplicity [5]. The key constraints that significantly impact the capacity of wireless systems are primarily bandwidth and signal power, as noted in source [6]. The limitations associated with bandwidth in modern wireless technology are important considerations that result in the unavailability of 5G networks in certain geographical areas. This disparity in access highlights how bandwidth constraints can hinder the overall performance and reach of such advanced technologies. Furthermore, it is important to note that the major driving factors for user preference lean heavily towards the adoption of FTTx and 5G networks.

Table 1. Wired Internet Technologies [5].

Technology	Upload/Download Speed (Mbps)	Coverage	Infrastructure	Advantages	Disadvantages
ADSL, ADSL2, ADSL2*	3/24	5 km	Public telephone network	- Using the existing telephone cable and network; - Fast installation; - Fixed crosstalk in Vectoring; - Increased frequency in G. Fast.	- High loss of copper cable.
VDSL, VDSL2, Vectoring	40/100	1 km			
G. Fast	1000	100 m			
CATV	100/200	2-100 km	Coaxial cable in streets and buildings, optical fiber in feeders, and the possibility of return channel	- Using the existing cable TV infrastructure; - Fast installation; - High bandwidth.	- High cost and installation time in places without cable TV infrastructure.
Optical Fiber	10,000/10,000	10-60 km	Light waves instead of radio waves and distribution to users with optical and electrical signals	- High bandwidth; - Low loss; - No interference.	- High investment cost; - Dependency of bandwidth on the type of FTTx.

Table 2. Wireless Internet Technologies [5].

Technology	Upload/Download Speed (Mbps)	Coverage	Infrastructure	Advantages	Disadvantages
HSPA HSPA+ (3G)	22.56/(56, 42.2) 22/(168, 84.4)	3 km	3G/4G mobile network and its accessories	- Wireless radio coverage; - Quick and easy setup.	- Interference; - Limited bandwidth.
4G-LTE (LTE-A)	30/100 (30/1000)				
5G	10,000/20,000	Home: 10 m Mobile: 3–6 km	5G base stations	- High data rate; - Low latency; - High reliability; - Higher frequency bands; - Multi-antenna structure.	- High investment to update base stations and equipment.
GEO satellite	10/30	Very high	Ground base stations and their accessories	- Suitable for dense users; - Optimum use of bandwidth; - Suitable for easy and quick coverage of distant areas.	- High latency; - High cost.
LEO satellites	Wi-Fi: 660/660 7000/7000	High		- Suitable for coverage of rural/remote areas; - Less latency and more economic access than GEO satellites.	- Controlling the asynchronous satellites from ground base stations.
Balloon	HSPA: 22.56/(56, 42.2) 22/(168, 84.4) LTE: 30/100 30/1000			- Suitable for coverage of rural/remote areas; - Improved coverage to ground Wi-Fi, LTE, and HSPA.	- Controlling balloons from ground base stations. - Being in the testing phase.
WiMAX (IEEE802.16e)	4/6 (70/70)	60 km			
Wi-Fi IEEE802.11n (IEEE802.11ad)	660/660 (7000/7000)	200 m (10 m)	Base stations and their accessories	- Inexpensive; - Quick and easy installation.	- Low range.
Li-Fi	Up to 224,000	A few meters		- Very high data rate; - Suitable for interference-sensitive applications; - Suitable for indoor usage.	- Low range; - Low reliability; - Clear sky; - High setup cost.

According to Table 3's technical comparison, FTTx technology outperforms 5G in terms of lower latency, higher data rates, and superior signal coverage. FTTH can achieve data speeds of up to 100 Gbps with minimal packet loss, ensuring strong signal quality [5]. The major challenges network provider facing in terms of evolving landscape of 5G networks and beyond is on improving spectral efficiency. In order to overcome this problem, wireless networks (cellular) is focusing on Non-Orthogonal Multiple Access (NOMA) model. NOMA model is used to allow multiple users of similar time and frequency [7].

Table 3. FTTx and 5G-based Comparison [5].

Criterion	Technology	
	FTTX-Based	5G-Based
Maximum achievable download (upload) speed	100 Gbs (100 Gbs)	20 Gbs (10 Gbs)
Average global latency	9	28 ms
Average global jitter	3 ms	9 ms
Packet loss	Insignificant	Average
Coverage radius	70 km	10 m (fixed)/3 km (mobile)
Convenience	High for fixed users	High for mobile users
Interference robustness	High	Low
Maintenance and inspection	High	Low
Security	High	High with cryptography
Establishment and startup time	High	Low

The increasing demand for wireless data and the necessity for faster online connectivity have significantly contributed to the advancement of next-generation mobile communication systems, particularly in the context of 5G technology. These developments are primarily aimed at achieving higher data transmission rates, reducing latency to enhance user experience, and providing more reliable connections overall. Enhanced

NOMA techniques have recently been proposed for these systems, particularly in downlink scenarios. NOMA enables multiple users to simultaneously share a single channel using various methods such as frequency, time, coding, or spatial techniques. However, the excessive multiple-access interference resulting from spectrum sharing among usersa key characteristic of NOMA has been a subject of investigation. This capability significantly enhances both spectrum and energy efficiency [8].

Fifth generation wireless systems aim for higher data rates, massive connectivity, and ultra-low latency, emphasizing the need for advanced communication technologies and the role of multi-carrier modulation in optimizing spectrum use. These techniques help mitigate the negative effects of frequency-selective fading, common in wireless communications. An in-depth analysis of several multi-carrier techniques employed in 5G systems places significant emphasis on critical performance metrics such as Bit Error Rate (BER), which indicates the number of error bits in a given amount of transmitted data, as well as Signal to Noise Ratio (SNR), which measures the level of a desired signal to the level of background noise, and Power Spectral Density (PSD), which represents the power distribution of a signal as a function of frequency [9].

An efficient and cost-effective transfer learning technique has been developed for recognizing various digital modulation formats in optical communication. This advanced method is capable of accurately classifying a variety of modulation formats, including QPSK, 8PSK, 16PSK, and 16QAM, effectively across a range of optical signal-to-noise ratios (SNRs). To achieve this, the technique employs two distinct transfer learning algorithms that have been trained using constellation diagrams. Through this innovative approach, the method successfully categorizes all identified formats with exceptional accuracy, reaching an impressive prediction accuracy rate of 100% [10]. The MFI Scheme effectively identifies various modulation formats, such as QPSK, 8QAM, and up to 128QAM, using the k-nearest neighbors (KNN) algorithm, making it suitable for digital coherent receivers. A key benefit is its ability to operate without prior knowledge of the optical signal-to-noise ratio (OSNR), increasing its practical applicability. Numerical simulations show that the MFI scheme achieves a 100% identification rate for all six formats. This success depends on the OSNR exceeding critical thresholds associated with a 20% forward error correction (FEC) rate, linked to a bit error rate (BER) of 2.4×0^{-2} [11].

FTTH networks may encounter scalability challenges as demand increases, with modulation techniques potentially facing bandwidth and interference limitations. Therefore, the impact of modulation schemes on network performance is analyzed. Integrating future technologies like 6G could mitigate these issues. Furthermore, advancements in materials and AI-driven network optimization may enhance the scalability of FTTH networks to meet rising demand. Moreover, as the landscape of digital communication evolves, the role of improved optical components will become increasingly critical.

Innovations in the field of photonic devices, along with advancements in fiber materials, hold the potential to significantly decrease signal loss and simultaneously boost data transmission rates. These improvements are crucial as they contribute to enhancing the overall capacity of FTTH infrastructures, which serve as the backbone for delivering internet services to residential areas. The enhancement of these core technologies is further complemented by sophisticated encoding techniques, such as orthogonal frequency-division multiplexing (OFDM), as well as WDM. These advanced methods create additional pathways that facilitate the expansion of bandwidth availability while effectively minimizing the levels of interference that can arise during data transmission.

The implementation of machine learning algorithms to predict and manage network traffic offers another significant leap forward. By utilizing real-time data analytics, networks can dynamically allocate resources based on user demand and usage patterns, ensuring that bandwidth is effectively utilized and reducing bottlenecks during peak usage periods. This proactive management approach not only enhances user experiences but also prolongs the lifecycle of existing infrastructure. Moreover, the seamless integration of IoT devices into FTTH networks can facilitate smarter homes and cities, demanding even greater bandwidth and reliability. As these devices proliferate, systems must adapt to provide consistent performance while maintaining security across increasingly complex connections. This integration will place additional strains on existing networks, but with the aforementioned technological advancements, it is possible to design FTTH systems that are not only scalable but resilient to future demands.

In the development and deployment of FTTH solutions, Active Optical Networks (AON) and Passive Optical Networks (PON) are essential components. Among these, PON stands out as a prominent and widely adopted optical access network technology. The architecture of a PON comprises several critical components, including multiple optical network units (ONUs), which are deployed at the customer premises, an Optical Distribution Network (ODN) that interconnects them, and a central office housing the OLT. This central office is essential to the overall functionality of the network as it connects to the 1 × N splitter through an optical cable, thereby facilitating efficient data distribution across the network. It offers services to end users and regulates data flow in both upstream and downstream directions over the ODN [12]. XGPON, also known as 10GPON, is the most recent iteration of GPON that has a bit rate of 10 gigabits.

PON is often used for low-capacity and short-distance transmission, yet it uses less power than AON [13]. Passive components are less expensive to use than active components since PON includes unpowered devices. As a result, it requires no electricity and has a reduced maintenance cost [13]. Figure 1 depicts the fundamental configuration of a WDM and PON combination designed to deliver high data rates and bandwidth in wireless communication [14]. Since in this journal TPS uses two types of different wavelengths, these wavelengths can be sent along the fiber simultaneously. The method of combining these two wavelengths (1490 nm & 1550 nm) onto the same fiber is known as WDM [14].

Triple Play Service (TPS) integrates voice, data, and video services, seamlessly delivering all three through a unified network infrastructure. Figure 2 illustrates 10G-PON simulation schemes with TPS. There are a few optimizations impacting the TPS system performance, which is modulation techniques, splitting ratio, multiplexing technique, Hybrid Optical Amplifier (HOA) as previous study [15–17]. However, in this research, we have chosen to use the modulation approach since it reduces the amount of noise present in the received signal, boosting its quality. Ccording to previous research, the goal of modulation approaches is to lessen the impact of Inter-Symbol Interference (ISI) by employing appropriate modulation techniques in both the transmitter and receiver sections [18].

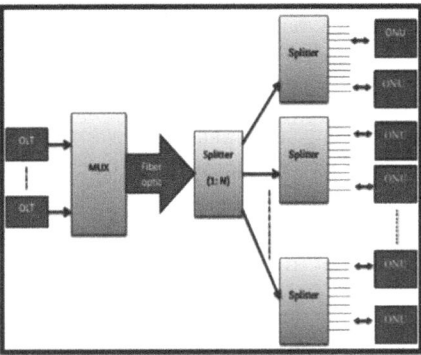

Fig. 1. Basic configuration of PON [13].

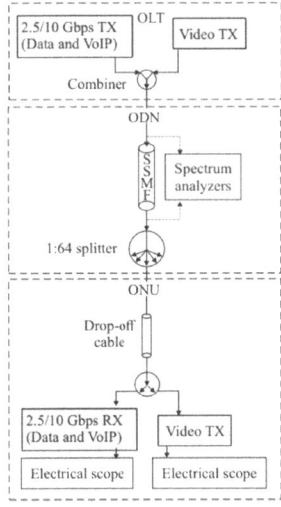

Fig. 2. GPON/XG-PON Simulation Schemes with Triple Play Services (TPS) [17].

Q-factor is used to assess the optical network, and it may be defined based on Eq. (1). μ_1 and μ_0 represents the average value of the received signal whereas σ_1 and σ_0 represent standard deviations. 1 and 0 represent logical operations. BER is used to determine the transmitter efficiency by indicating the number of bits that have errors to the received bits. Non-Return-to-Zero On-Off Keying (NRZ-OOK), Non-Return-to-Zero Differential Phase Shift Keying (NRZ-DPSK), Chirp-Return-to-Zero Differential Phase Shift Keying (CRZ-DPSK), Duobinary, and 16 Quadrature Amplitude Modulation (QAM) were among the modulation schemes used in this study to assess Inter-Symbol Interference (ISI). These schemes were selected to assess their performance under different transmission conditions and their impact on signal quality. Analysis of the BER and Q-factor is used to tabulate the data. This study develops four 10G-PON designs based on various modulation approaches, representing unidirectional or single directions.

$$Q = (\mu_1 - \mu_0)/(\sigma_1 - \sigma_0) \tag{1}$$

Ultimately, collaboration among stakeholders,network operators, technology developers, and policymakers it will be essential to create a holistic strategy for the future of FTTH networks. By fostering innovation and investing in robust infrastructure, we can pave the way for a more interconnected society that can readily meet the challenges posed by an ever-expanding digital frontier. 6G-based technology will be the successor to 5G-based technology by 2031, refers to sixth generation of wireless communication technology as shown in Fig. 3. 6G technology can effectively support extensive coverage and AI services with high data rates. However, challenges like network security, data privacy, deployment costs, and mobile battery life must be addressed before its implementation [19].

Fig. 3. Timeframe of 6G-based Technology [20].

The performance study of single-ONU customer access networks adopting different modulation algorithms gives substantial insights into efficiency and reliability. Different modulation systems, such as Differential Pulse Position Modulation (DPPM) and Index Modulation (IM), display diverse capacities in boosting network performance, notably in terms of spectrum efficiency and error rates. DPPM approach increases frame error probability and network performance by employing generalized optical orthogonal codes, allowing for greater simultaneous user transmissions [21]. IM-OFDM with single subcarrier activation provides low peak-to-average power ratio (PAPR) and beats classical OFDM in spectral efficiency, especially when high power amplifiers are utilized. DPPM and IM-OFDM both concentrate on decreasing BER, critical for ensuring data integrity in high-speed networks [21].

This research aims to enhance signal quality and service delivery through the use of various modulation techniques and optimized network design. Additionally, it includes a comparative analysis of these techniques in terms of speed, reliability, and scalability.

This study aims to identify the best modulation strategy to guarantee that end users receive signals from the service provider without sacrificing the quality of the data provided to them. The limiting factors in this research are number of access users and upstream direction.

2 Methodology

The Internet+VoIP (1490 nm) service was assessed using Phase Shift Keying (PSK) and On-Off Keying (OOK) modulation techniques, while QAM was used for analyzing the Broadcasting video (1550 nm) service. The BER analyzer component's eye pattern is a way to look at how Inter-Symbol Interference (ISI) impacts BER and Q-factor. OLT, ODN and ONT make up this architecture. This simulation model facilitates the transmission of data from the OLT at the operator's end to the Optical Network Terminal (ONT) at the user's end, implementing a downlink transmission process. The sequence of steps involved in designing the FTTH access network, as outlined in the methodology, is summarized in Fig. 4, providing a clear roadmap for the simulation and network design process.

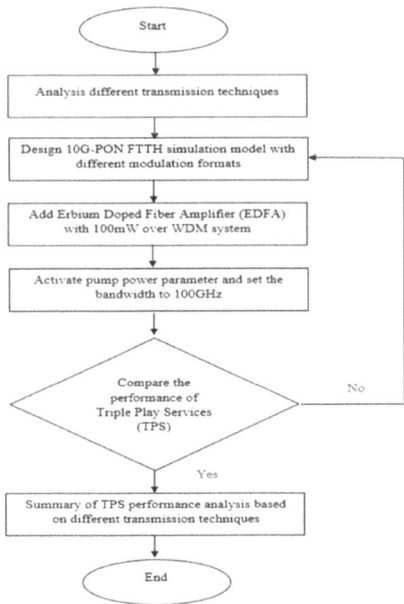

Fig. 4. GPON FTTH Assess Network Design Steps.

Figure 5 illustrates the schematic diagram of a unidirectional structure designed to connect two OLTs to two ONUs. In this configuration, the OLT plays a primary role in managing and controlling the information flow across the ODN. On the receiving end, the ONU serves as the receiver unit, equipped with an optical filter and a photodetector to process the incoming optical signals. The connection between the OLT and ONU is established using Single Mode Fiber (SMF), ensuring high-speed and reliable data transmission over long distances [22].

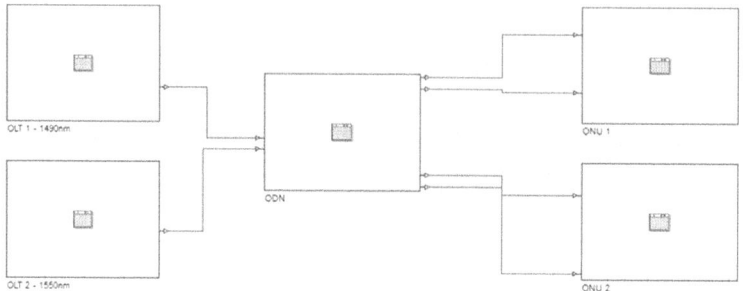

Fig. 5. Schematic Diagram for Unidirectional Scheme.

The downstream transmission in this study was designed using a wavelength of 1575 nm, providing efficient signal delivery. In Fig. 6, the term Optical Distribution Network (ODN) encompasses components such as power splitters, connectors, and Single Mode Fiber (SMF), which collectively form the backbone of the network. The 10G-PON unidirectional model was constructed and simulated using the OptiSystem software, a powerful tool for designing and analyzing optical communication systems. The detailed design of this model is depicted in Fig. 6.

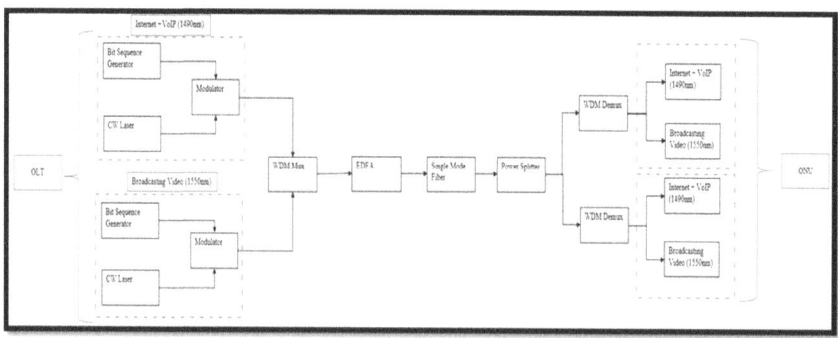

Fig. 6. Block Diagram for whole system.

There are three components to a 10 G-PON FTTH system: transmitter, channel, and receiver. The transmitter section's component configuration varies depending on the modulation type, since each one requires a distinct transmitter structure. A Continuous-wave (CW) lasers, pulse generators, modulators, and pseudo random binary sequence generators (PRBS) make up the transmitter component. An EDFA is created by merging channels using a WDM multiplexer. All transmitter configurations utilize a CW laser with a transmitter power of −10 dBm, ensuring consistent and high-quality signal generation. Data traffic is emulated via a pseudo-random binary sequence (PRBS), generating an input signal encoded as either non-return-to-zero (NRZ) or return-to-zero (RZ). The input signal is modulated by the continuous wave (CW) laser using a Mach-Zehnder Modulator (MZM), with the CW laser acting as a dependable source of optical signals. The continuous wave laser operates at wavelengths of 1490 nm and 1550 nm with a power level of −10 dBm, facilitating modulation at 10 Gbps across many modulation schemes. The choice between intensity modulation and phase modulation depends on the type of modulator used, whether it is a Mach-Zehnder Modulator for intensity modulation or a phase modulator for phase modulation [23].

In the transmitter section, the configuration of components varies for each modulation format, as each requires a unique transmitter structure. The transmitter setup consists of a CW laser, a Pseudo-Random Binary Sequence (PRBS) generator, pulse generators, modulators, and a WDM multiplexer. The EDFA)amplifies the channels after they are combined using the WDM multiplexer.. All transmitter configurations rely on a CW laser operating at a power of −10 dBm, providing a stable optical signal source. PRBS is utilized to simulate data traffic, generating NRZ or RZ input signals in the simulation setup, ensuring realistic and flexible testing conditions. A scheme with the 10 GPON FTTH system implemented by the different modulation techniques' full architecture is shown in Fig. 7.

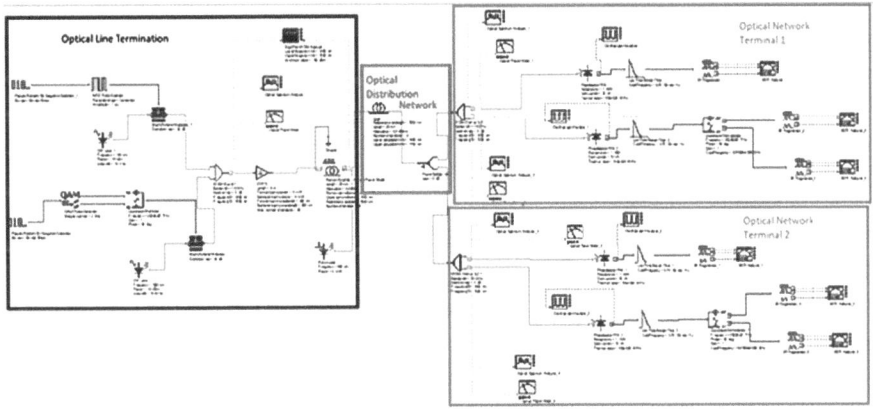

Fig. 7. Simulation Model of 10G-PON FTTH System.

In the channel section, an EDFA with Single-Mode Fiber (SMF) transmits signals over 25 km. Wavelengths of 1490 nm and 1550 nm are multiplexed into a WDM with zero insertion loss and passed through a 5-m Erbium-Doped Fiber (EDF) for amplification. The EDFA operates with a 100 mW forward pump power at 980 nm. The 25 km SMF efficiently delivers the amplified signals to the receiver.[24]. A 1 × 2 splitter allows a SMF to be shared between two ONTs by dividing the signal power, ensuring consistent signal quality and performance across both connections. In the receiver section, optical pulses are converted into electrical bits through a series of interconnected components designed for efficient signal processing. The receiver configuration consists of a 3R regenerator, a low-pass Bessel filter, and a PIN photodiode that are connected to a BER analyzer for VoIP and Internet performance analysis. For intensity-modulated signals, a PIN photodetector is used. For phase-modulated signals, a delay interferometer is placed before the PIN photodetector. This setup ensures direct detection by enabling the PIN photodetector to process the optical outputs of the interferometer, with the electrical signal output derived from the difference in PIN-detected currents. A WDM demultiplexer splits the multiplexed optical signals into separate channels, and the bandwidth for each Optical Network Terminal (ONT) is set to 100 GHz [25]. T The low-pass Bessel filter, set at 0.75 times the bit rate, effectively rejects high-frequency noise while allowing low-frequency signals through. The 3R regenerator at the output reconstructs the bit sequence and produces a modulated electrical signal for precise BER analysis, ensuring signal fidelity and performance.

3 Results and Discussion

There are two primary parts to the FTTH simulation with TPS via 10G-PON. The first section evaluates four optical modulation methods, as illustrated in Figs. 8, 9, 10 and 11. A CW laser emits at 1490 nm with −10 dBm (0.1 mW) output power, ensuring a stable source for the modulation tests. Refer to Figs. 8, 9, 10 and 11 represent Eye Diagrams and associated Q-Factor and BER values for different modulation formats. For Duobinary it shows moderate quality, high ISI andless suitable for high-speed environments while for NRZ-OOK shows high quality, simple and effective and also good for low complexity systems. Modulation format NRZ DPSK shows strong noise resilience and slightly more complex than OOK. The best performance overall and suitable for high-speed for long-distance links is CRZ DPSK modulation format. Thus, CRZ DPSK offers the best performance in terms of Q-factor and BER, making it ideal for advanced communication systems. Duobinary has the weakest performance, making it less suitable for high-demand applications but potentially useful in bandwidth-limited scenarios.

The relevant BER and Q-factor values for a 100 GHz bandwidth are shown in Table 4. Due to its lower BER and Q-factor when paired with a reference pump power of 100 mW, duobinary modulation was chosen for the 1490 nm channel based on Table 4. Randomly generated data was divided into two arms for the duobinary transmission approach. One of the two arms uses a logical NOT component to flip the data's logical state. Both signals are modified using an optical Mach-Zehnder modulator after passing via an NRZ modulator driver. This configuration ensures precise modulation of the optical signals, leveraging the DDMZM's ability to handle dual-input signals for enhanced performance and flexibility in the modulation process.

Higher and lower values of the BER and Q-factor are caused by the greatest degree of Inter-Symbol Interference (ISI). The Q-factor for this network, which supports about 2 users, is 35.8226, exceeding the required Q-factor. For a signal to be considered to have an acceptable SNR, the signal must match or exceed the minimum Q-factor of 6. The minimum requirement for the BER is $1 \times 10^{(-9)}$. Our simulation shows that our proposed network with two users has less than $1 \times 10^{(-9)}$ for all modulation techniques.

The simulation stage focuses on the 1550 nm channel for video broadcasting, using 16-QAM modulation to achieve 10 Gbps data rates for high-definition quality. The

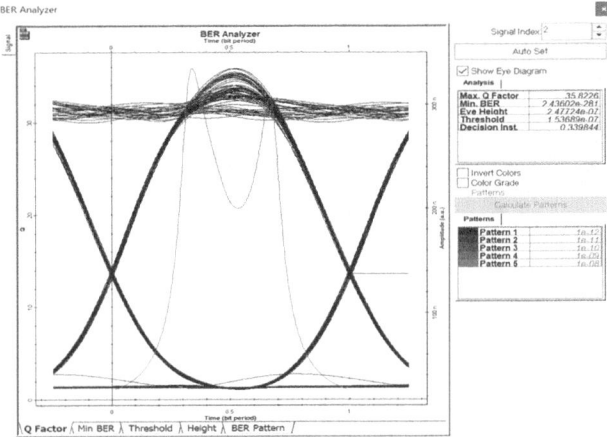

Fig. 8. Eye diagram, Q factor, and BER – Duobinary.

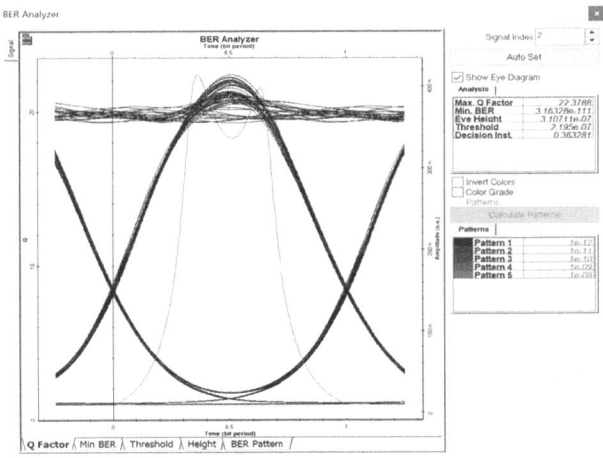

Fig. 9. Eye diagram, Q factor and BER – NRZ OOK.

research did not test 64-QAM. An MZM modulates the electrical signal to an optical carrier. 16-QAM OOK recorded acceptable values. The eye diagrams of the received signals in Figs. 12 and 13 offer valuable insights into signal quality. In Fig. 12, the width of the eye diagram represents the time window available for accurate sampling. A clear and wide eye opening indicates minimal inter-symbol interference (ISI), reflecting good signal integrity. Although some noise is visible in the eye diagram, the signal levels are well-separated at the decision threshold, ensuring reliable decoding. The symmetry of the eye further supports consistent decoding performance.

Figure 12 shows that value for Q-Factor is 15.7 which is a very high value, indicating excellent signal quality and a low likelihood of bit errors. Similar to Fig. 12, the eye is well-formed but slightly affected by noise or distortion as shown in Fig. 13. Compared to

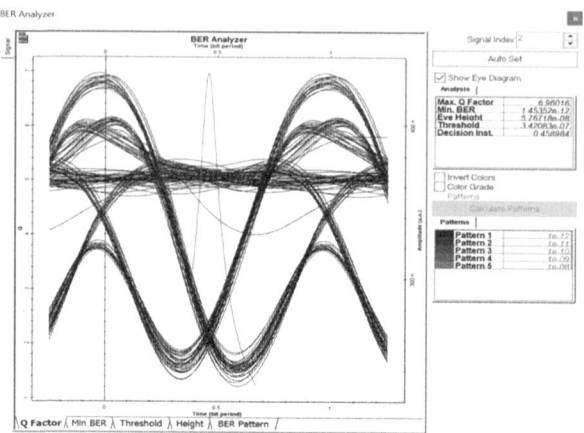

Fig. 10. Eye diagram, Q factor and BER – NRZ DPSK.

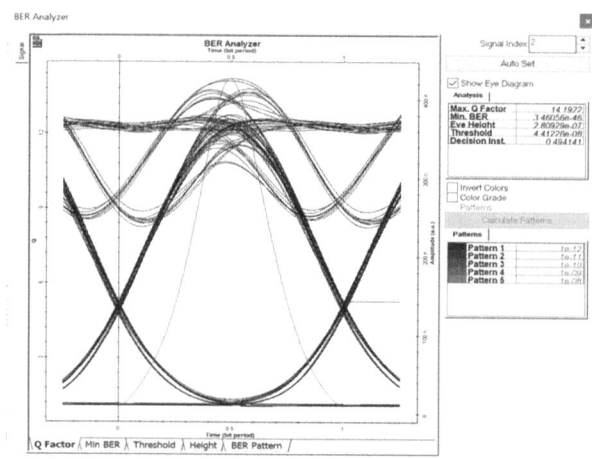

Fig. 11. Eye diagram, Q factor and BER – CRZ DPSK.

the in-phase diagram, the Quadrature signal shows a narrower eye opening, potentially making it more susceptible to noise and errors. The signal remains clearly distinguishable above the threshold, ensuring reliable detection. As can be seen in Fig. 13, the quadrature component may be subject to somewhat greater distortion or noise, as the Q-Factor value of 14.8 is marginally lower than that of the in-phase component. However, a Q-factor of 14.8 nevertheless remains excellent and indicates consistent performance.

These metrics validate the use of 16-QAM modulation for FTTH networks, balancing high data rates with excellent signal integrity. The findings suggest the network is robust enough to handle real-world conditions, ensuring reliable high-speed communication for end-users. The study investigates modulation techniques to enhance signal quality for high-demand applications in FTTH networks, such as video streaming, telemedicine, and remote work. By comparing various modulation methods, the research identifies

Table 4. Characteristics of Q factor following Triple Play Service (TPS) system.

Service	Modulation	BER	Q-factor [dB]
Internet+VoIp (1490 nm)	Duobinary	$2.4360 \times e^{-281}$	35.8226
	NRZ OOK	$3.1633 \times e^{-111}$	22.3788
	NRZ DPSK	$1.4535 \times e^{-12}$	6.9602
	CRZ DPSK	$3.4606 \times e^{-46}$	14.1922
Broadcast video (1550 nm)	16-QAM OOK	NRZ DPSK	
	In phase	$1.3274 \times e^{-45}$	In phase
	In quadrature	$3.1483 \times e^{-45}$	14.0642
	16-QAM OOK	Duobinary	
	In phase	$4.5797 \times e^{-41}$	In phase
	In quadrature	$2.7858 \times e^{-39}$	In quadrature
	16-QAM OOK	NRZ OOK	
	In phase	$1.3674 \times e^{-53}$	In phase
	In quadrature	$6.4291 \times e^{-50}$	In quadrature
	16-QAM OOK	CRZ DPSK	
	In phase	$1.7137 \times e^{-39}$	In phase
	In quadrature	$1.5003 \times e^{-37}$	12.7530

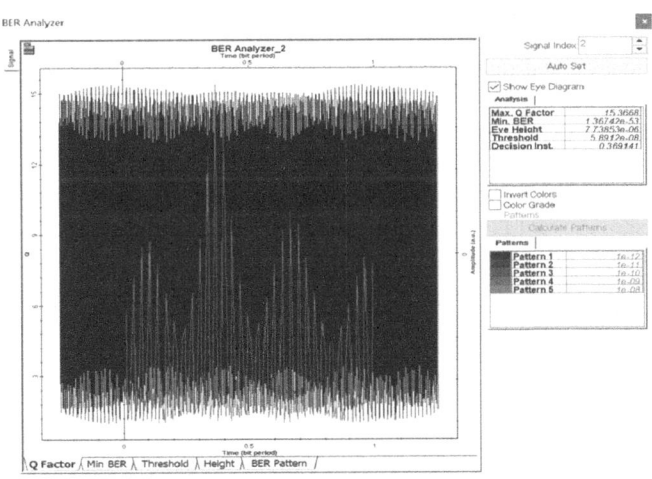

Fig. 12. Eye diagram, Q factor and BER – In Phase (16 QAM).

optimal strategies for different scenarios. Effective modulation techniques can reduce power consumption and costs for network providers. The findings demonstrate how FTTH networks can scale to meet growing user demands and bandwidth requirements.

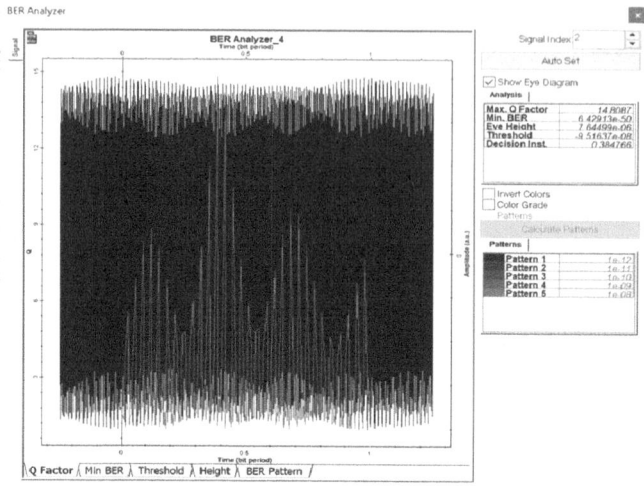

Fig. 13. Eye diagram, Q factor, and BER – In Quadrature (16 QAM).

As 6G networks demand a robust backbone infrastructure, this research ensures that FTTH networks do not become a bottleneck.

4 Conclusion

This study outlines how to ensure that end users receive high-quality signals that meet established specifications for Q factor and BER. The investigation focused on modulation techniques, examining eye diagrams, Q factors, and BER data. Our analysis of network architecture and modulation techniques revealed that 1490 nm and 1550 nm channels offer the best performance, evidenced by their lower BER values (below $1 \times 10^{(-9)}$) compared to standard optical communication systems. Consequently, the system meets the requirements for Internet+VoIP (1490 nm) and video broadcasting (1550 nm) services, utilizing a FTTH network with a 10 Gbps data transfer rate and PON architecture. Contemporary FTTH systems may have challenges with scalability and maintenance expenses as demand increases. The study delineates deficiencies and obstacles in existing FTTH technology, including interference management and bandwidth optimization. This establishes a framework for forthcoming research, promoting innovation in communication technologies.

5 Recommendation

The research should consider compatibility with evolving technologies, such as integrating machine learning for dynamic modulation and optimization, as well as utilizing findings to develop hybrid networks. Next-Generation PON 2 (NG-PON2) and other greater capacity PON technologies will ensure that the system is capable of managing the higher bandwidth needs (200 GHz, 300 GHz, etc.) to enable a bandwidth consumption above 100 GHz in a 10G-PON FTTH system. Higher bandwidth allows for the simultaneous transmission of more signals or data.

Acknowledgment. This research was funded by the Ministry of Higher Education through the Fundamental Research Grant Scheme (FRGS) code number FRGS/1/2023/TK07/UNISEL/02/1 and received support from the University of Selangor, along with contributions from the Faculty of Engineering and Built Environment at Universiti Kebangsaan Malaysia (UKM).

References

1. Prayoga, W.M., Sani, A.: Perancangan Jaringan Fiber to the Home (FTTH) Menggunakan Teknologi Gigabit Passive Optical Network (GPON). EBID: Ekonomi Bisnis Digit. **1**(2), 179–188 (2023). https://doi.org/10.37365/EBID.V1I2.220
2. Adila Asril, A., Maria, P., Antonisfia, Y., Hadi, R.: Fiber to the Home (FTTH) network design with addition of Optical Distribution Point (ODP) using the branching method. Int. J. Adv. Sci. Comput. Eng. **5**(2), 95–107 (2023)
3. Shibghatullah, A.S., Mohammed, M.M., Doheir, M., Majed, A.N.: Fiber-to-the-Home (FTTH) architecture for Mosul, Iraq. Int. J. Hum. Technol. Interact. **1** (2017)
4. Harstead, E., Sharpe, R.: Future fiber-to-the-home bandwidth demands favor time division multiplexing passive optical networks. IEEE Commun. Mag. **50**(11), 218–223 (2012). https://doi.org/10.1109/MCOM.2012.6353704
5. Shirvani Moghaddam, S.: The past, present, and future of the internet: a statistical, technical, and functional comparison of wired/wireless fixed/mobile internet. Electron. (Switz.) **13**(10), (2024). https://doi.org/10.3390/electronics13101986
6. Ezenugu, I.A., Ezeh, I.H.: Challenges of bandwidth and power limitations in cellular communication: a review **7**(4), 1–11 (2020). https://doi.org/10.9790/0050-07040111
7. Al Ghafri, Y., Asif, H.M., Tarhuni, N., Nadir, Z.: NOMA-empowered technologies for non-line of sight environments: a comprehensive review, 02 July 2024. https://doi.org/10.20944/preprints202407.0219.v1
8. Yonis, A.Z., Nawaf, A.: Investigation of evolving multiple access technologies for 5G wireless system. In: 8th IEC 2022 - International Engineering Conference: Towards Engineering Innovations and Sustainability, Institute of Electrical and Electronics Engineers Inc., pp. 118–122 (2022). https://doi.org/10.1109/IEC54822.2022.9807471
9. Dhruvakumar, T., Manasa, R., Monika, P., Gowthami, N., Thunga, U.M.: Performance analysis of multi-carrier scheme for 5G communication system. In: 2024 International Conference on Smart Systems for Applications in Electrical Sciences (ICSSES), pp. 1–6 (2024)
10. Jha, D.K., Mishra, J.K.: Transfer learning-assisted modulation format identification for low OSNR. In: 2023 IEEE IAS Global Conference on Emerging Technologies (GlobConET), pp. 1–4 (2023)
11. Hao, M., et al.: Modulation format identification based on multi-dimensional amplitude features for elastic optical networks. Photonics **11**(5), (2024). https://doi.org/10.3390/photonics11050390
12. Kumari, M., Sharma, R., Sheetal, A.: Passive optical network evolution to next generation passive optical network: a review. In: 2018 6th Edition of International Conference on Wireless Networks & Embedded Systems (WECON), p. 102. IEEE, November 2018
13. Abdulwahab, M.M., Eldaw, M.M.: Performance evaluation of symmetric and asymmetric XG passive optical networks. Int. J. Electr. Eng. Appl. Sci. (IJEEAS) **4**(1), (2021). https://ijeeas.utem.edu.my/ijeeas/article/view/6035/4011. Accessed 29 Oct 2024
14. Pandey, V.K., Gupta, S., Chaurasiya, B.: Performance analysis of WDM PON and ROF technology in optical communication based on FBG. Int. J. Eng. Res. **3**(10), (2014)
15. Al-Quzwini, M.M.: Design and implementation of a fiber to the home FTTH access network based on GPON. Int. J. Comput. Appl. **92**(6), 0975–8887 (2014)

16. Alobaidan, H.: Current and future FTTH technologies. J. Wirel. Netw. Commun. **7**(2), 35–40 (2017). https://doi.org/10.5923/J.JWNC.20170702.02
17. Agalliu, R., Burtscher, C., Lucki, M., Seyringer, D.: Optical splitter design for telecommunication access networks with triple-play services. J. Electr. Eng. **69**(1), 32–38 (2018). https://doi.org/10.1515/jee-2018-0004
18. Beyene, W.T., Amirkhany, A.: Controlled intersymbol interference design techniques of conventional interconnect systems for data rates beyond 20 Gbps. IEEE Trans. Adv. Packag. **31**(4), 731–740 (2008). https://doi.org/10.1109/TADVP.2008.924224
19. Alsharif, M.H., Kelechi, A.H., Albreem, M.A., Chaudhry, S.A., Sultan Zia, M., Kim, S.: Sixth Generation (6G) wireless networks: vision, research activities, challenges and potential solutions. Symmetry (Basel) **12**(4), 676 (2020). https://doi.org/10.3390/SYM12040676
20. Alsharif, M.H., Kelechi, A.H., Albreem, M.A., Chaudhry, S.A., Sultan Zia, M., Kim, S.: Sixth generation (6G) wireless networks: vision, research activities, challenges and potential solutions. MDPI AG, 01 April 2020. https://doi.org/10.3390/SYM12040676
21. Khazraei, S., Shoaie, M.A., Pakravan, M.R.: Efficient modulation technique for optical code division multiple access networks: differential pulse position modulation. IET Optoelectron. **8**(5), 181–190 (2014). https://doi.org/10.1049/IET-OPT.2013.0076
22. Mohamed, I.M.M., Bin Ab-Rahman, M.S.: Options and challenges in next-generation optical access networks (NG-OANs). Optik (Stuttg) **126**(1), 131–138 (2015). https://doi.org/10.1016/J.IJLEO.2014.08.131
23. Ab-Rahman, M.S., Rahman, J.A., Arifin, N.F.M., Kaharudin, I.H., Hwang, I.S.: Upscaling customer access network using spectrum conversion–slicing–duplication technique. Photonics **10**(11), (2023). https://doi.org/10.3390/photonics10111271
24. Aldouri, M.Y., Aljunid, S.A., Anuar, M.S., Fadhil, H.A., Ahmed, N.: One EDFA loop in 16 channels spectrum slicing WDM for FTTH access network. In: 2010 International Conference on Photonics, ICP 2010, p. 1. IEEE, July 2010. https://doi.org/10.1109/ICP.2010.5604414
25. Ikhsan, R., Syahputra, R.F., Suhardi, Saktioto, Husein, N.A., Okfalisa: Development of WDM system in optical amplifiers by manipulating fiber length and bandwidth for telecommunication system. In: Saeed, F., Mohammed, F., Gazem, N. (eds.) Emerging Trends in Intelligent Computing and Informatics, IRICT 2019. AISC, vol. 1073. Springer, Cham (2020). https://doi.org/10.1007/978-3-030-33582-3_58

Extracting Narrative Events in Andersen's Fairy Tales Using a Hybrid BERT-LSTM Model

Erna Daniati, Aji Prasetya Wibawa[✉], and Wahyu Sakti Gunawan Irianto

Universitas Negeri Malang, Malang, Indonesia
`aji.prasetya.ft@um.ac.id`

Abstract. This research investigates the extraction of narrative events from Andersen's fairy tales by employing a hybrid BERT-LSTM model to enhance the understanding and analysis of classical literary works through natural language processing (NLP). The model merges the robust language representation capabilities of BERT with the sequential processing strengths of LSTM to capture both the semantic and contextual subtleties of narrative events throughout Andersen's tales. By leveraging this hybrid approach, the study effectively identifies key narrative components such as character actions, plot developments, and crucial events. The model's performance was rigorously assessed, achieving a precision of 0.88, a recall of 0.82, and an F1-score of 0.75, demonstrating effective accuracy and a solid balance between detecting true positives and reducing false negatives. The results contribute to computational literary analysis, providing a dependable method for analyzing complex narrative structures in fairy tales and similar texts. This method not only enriches literary research with quantitative insights but also has potential applications in automated story generation, educational tools, and digital humanities. Future research will aim to improve recall rates and further refine the model for broader application across various literary genres.

Keywords: Andersen's Fairy Tales · BERT-LSTM Model · Computational Literary Analysis · Narrative Event Extraction · Natural Language Processing

1 Introduction

Recent advancements in natural language processing, such as the use of transformer-based language models like BERT (Bidirectional Encoder Representations from Transformers) [4], have demonstrated remarkable abilities in understanding and representing text. Additionally, convolutional neural networks (CNN) [1, 2], particularly LSTM (Long Short-Term Memory) [3, 4] have proven effective in modeling sequential data like text. This approach enables models to grasp text context more deeply. As a transformer-based language model, BERT has shown its capability to comprehend the entire context of text by utilizing contextual word representations [5]. This allows BERT to identify events in fairy tales while considering the broader narrative context, offering deeper insights into event patterns and character relationships [6]. On the other hand, LSTM, a recurrent neural network, effectively models the order of information in text by considering temporal dependencies between words. The use of LSTM in event extraction enables the model to capture the chronological connections between events in fairy tales, significantly improving accuracy and narrative understanding [7]. Furthermore, CNNs have become popular for feature extraction in text data due to their ability to detect both local and global patterns. By applying deep CNN for event extraction, the model can identify crucial features in fairy tale texts that reflect key narrative events [8]. Therefore, the combination of BERT, LSTM, and CNN offers a comprehensive and effective approach to event extraction in fairy tales, resulting in a deeper understanding of narrative structure and event dynamics in literary works [9].

Despite the widespread use of BERT and LSTM in various NLP applications, their application in fairy tale analysis remains limited [10]. Previous studies have primarily focused on news texts, product reviews, or other formal documents. The use of BERT and LSTM for event extraction in fairy tales reflects the combined strengths of deeply understanding text context and modeling sequential information [4]. BERT, with its contextual understanding, can identify and interpret relationships between words in a text, including the events occurring. In contrast, LSTM excels in modeling information sequences, allowing it to recognize complex patterns in story narratives. Combining these two methods addresses some of the challenges each faces individually [11]. For example, while BERT excels at understanding the broader context of a text, it may be less effective at modeling the sequence of information. On the other hand, LSTM is adept at modeling sequences but may be less sensitive to the overall context. By combining BERT and LSTM, these limitations are overcome, allowing the model to understand text context more thoroughly and better recognize patterns in the sequence of information. As a result, event extraction becomes more accurate, offering richer insights into the narrative structure of fairy tales.

This study aims to develop and evaluate a BERT-LSTM hybrid model for extracting narrative events from Andersen's fairy tales. The primary objectives are to identify narrative elements, such as characters, actions, and events, with high accuracy and to compare the performance of this hybrid model with that of individual BERT and LSTM models [9]. Additionally, the study seeks to measure the model's effectiveness in understanding story structure using performance metrics, including precision, recall, and F1-score, and to explore the potential applications of this model in digital literary studies and automated story generation. This research is expected to contribute to computational literary analysis and open up new opportunities for applying this method in various digital literacy contexts.

In this study, we aim to develop an approach combining BERT and LSTM models for event extraction in fairy tales. This includes identifying key events in fairy tales and analyzing the emerging event patterns [12]. The performance of the proposed approach in understanding and extracting events will be evaluated and compared with existing event extraction methods.

2 Method

2.1 Research Flow

Datasets used for event extraction in closed domains exhibit several distinct characteristics when compared to datasets from open domains. A closed domain refers to a dataset derived from specific areas or subjects, such as medical, financial, legal, or technical fields. Below are some key characteristics of datasets for event extraction in closed domains:

1. **Domain-Specific:** These datasets are closely tied to a particular domain or industry. The information extracted usually pertains to events or entities that are relevant to the specific subject area. For instance, in the medical field, an event might be a patient's diagnosis, a medical procedure, or the side effects of a drug.
2. **Scope Limitations:** Closed domain datasets generally have a more restricted scope than open domain datasets. The extracted information primarily focuses on topics within that specific domain, excluding information from other domains.
3. **Imbalance:** There is often an imbalance in the class or type distribution of events within closed domain datasets. Certain types of events may occur more frequently than others, which can impact the performance of deep models in recognizing less common events.
4. **Limited Data Availability:** Datasets in closed domains can be hard to obtain or may not be widely available due to the sensitivity or specificity of the domain. This limitation can hinder the development and evaluation of event extraction models in closed domains.
5. **Annotation Challenges:** The annotation process in closed domains can be more complex, as it requires in-depth, expert knowledge of the subject matter. This can lead to limitations in the volume of data that can be annotated accurately.
6. **Difficult Domain Expansion:** Expanding event extraction models from one closed domain to another often requires significant adaptation and updates. This is due to differences in terminology, sentence structure, and domain-specific rules.

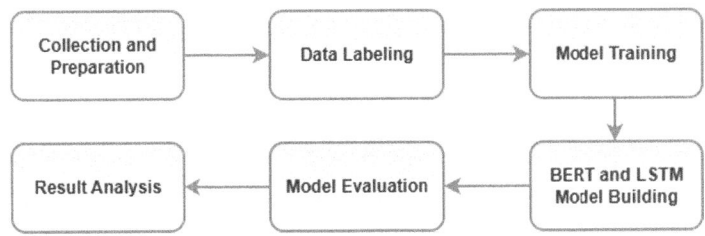

Fig. 1. Research Flow

The research process for event extraction using BERT and LSTM is illustrated in Fig. 1. The approach applied to fairy tale datasets involves several important stages, outlined as follows:

1. **Collection and Preparation**

   The research begins with the collection of a fairy tale dataset for event extraction. This dataset typically consists of various fairy tales formatted as text. The next step involves preparing the data, which includes cleaning the text, tokenization, and other preprocessing tasks such as removing punctuation, converting letters to lowercase, and splitting the text into sentences.

2. **Data Labeling**

   Before training the model, the data must be labeled with the events present in the fairy tales. This step involves identifying key events within the text, including the introduction of trigger events, event types, and event arguments.

3. **Model Training**

   After the data is labeled, the BERT and LSTM models are trained using the labeled dataset. The training process optimizes the model parameters to understand the context and patterns within the fairy tale dataset, enabling the model to identify events with high accuracy.

4. **BERT and LSTM Model Development**

   The next step is the development of an event extraction model using the BERT and LSTM approaches. The BERT model is employed to deeply understand the text's context, while LSTM is used to model the order of words within the text. Both models are integrated into a single architectural framework.

5. **Model Evaluation**

   After training, the model is evaluated using relevant metrics such as accuracy, precision, and recall. The evaluation is performed on a separate test dataset, which was not used during training, to ensure the model's ability to extract events accurately from new data.

6. **Analysis of Results**

   The final step involves analyzing the results of the trained model. This includes assessing the model's performance, comparing it with existing event extraction approaches, and analyzing the patterns of events extracted from the fairy tale dataset. This analysis offers insights into the model's ability to understand the narrative of fairy tales and its success in extracting relevant event information.

By following these steps, it is expected that this line of research will develop an effective approach to event extraction in fairy tales using the BERT and LSTM models, which will provide valuable insights into the narrative structure and event patterns in literary works.

Consistency and uniformity in the use of terminology, formats, and data structures are crucial in closed domain datasets to ensure accuracy and consistency in event extraction. Both LSTM (Long Short-Term Memory) and BERT (Bidirectional Encoder Representations from Transformers) methods are used for event extraction, employing different approaches but sharing the common goal of understanding and extracting information from text.

2.2 Long Short-Term Memory (LSTM)

LSTM is a type of Recurrent Neural Network (RNN) specifically designed to address problems involving long-range dependencies in sequential data, such as text [13]. LSTM utilizes complex memory units to capture intricate temporal relationships within text.

In event extraction, LSTM can be employed to model and extract recurring patterns found in descriptive event text [14]. LSTM models can learn from labeled training data containing event-related information and can subsequently be used to recognize and extract similar events from new texts.

One of the key advantages of LSTM is its ability to handle long temporal dependencies and understand a broader context within the text, enabling it to generate rich and informative representations of the events described in the text [9].

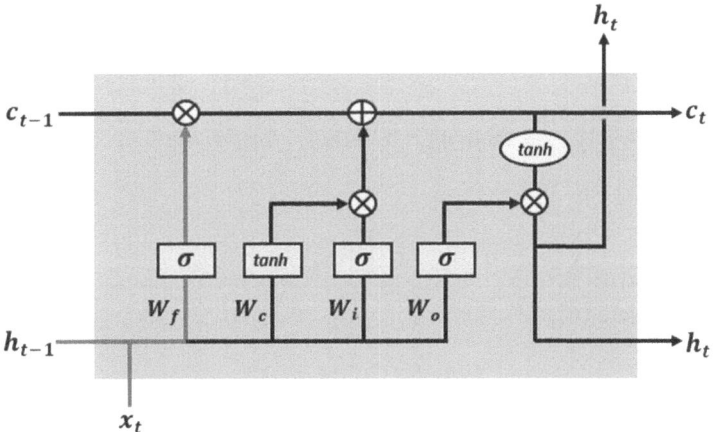

Fig. 2. Process Flow in LSTM

The flow of the LSTM (Long Short-Term Memory) method is illustrated in Fig. 2. It is used to process data sequences, such as text or time-series data, where the relationship between elements in a sequence is crucial. Below is the general flow of the LSTM method:

1. **Data Input**
   Sequential data (for example, word order in a sentence) is input into the LSTM model. Each element in the sequence is represented as a vector in a feature space.
2. **Forward Pass**
   At each time step, the LSTM computes the output for the current time step based on both the input at that step and the stored information from the previous time step. This is achieved through a series of complex mathematical operations within the LSTM [15].
3. **State Updates**
   LSTM has two types of internal memory: cell memory (for long-term information) and hidden state (for short-term information). These memories are updated and filtered through a set of internal gates at each time step.

4. Backpropagation Through Time (BPTT)

After completing the iteration for all data sequences, the LSTM model is evaluated using an appropriate loss function (e.g., cross-entropy loss for text classification) [16]. The gradient of the loss function with respect to the model parameters is then calculated using the backpropagation algorithm, enabling the model to "learn" from the data and update its parameters. Based on this, the LSTM method can be expressed as follows [17].

Update Gate (i)

$$i_t = \sigma(W_{xi} \cdot x_t + W_{hi} \cdot h_{t-1} + b_i) \tag{1}$$

Forget Gate (f)

$$f_t = \sigma(W_{xf} \cdot x_t + W_{hf} \cdot h_{t-1} + b_f) \tag{2}$$

Input Gate (g)

$$g_t = \tanh(W_{xg} \cdot x_t + W_{hg} \cdot h_{t-1} + b_g) \tag{3}$$

Output Gate (o)

$$o_t = \sigma(W_{xo} \cdot x_t + W_{ho} \cdot h_{t-1} + b_o) \tag{4}$$

Cell State Update (C_t)

$$C_t = f_t \cdot C_{t-1} + i_t \cdot g_t \tag{5}$$

Hidden State (h_t)

$$h_t = o_t \cdot \tanh(C_t) \tag{6}$$

In above equation, x_t is the input at time step t, h_{t-1} is the hidden state of the previous time step, W is weight matrix, b is bias vector, and σ is the sigmoid function. i_t, f, g_t, and o_t is controlling vector *gate*, meanwhile C_t are memory cells and h_t is forest memory generated at time step t.

2.3 BERT (Bidirectional Encoder Representations from Transformers)

BERT is a language model created by Google that employs Transformer architecture to understand the contextual relationships between words in text. BERT is well-known for its ability to comprehend context and handle a wide range of language task [7]. Document-level event extraction aims to identify event mentions and extract events that include event arguments and their corresponding roles from text. This paper proposes an end-to-end model for closed-domain tasks based on BERT. We introduce event type and entity node embeddings into the subsequent layer for event argument and role identification, representing the relationships between events, arguments, and roles, which enhances the accuracy of classifying multi-event arguments.

In event extraction, BERT can be used to comprehend and extract event-related information from text, considering the surrounding context [8]. BERT can be trained on various text data that contains events and their contexts, and then used to identify events and their relationships with entities and other information in the text.

One of the main advantages of BERT is its ability to understand contextual relationships between words, even in long and complex sentences. By leveraging its rich text representation, BERT can produce more accurate and informative event extraction results.

The second method can also be effectively used for event extraction, depending on the specific goals and characteristics of the available text data. LSTM excels in handling long temporal dependencies and identifying recurring patterns in text, while BERT is better at understanding the context and relationships between words in the text. Combining or using both methods together can also yield favorable results in event extraction. One of BERT's main features is its ability to understand the context of text as a whole. BERT uses an "unsupervised pre-training" strategy, where the model is trained on large text corpora, such as Wikipedia or books, to develop a deeper understanding of word meanings in their appropriate contexts. The word representation in BERT is described by Eq. 1 [18].

$$\text{BERT}(\omega_i) = \text{embed}(\omega_i) + \text{Pos}(i) + \text{Seg}(i) \qquad (7)$$

where:

ω_i is the i -th word in sentence

embed (ω_i) is a word embedding representation,

Pos(i) is the embedding of the word position in sentence,

Seg(i) is an embedding segment for marking sentence first or second in partner sentence (BERT can take two sentences as input for some task).

BERT has demonstrated remarkable capabilities in understanding text context and has had a significant impact on various NLP applications, including event extraction [19]. The following outlines the general flow of the BERT method:

1. **BERT Pretraining**

   BERT is trained on two main tasks: Masked Language Model (MLM) and Next Sentence Prediction (NSP). In MLM, some tokens in the input text are masked, and the model is tasked with predicting the missing tokens. In NSP, the model is given two sentences and must predict whether the second sentence logically follows the first in the original text.

2. **Fine-Tuning BERT**

   After pretraining, the BERT model can be fine-tuned for specific tasks, such as classification, sequence tagging, or information extraction, including event extraction. During fine tuning, the final layers of the BERT model may be modified or additional layers added to adapt the model to the desired task.

3. **Input Encoding**

   The input text is fed into the BERT model through a process called tokenization, specifically using WordPiece tokenization. Each token is converted into a word-sized vector representation (word embedding), which is then passed into the transformer layer.

4. **Transformer Blocks**

The BERT model consists of multiple transformer blocks, which are responsible for processing and understanding the text's context. Each transformer block has two sub-layers: the multi-head self-attention layer and the feedforward neural network [20].

5. **Output Layer**

In the final stage, the output of the BERT model can be obtained in various ways, depending on the task to be solved. For instance, for text classification, the output from the first token ("[CLS]") can be used as input for an additional classification layer. The flow of these stages is illustrated in Fig. 2.

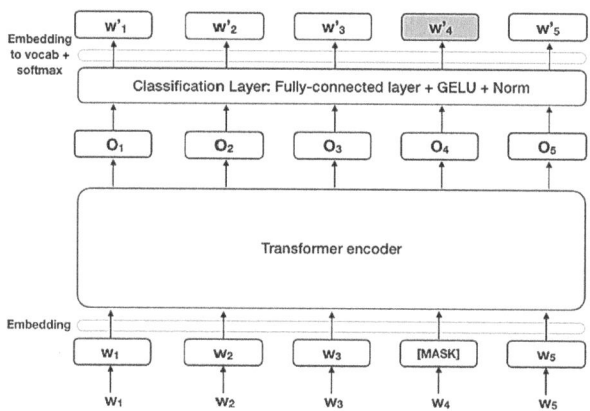

Fig. 3. BERT Architecture

Figure 3 represents the architecture and flow of a masked language model (MLM) using a Transformer encoder, which is commonly used in models like BERT (Bidirectional Encoder Representations from Transformers). Here's a detailed explanation:

1. Input Sequence:

   The input sequence consists of tokens W_1, W_2, W_3, [MASK], W_5, where MASK indicates a masked token that the model aims to predict [21].

2. Embedding Layer:

   Each token in the input sequence is converted into an embedding vector W_1, W_2, W_3, [MASK], W_5 through an embedding layer [22].

3. Transformer Encoder:

   The sequence of embeddings is passed through a Transformer encoder, which processes the entire sequence to capture contextual information for each token. The output of the encoder is a sequence of context-aware representations O_1, O_2, O_3, O_4, O_5 [23].

4. Classification Layer:

   Each context-aware representation O_1, O_2, O_3, O_4, O_5 is fed into a classification layer, which typically includes a fully-connected (dense) layer, followed by a GELU (Gaussian Error Linear Unit) activation function and normalization [24].

5. Output Sequence:

The classification layer outputs a sequence of vectors W'_1, W'_2, W'_3, W'_4, W'_5, where each vector corresponds to the predicted token for each position in the input sequence.

The output corresponding to the masked token *MASK* is W'_4, which is the model's prediction for the original token at that position.

6. Softmax Layer:

Finally, each output vector is passed through a softmax layer to convert it into a probability distribution over the vocabulary, allowing the model to predict the most likely token for each position.

The process begins with input tokenization, where the given text is split into tokens such as W_1, W_2, W_3, W_4, and W_5, with W_4 being masked [25]. Next, these tokens are converted into their corresponding embeddings through the embedding layer, resulting in vectors like W_1, W_2, W_3, *[MASK]*, and W_5. These embedding vectors are then processed by the Transformer encoder, which uses self-attention mechanisms to understand the context and relationships between tokens. The output of the encoder is a sequence of context-aware representations, such as O_1, O_2, O_3, O_4, and O_5 [26]. Each context-aware representation O_i is then passed through a classification layer to produce output vectors like W'_1, W'_2, W'_3, W'_4, *and* W'_5. Finally, the output vector corresponding to the masked token (W'_4) is used for prediction, where it is passed through a softmax layer to obtain a probability distribution over the entire vocabulary, ultimately predicting the original token that was masked.

The goal of this architecture is to predict the original tokens in the input sequence, especially the masked ones, using the context provided by the entire sequence. This process is a key component of pre-training language models like BERT, which can then be fine-tuned for various natural language processing (NLP) tasks.

2.4 Proposed Hybrid Model

The combination/hybrid between BERT and LSTM methods for event extraction offers substantial advantages in NLP, particularly in the analysis of narratives such as fairy tales. This approach leverages the strengths of both models: BERT's ability to deeply understand context and LSTM's proficiency in modeling the sequence of information in text. By utilizing BERT's robust contextual representation, the model can comprehend the connections between words and their overall context. Subsequently, LSTM processes the sequence of information, enabling the model to capture complex patterns in word order, including relationships between events, characters, and the broader story context. This combination enhances event extraction by improving both contextual awareness and the accuracy of sequence modeling. The anticipated outcome is a deeper understanding of the narrative structure of fairy tales and enhanced model performance in identifying and extracting significant events in literary works.

The architecture shown in Fig. 4 for event extraction in Andersen's fairy tales integrates two powerful models, BERT and LSTM, to identify and extract key events from the narrative. This method is tailored to capture both the contextual meaning of words and the temporal relationships between events, which are essential for comprehending the structure of stories like those by Andersen. The process begins with the Input Text, where the fairy tale is entered into the system. The first step is Text Preprocessing, where the raw text is tokenized and normalized. Tokenization splits the text into smaller units (tokens), such as words or subwords, while normalization ensures the text is standardized, facilitating easier processing by the model. This preparation sets the stage for more advanced analysis.

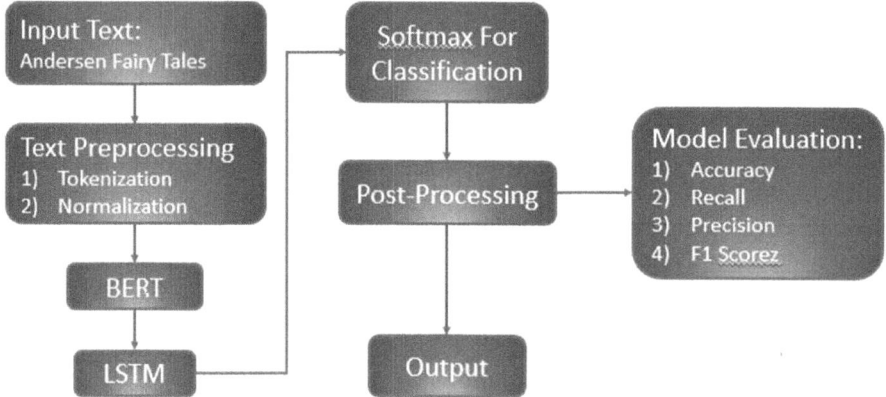

Fig. 4. Proposed model

The text is then input into the BERT model, which is based on transformer architecture and generates rich, context-aware vector representations for each word or token in the text. BERT achieves this by considering the surrounding words in both directions (bidirectional), enabling it to fully grasp the meaning of each token within its specific context. This step is essential because words in a narrative can carry different meanings depending on the surrounding context, and BERT effectively captures these nuances. The contextualized vectors generated by BERT are subsequently passed into an LSTM (Long Short-Term Memory) network. LSTM, a type of recurrent neural network (RNN), is renowned for its ability to capture dependencies over time or across sequences. In this setup, LSTM identifies the temporal relationships between events in the fairy tale, aiding in the understanding of the sequence of events and their interconnections within the narrative.

After passing through the LSTM, the output is directed to a Classification Layer, which employs a Softmax function to classify each token as either part of an event or not. This step enables the model to differentiate between significant events and merely supporting details. Finally, in the Post-Processing stage, the extracted events are grouped and arranged into a coherent sequence that reflects the unfolding events of the story. The end result is a structured sequence of events extracted from the fairy tale, which can be further used for analysis or applications like summarization, question answering, or automated storytelling. By integrating BERT and LSTM, this architecture harnesses both deep contextual comprehension and sequence modeling, making it especially effective for complex narratives such as Andersen's fairy tales.

The pseudocode in Fig. 5 describes a procedure for extracting narrative events from Hans Christian Andersen's fairy tales using a combined model that merges the advantages of both BERT and LSTM. This hybrid approach capitalizes on BERT's ability to

```
# Initialize BERT model for feature extraction
initialize BERT_model

# Load Andersen's fairy tales dataset
dataset = load_fairy_tales("Andersen_Fairy_Tales")

# Preprocess the text data
for each tale in dataset:
    preprocess tale
    split tale into sentences
    tokenize each sentence

# Represent each sentence with BERT embeddings
sentence_embeddings = []
for each sentence in tale:
    embedding = BERT_model.encode(sentence)
    sentence_embeddings.append(embedding)

# Initialize LSTM model for sequence learning
initialize LSTM_model with input_size = BERT_embedding_size

# Define the event extraction function
define extract_events(sentence_embeddings):
    lstm_output = LSTM_model.process(sentence_embeddings)
    events = []

    # Identify narrative events
    for each output in lstm_output:
        if output represents an event:
            events.append(output)
    return events

# Extract events from each tale
all_events = []
for each tale in dataset:
    sentence_embeddings = [BERT_model.encode(sentence) for sentence in tale]
    events = extract_events(sentence_embeddings)
    all_events.append(events)

# Output extracted events
for each tale, events in zip(dataset, all_events):
    display("Tale:", tale)
    display("Extracted Events:", events)
```

Fig. 5. BERT-LSTM in pseudocode

generate contextual sentence representations and LSTM's strength in sequence processing, enabling the identification and extraction of relevant narrative events across multiple sentences in the stories.

The BERT model in Fig. 5 is first initialized to generate sentence embeddings. The Andersen fairy tales dataset is then loaded and preprocessed, with each story being divided into individual sentences. Each sentence is tokenized to prepare it for BERT's encoding process. Using BERT, each sentence is converted into a numerical representation (embedding) that captures both its meaning and contextual information. This step is crucial for understanding the intricate relationships between words and phrases within each sentence, which is essential for accurate event extraction.

After generating the sentence embeddings, the LSTM model is initialized. LSTM is highly effective for sequence learning tasks, allowing it to handle the sequential dependencies within the sentence embeddings produced by BERT. The LSTM processes the full sequence of sentence embeddings for each tale, learning the narrative flow and the contextual links between sentences. This step is important for identifying narrative events, as they are often defined by not just individual sentences, but by their relationships within the larger narrative context.

The primary event extraction task is managed by the extract_events function. For each output produced by the LSTM, it determines whether the output represents an event. These events are then organized and stored for each fairy tale. Once all tales are processed, the extracted events are presented as output, providing a structured view of the narrative sequence. This methodology allows for a thorough understanding of the story structure and the specific narrative elements within each fairy tale, which is useful for further analysis or applications in areas like storytelling, literature analysis, and automated narrative generation.

In summary, this hybrid BERT-LSTM approach successfully combines the strengths of both models to address the complexities of narrative event extraction. By utilizing BERT for contextually rich sentence embeddings and LSTM for sequence modeling, this combined approach improves event extraction, facilitating a systematic analysis and categorization of the narrative flow in Andersen's fairy tales.

3 Results and Discussion

In the event extraction results using the LSTM method, the model demonstrates its capability to capture repetitive patterns and long-term dependencies in text. By employing a complex memory unit, LSTM is able to generate detailed and informative representations of the events described in the text. The extraction results from LSTM typically emphasize the temporal relationships between entities or words in the text, offering a deeper understanding of the sequence and interconnections between events in the narrative.

Conversely, the event extraction results using the BERT method highlight the model's ability to comprehend context and relationships between words in the text which is shown in Fig. 6, considering the broader context. By leveraging rich, multidimensional representations of the text, BERT is able to produce more accurate and contextually relevant event extractions. The results from BERT tend to offer a more comprehensive view of events, including the entities involved, their context, and their relationships with other elements in the text.

Both methods have their unique advantages and characteristics when it comes to event extraction from text. While LSTM excels in handling temporal distances and iterative patterns, BERT shines in understanding context and the relationships between words. When used together, these methods can provide more comprehensive and informative event extraction results, drawing on the strengths of each approach to better understand and extract information from the text.

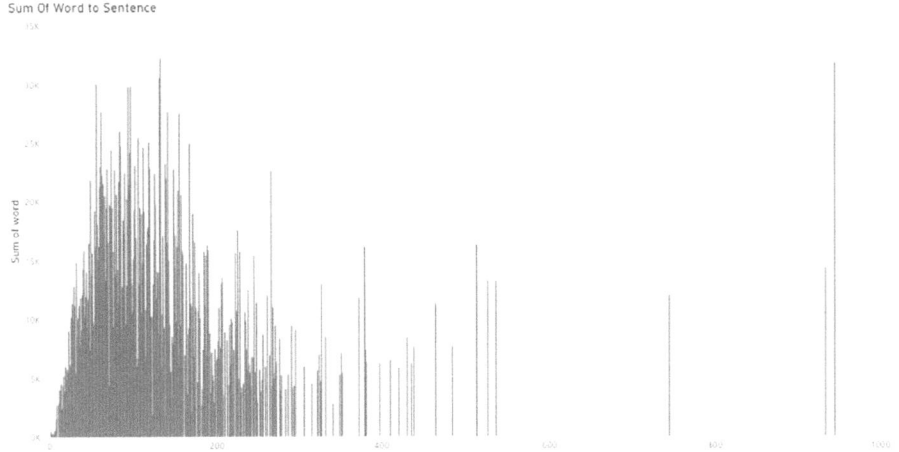

Fig. 6. Sum Of Words in Sentences

In event extraction analysis, understanding the relationship between the number of words and sentences in each story can be valuable in several ways which is shown in Fig. 7. First, knowing the word and sentence count helps determine the complexity and length of the story, which can influence the number of events that can be extracted. Second, the word-to-sentence ratio provides insight into the density of information within the story, which can affect the difficulty of event extraction which is shown in Fig. 8. Therefore, analyzing the relationship between the number of words, sentences, and the word-to-sentence ratio in the dataset can assist in refining the extraction strategy, evaluating the story's complexity, and estimating the expected outcomes of the extraction process.

In preparing for event extraction using the LSTM method, the first step is to collect and organize the text data to be used for training. This data typically consists of a corpus containing event-related information along with its context. Next, the text is processed and converted into a format suitable for the LSTM model, including tasks like tokenizing words, vectorizing, and splitting the data into training and test sets. The LSTM training process involves feeding the text data into a randomly initialized model, and then adjusting the model parameters through a learning process that uses the backpropagation algorithm. This cycle repeats until the model reaches the desired performance level. After training is complete, the LSTM model is ready for use in the event extraction phase. In contrast, preparing for event extraction with the BERT method

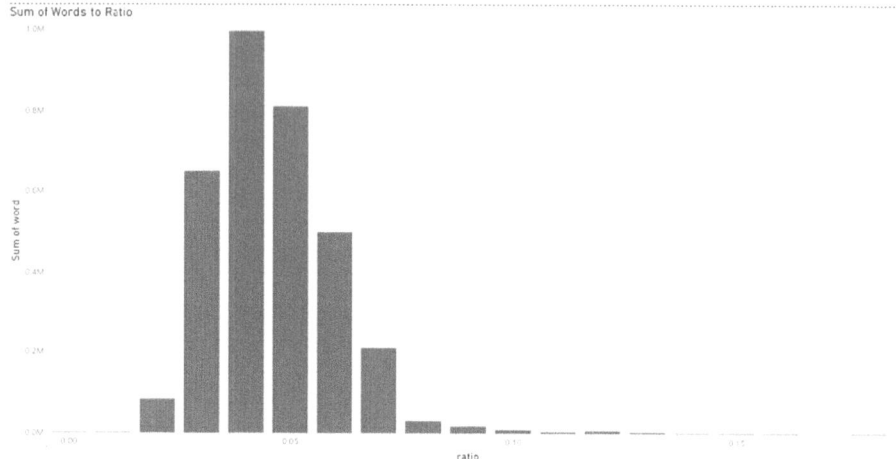

Fig. 7. Sum Of Words to Ratio

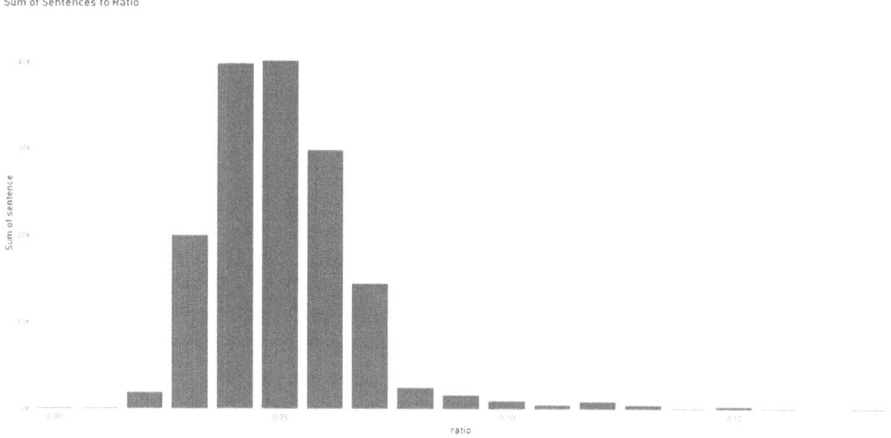

Fig. 8. Sum Of Sentences to Ratio

involves similar steps in data collection and preparation, but with a focus on formatting the data for BERT's input requirements.

The text data is converted into a vector format compatible with BERT, and split into training and test sets. The BERT training process consists of feeding the text into a pre-trained BERT model and fine-tuning its weights to suit the specific event extraction task. This process is also repeated until the model achieves the desired performance. Throughout this process, both LSTM and BERT models are evaluated using separate test data to assess their event extraction performance which is shown in Table 1. Evaluation metrics may include qualitative measurements such as accuracy, precision, recall, and F1-score, as well as testing on new datasets to ensure the models' ability to generalize well.

The discussion of the event extraction results from the LSTM method will highlight the model's ability to detect temporal patterns and relationships between entities in the text, focusing on the sequence of events. In contrast, the BERT-based discussion will emphasize the model's capacity to understand context and word relationships, providing a more comprehensive understanding of events, including the entities involved, their context, and their relationships with other information in the text. A comparative analysis between the two methods will offer valuable insights into the strengths and weaknesses of each, as well as the potential benefits of using them together for more effective event extraction results.

Table 1. Evaluation using comparison with other methods

Model	Precision	Recall	F1-Score
BERT-LSTM (Hybrid)	0.88	0.82	0.75
BERT (without LSTM)	0.82	0.76	0.79
LSTM (without BERT)	0.8	0.7	0.74
CNN	0.78	0.65	0.71
CRF	0.76	0.68	0.72

The results of the study compared with other models shown in Table 1 and Fig. 9 show that the BERT-LSTM hybrid model has superior performance compared to other models, both in terms of precision, recall, and F1-score. This model is designed to combine the advantages of BERT, which excels in understanding semantic context in depth, with LSTM, which is able to capture sequential relationships in text. Thus, BERT-LSTM successfully handles two important aspects in narrative analysis, namely understanding global context and processing event sequences.

In the performance evaluation, the BERT-LSTM hybrid model achieved a precision of 0.88, demonstrating its high accuracy in identifying relevant narrative events. When compared to other models, such as BERT without LSTM, which achieved a precision of 0.82, or LSTM without BERT, which had a precision of 0.80, the hybrid model proved more effective in minimizing false positives. This highlights that combining the BERT and LSTM architectures results in more precise event recognition.

Furthermore, the BERT-LSTM model exhibited a recall of 0.82, indicating its ability to capture most narrative events in the dataset. While BERT without LSTM reached a recall of 0.76, and CRF had a recall of 0.68, the BERT-LSTM model still outperformed these models in ensuring a more comprehensive event capture. This high recall is particularly crucial in narrative extraction, where the omission of relevant events can affect the understanding of the overall story structure.

Regarding the F1-score, which balances precision and recall, BERT-LSTM achieved a value of 0.75, surpassing other models like LSTM without BERT (0.74), BERT without LSTM (0.79), and CNN (0.71). While the F1-score of BERT alone is similar to that of BERT-LSTM, the hybrid model's overall performance is more consistent across all evaluation metrics, making it a better solution for narrative event extraction tasks.

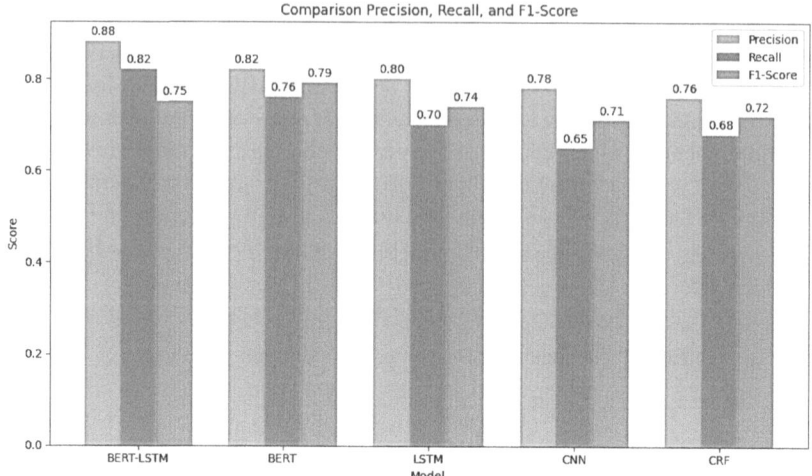

Fig. 9. Visualization for comparison with other methods

This comparison demonstrates that the hybrid approach has clear advantages over models based on a single architecture. By utilizing BERT for semantic understanding and LSTM for managing temporal relationships, the BERT-LSTM model overcomes the limitations of each individual model. This advantage significantly contributes to the analysis of classic literature, such as Andersen's fairy tales, while also offering potential for further applications in digital humanities, automated story creation, and technology-driven educational tools.

4 Conclusion

Datasets for event extraction in specific domains have distinct characteristics when compared to datasets in open domains. A closed domain refers to datasets sourced from a specific field or subject area, such as medical, financial, legal, or technical sectors. These datasets are highly relevant to particular industries or domains. The extracted information typically focuses on events or entities that are pertinent to the domain's subject matter. In event extraction, LSTM can be used to model and extract recurring patterns within descriptive text about incidents. LSTM models can learn from training data, which consists of labeled text containing event-related information, and then apply this knowledge to recognize and extract similar incidents from new texts. In event extraction, BERT can be employed to understand and extract event-related information from text, taking into account the surrounding context. BERT can be trained on various text datasets containing events and their contexts and then used to identify events and their connections to entities and other information in the text. A key advantage of BERT is its ability to understand contextual relationships between words, even in long and complex sentences.

References

1. Shuo, W., Yuan, R., Xiaobing, F., Jiangnan, Q.: Joint event extraction model based on multi-feature fusion. Procedia Comput. Sci. **174**, 115–122 (2020). https://doi.org/10.1016/j.procs.2020.06.066
2. Wibawa, A.P., et al.: Decoding and preserving Indonesia's iconic Keris via A CNN-based classification. Telemat. Inform. Rep. **13**, 100120 (2024). https://doi.org/10.1016/j.teler.2024.100120
3. Scaboro, S., Portelli, B., Chersoni, E., Santus, E., Serra, G.: Extensive evaluation of transformer-based architectures for adverse drug events extraction. Knowl.-Based Syst. **275**, 110675 (2023). https://doi.org/10.1016/j.knosys.2023.110675
4. Utama, A.B.P., et al.: Improving time-series forecasting performance using imputation techniques in deep learning. In: 2024 International Conference on Smart Computing, IoT and Machine Learning (SIML), pp. 232–238. IEEE (2024). https://doi.org/10.1109/SIML61815.2024.10578273
5. Wang, Z., Liu, Z., Luo, L., Chen, X.: A multi-neural network fusion based method for financial event subject extraction. In: 2020 3rd International Conference on Advanced Electronic Materials, Computers and Software Engineering (AEMCSE) (2020). https://doi.org/10.1109/aemcse50948.2020.00084
6. Guda, V., Sanampudi, S.K.: A hybrid method for extraction of events from natural language text. In: Advances in Intelligent Systems and Computing, pp. 301–307. Springer, Singapore (2017). https://doi.org/10.1007/978-981-10-3223-3_28
7. Yu, X., Rong, W., Liu, J., Zhou, D., Ouyang, Y., Xiong, Z.: LSTM-based end-to-end framework for biomedical event extraction. IEEE/ACM Trans. Comput. Biol. Bioinforma. **17**(6), 2029–2039 (2020). https://doi.org/10.1109/TCBB.2019.2916346
8. Kuila, A., Bussa, S.C., Sarkar, S.: A neural network based event extraction system for Indian languages. Fire (2018)
9. Purnawansyah, P., et al.: Congestion predictive modelling on network dataset using ensemble deep learning. J. Appl. Data Sci. **5**(4), 1597–1613 (2024). https://doi.org/10.47738/jads.v5i4.333
10. Yazdi, M.A., Farhadi Ghalatia, P., Heinrichs, B.: Event log abstraction in client-server applications. In: Proceedings of the 13th International Joint Conference on Knowledge Discovery, Knowledge Engineering and Knowledge Management, SCITEPRESS - Science and Technology Publications, pp. 27–36 (2021). https://doi.org/10.5220/0010652000003064
11. Nada, A.Q., Wibawa, A.P., Putri Syarifa, D.F., Fajarwati, E., Putri, F.I.: Optimizing Indonesian-sundanese bilingual translation with adam-based neural machine translation. J. RESTI (Rekayasa Sist. dan Teknol. Informasi) 8(6), 690–700 (2024). https://doi.org/10.29207/resti.v8i6.6116
12. Juárez Gambino, O., Ortega-Pacheco, J.-D., García-Mendoza, C.-V., Felix-Mata, M.: Automatic detection and registration of events by analyzing email content. Res. Comput. Sci. **130**(1), 35–43 (2016). https://doi.org/10.13053/rcs-130-1-3
13. Yu, W., Yi, M., Huang, X., Yi, X., Yuan, Q.: Make it directly: event extraction based on tree-LSTM and bi-GRU. IEEE Access **8**, 14344–14354 (2020). https://doi.org/10.1109/access.2020.2965964
14. Chen, Z., Ji, W., Ding, L., Song, B.: Fine-grained document-level financial event argument extraction approach. Eng. Appl. Artif. Intell. **121** (2023). https://doi.org/10.1016/j.engappai.2023.105943
15. Ding, R., Li, Z.: Event extraction with deep contextualized word representation and multi-attention layer. In: Advanced Data Mining and Applications, pp. 189–201. Springer, Cham (2018). https://doi.org/10.1007/978-3-030-05090-0_17

16. Song, D., Xu, J., Pang, J., Huang, H.: Classifier-adaptation knowledge distillation framework for relation extraction and event detection with imbalanced data. Inf. Sci. (Ny) **573**, 222–238 (2021). https://doi.org/10.1016/j.ins.2021.05.045
17. Vo, T.: SynSeq4ED: a novel event-aware text representation learning for event detection. Neural. Process. Lett. **54**(1), 227–249 (2022). https://doi.org/10.1007/s11063-021-10627-2
18. Liu, J., Chen, Y., Liu, K., Bi, W., Liu, X.: Event extraction as machine reading comprehension. In: Proceedings of the 2020 Conference on Empirical Methods in Natural Language Processing (EMNLP), Stroudsburg, PA, USA, pp. 1641–1651. Association for Computational Linguistics (2020). https://doi.org/10.18653/v1/2020.emnlp-main.128
19. Supriyono, Wibawa, A.P., Suyono, Kurniawan, F.: Advancements in natural language processing: implications, challenges, and future directions. Telemat. Inform. Rep. **16**, 100173 (2024). https://doi.org/10.1016/j.teler.2024.100173
20. Suryanto, T.L.M., Wibawa, A.P., Hariyono, H., Nafalski, A.: Comparative performance of transformer models for cultural heritage in NLP tasks. Adv. Sustain. Sci. Eng. Technol. **7**(1), 0250115 (2025). https://doi.org/10.26877/asset.v7i1.1211
21. Wang, Y., Luo, S., Hu, Z., Han, M.: A study of event elements extraction on Chinese bond news texts. In: 2018 IEEE International Conference on Progress in Informatics and Computing (PIC), pp. 420–424. IEEE (2018). https://doi.org/10.1109/PIC.2018.8706322
22. Yang, G., Li, M., Zhang, J., Lin, X., Chang, S.-F., Ji, H.: Video event extraction via tracking visual states of arguments (2022). https://doi.org/10.48550/arXiv.2211.01781
23. Permataning Tyas, S.M., Sarno, R., Haryono, A.T., Rossa Sungkono, K.: A robustly optimized BERT using random oversampling for analyzing imbalanced stock news sentiment data. In: 2023 International Conference on Computer Science, Information Technology and Engineering (ICCoSITE), Jakarta, Indonesia, pp. 897–902. IEEE (2023). https://doi.org/10.1109/ICCoSITE57641.2023.10127725
24. Kan, Z., Qiao, L., Yang, S., Liu, F., Huang, F.: Event arguments extraction via dilate gated convolutional neural network with enhanced local features. IEEE Access **8**, 123483–123491 (2020). https://doi.org/10.1109/ACCESS.2020.3004378
25. Shreyas, A., Priyanka, G., Pearl, M., Swapnal, S.: Event information extraction from e-mail and updating event in calendar. Int. J. Adv. Res. Innov. Ideas Educ. **4**(3) (2018)
26. Kar, D., Sarkar, S., Goyal, P.: Event argument extraction using causal knowledge structures. In: 17th International Conference on Natural Language Processing (ICON 2020) (2021). https://doi.org/10.48550/arXiv.2105.00477

Galvanic Corrosion Progression Analysis at Aluminium-Carbon Fiber Interfaces in ACCC/TW Conductors

Mohamad Izzat Nawawi[1](✉), Muhammad Iqbal Shafei[1], Shahnurriman Abdul Rahman[1], and Konstantinos Kopsidas[2]

[1] Faculty of Engineering and Built Environment, Universiti Sains Islam Malaysia Nilai, Nilai, Malaysia
izzatnawawi@raudah.usim.edu.my

[2] The University of Manchester, Manchester, UK

Abstract. Galvanic corrosion poses a large risk particularly to bimetallic overhead conductors as the phenomenon may potentially degrade the conductors' structural integrity, electrical performance, and overall service life. Various environmental and operational factors may accelerate the galvanic corrosion development within the conductors which worsen the situation if not properly addressed. This paper focuses on analyzing the impact of galvanic corrosion on the total area of the aluminum strands hence the ampacity of the Aluminum Conductor Composite Core (ACCC/TW) conductors. Using the numerical modeling approach of COMSOL Multi-physics, the galvanic corrosion occurring at the aluminum-carbon fiber core interfaces within ACCC/TW conductors are simulated. The impacts of galvanic corrosion development are observed for both perfect and imperfect interfaces which later demonstrate the existence of 25% and 50% artificial mechanical crack in the fiberglass layer. The simulated results reveal a relationship between increased in mechanical crack and accelerated aluminum strand area loss, with a corresponding reduction in ampacity. For example, after 25 months of exposure to galvanic corrosion, the ACCC/TW with 25% crack would lose 4.6% of aluminium strands which reduce the conductor ampacity by 2.3%. The ampacity reduction is found to be worse at 5.03% for the ACCC/TW with 50% of crack. These simulations reveal that as the extent of cracking increases, the rate of aluminum corrosion increases significantly, directly impacting the conductor's ampacity.

Keywords: Galvanic corrosion · conductors · modelling

1 Introduction

The overhead transmission lines serve as the basis of modern electrical power system networks. They are vital to transfer electricity from the power plants to the consumers, and they are holding a crucial position within the overall transmission line [1]. With increasing demands on power grids and the push toward integrating renewable energy sources, the durability and efficiency of these conductors are more critical than ever. However, corrosion poses operational challenges to the power utilities as the phenomenon may degrade the structural integrity and performance of these conductors, potentially shortening their service life and diminishing their electrical efficiency [2]. Galvanic corrosion is a particular concern for bimetallic conductors of Aluminum Conductor Composite Core (ACCC) that area may continuously exposed to moisture and other corrosive elements during their operation [3]. The galvanic corrosion creates the electrochemical degradation process when the aluminum and carbon fiber, are in contact within a conductive environment which eventually creates an electrochemical cell that accelerates the loss of aluminium atomic bonding. The resulting degradation can lead to a reduction in cross-sectional area, increasing resistance, and eventually compromised mechanical and electrical performance if not properly addressed [4].

The Aluminum Conductor Composite Core (ACCC) has shown to offer various benefits over the other traditional aluminium alloy conductors [5]. For example, the ACCC conductors are lighter in weight, less sag at high temperature, and exhibit improved resistance to galvanic coupling and corrosion [3]. Nevertheless, the ACCC/TW conductors may still be susceptible to corrosion under certain conditions, especially when protective barriers, such as the fiberglass layer around the core, experience mechanical damage or cracking. Mechanical crack may happen to the fiber glass layer when the conductor is installed in harsh environmental conditions such as strong winds, high temperatures and extreme thermal cycling and causing high tension during operation [6]. The crack in the fiberglass layer exposes the carbon fiber core to aluminum strands, potentially accelerating galvanic corrosion and amplifying ampacity loss over time.

There are studies [7, 8] focusing on the benefits of composite core conductors from a mechanical and structural view, but the effect of galvanic corrosion on the performance and durability of the "imperfect" ACCC/TW conductors remains underexplored. This study addresses this knowledge gap by analyzing the impact of galvanic corrosion on two perspectives which are the loss of area in the aluminium strands and the ampacity. Through COMSOL numerical modelling approach, the electrochemical processes occurring at the aluminum-carbon fiber interface are simulated which then allows for comparing its impact on the perfect conductor model and imperfect conductor model that has 25% or 50% artificial mechanical crack on its fiberglass layer. The analyses are aimed to evaluate the relationship between the severity of cracks and the corrosion rate. The results would offer an important understanding of the structural strength of ACCC/TW conductors, emphasizing the significance of maintaining fiberglass layer integrity to preserve ampacity and prolong service life in corrosive conditions.

Section 2 and Sect. 3 of this paper will provide the fundamental understanding on the design and materials of composite ACCC/TW conductor and the development of galvanic corrosion mechanism within the conductors respectively. Section 4 explains the methodology used to model the galvanic corrosion phenomena, while the results are presented and analyzed in Sect. 5. Following that, the study is concluded in Sect. 6.

2 Composite Core Conductors

2.1 The Advanced ACCC/TW Conductors

Composite core conductors of Aluminum Conductor Composite Core with Trapezoidal Wire (ACCC/TW), represent an advancement in conductor technology aimed at enhancing performance and durability under various operational stresses. The conductors incorporate a composite core made of carbon fiber encased in a fiberglass layer. The core of ACCC conductors is a central element in their design, contributing to both mechanical stability and corrosion resistance. Made from high-strength carbon fiber, the composite core provides exceptional tensile strength while remaining significantly lighter than steel. This reduction in weight allows for a greater proportion of aluminum to be included in the conductor without increasing its overall weight, which is essential for enhancing electrical conductivity and ampacity. Moreover, the use of a carbon fiber core reduces the overall conductor sag at high operating temperatures—a critical factor for overhead lines that need to maintain adequate ground clearance under load conditions [9].

The outer layer of the composite core is typically layered in fiberglass to provide an additional protective barrier against environmental and mechanical stresses. This fiberglass layer serves as an insulator that prevents direct contact between the aluminum strands and the carbon fiber core, which is essential for reducing galvanic corrosion risk. In ACCC/TW conductors as illustrated in Fig. 1 (right) the aluminum is shaped into trapezoidal wire (TW) strands, which enables a denser packing arrangement around the core. This configuration increases the aluminum cross-sectional area, enhancing the conductor's ampacity and overall conductivity [10].

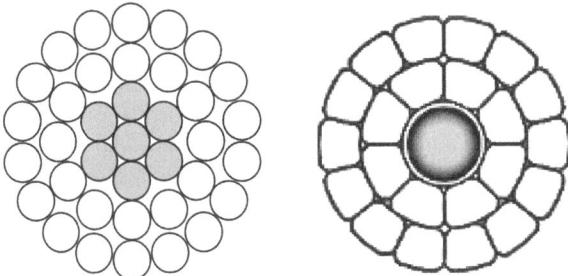

Fig. 1. Visualization of round (left) and trapezoidal (right) conductors

In comparison, Fig. 1 (left) shows the Aluminium Conductor Steel Reinforced (ACSR) conductors which rely on a central steel core that, while providing strength, is also susceptible to galvanic corrosion when in contact with aluminum. Steel and aluminum form a galvanic couple in the presence of an electrolyte, leading to accelerated corrosion of the aluminum strands over time, especially in environments with high moisture levels. This corrosion degrades the aluminum area, reducing the conductor's current-carrying capacity (ampacity) and structural stability. Although ACSR conductors are widely used and cost-effective, their susceptibility to corrosion requires regular maintenance and can reduce their operational lifespan in harsh environmental conditions [11].

The design and material choices in ACCC conductors offer enhanced corrosion resistance compared to ACSR, as the non-metallic composite core eliminates the primary bimetallic interface where galvanic corrosion typically occurs. Additionally, the composite core's higher strength-to-weight ratio allows for greater ampacity without the thermal expansion issues associated with steel and designed to operate continuously at 180 °C and in emergencies up to 200 °C [12]. However, while the ACCC design mitigates many of the challenges seen in ACSR conductors, it is not immune to damage. For instance, if the fiberglass layer surrounding the carbon fiber core becomes cracked or damaged, the risk of galvanic corrosion at the aluminum interface increases, potentially accelerating conductor degradation. This study explores how such damage impacts corrosion progression and ampacity, under varying operational and environmental conditions.

2.2 The Use of Aluminium Trapezoidal Wire in ACCC/TW Conductors

The ACCC/TW conductors use the trapezoidal shape wire in its aluminium strands. The analysis involving the trapezoidal wire in this study is particularly relevant because the increased aluminum area is directly exposed to environmental factors that can initiate and accelerate corrosion, especially when galvanic coupling with the composite core occurs. The larger aluminum surface area, coupled with a more densely packed structure, amplifies the impact of any corrosion-related aluminum loss on ampacity performance. Additionally, the dense packing of trapezoidal strands reduces gaps between wires, affecting how moisture or corrosive agents penetrate the conductor which becomes the key factors that influence the corrosion process [13].

The use of trapezoidal wire in the ACCC/TW conductor also allows for a more controlled and comparative analysis of corrosion progression. The enhanced aluminum area can provide clearer understandings into how varying degrees of fiberglass cracking influence ampacity loss, as even small amounts of corrosion in TW conductors can cause noticeable ampacity reductions. This makes trapezoidal wire an optimal choice for quantifying the impacts of galvanic corrosion on conductor performance in this study.

3 Galvanic Corrosion in Composite Core Conductors

The galvanic corrosion process in ACCC/TW conductors begins at the aluminum-carbon fiber interface when a path is created for electrochemical reactions. Theoretically, under the ideal condition, the fiberglass layer surrounding the carbon fiber core acts as an insulator, preventing direct electrical contact between the aluminum strands and the carbon fiber. However, when the fiberglass layer sustains damage, such as cracks due to thermal cycling, mechanical stress, or other environmental factors, the barrier may be compromised. This damage allows electrolyte access, enabling galvanic coupling between the aluminum and carbon fiber, which accelerates corrosion.

3.1 Galvanic Series: Electrochemical Potential and Nobility of Metal

Polymer matrix composites, utilized in the ACCC/TW core and reinforced with carbon fibers or similar conductive fibers, can contribute to metal corrosion under specific conditions. Carbon fibers, often referred to as graphite fibers, may not be entirely encased within the polymer matrix, allowing them to engage in galvanic corrosion when an electrolyte is present. Carbon fiber is more noble than most metals (see the galvanic series in Fig. 2) and can cause a strong corrosive attack of metallic components. For ACCC/TW conductors, carbon fibers may act as a noble electrode if the protective galvanic barrier becomes compromised. Galvanic corrosion arises due to the electrochemical potential difference between aluminum, which is more anodic, and carbon fiber, which is more cathodic. This potential difference is governed by their respective positions in the galvanic series as shown in Fig. 2. Metals and carbon fiber each have their own unique electrode potential, which is essentially a measure of how much a material wants to give up or receive electrons, and this is measured in volts. Materials that are further apart from one another in electrode potential, and thus further apart in the galvanic series, have a higher potential for galvanic corrosion when they come in contact. Carbon fiber and aluminum are far apart in the galvanic series, so they have a high potential for galvanic corrosion if they are used together [14].

Fig. 2. Galvanic Series [4]

3.2 Electrochemical Reactions within ACCC/TW Conductors

In the context of ACCC (Aluminum Conductor Composite Core) conductors, galvanic corrosion typically involves the aluminum strands acting as the anode (where oxidation occurs) and the carbon fiber core acting as a cathode (where reduction occurs) only if the fiberglass barrier is compromised. Although carbon fiber itself is not a metal, it is conductive and can promote galvanic reactions in contact with metals like aluminum, especially in the presence of an electrolyte (e.g., moisture, salt water). The primary anodic reaction occurs at the aluminum, where aluminum atoms lose electrons and form aluminum ions. These ions then react with hydroxide ions in the electrolyte to produce aluminum hydroxide, a corrosion product. The anodic reaction reduces the aluminum's structural integrity, which can lead to localized thinning and pits in the conductor strands [15]. When both anodic and cathodic reactions occur, the complete reaction results in the consumption of aluminum and the formation of aluminum hydroxide as in (1):

$$Overall: 4Al + 3O_2 + 6H_2O \rightarrow 4Al(OH)_3 \qquad (1)$$

These corrosion products can accumulate over time, leading to a buildup that effectively reduces the cross-sectional area of the conductor. The reduction in cross-sectional area decreases the electrical conductivity of the conductor and increases its resistance, potentially affecting the efficiency and performance of the transmission line.

4 Methodology

The analysis of galvanic corrosion on ACCC/TW overhead conductor is studied using COMSOL Multi-physics and the methodological flowchart is presented in Fig. 3. The simulation process begins with 2D model preparation, where conductor geometry and material properties are defined for three scenarios: Perfect with no crack, 25% crack, and 50% crack at the fiberglass layer. The next phase involves setting up the galvanic corrosion simulation, specifying electrochemical properties, boundary conditions, and electrolyte parameters to accurately model the corrosion reactions at the aluminum-carbon fiber interface. Following this, a corrosion simulation is executed to analyze aluminum loss over time in each cracking scenario. The resulting ampacity reduction is then calculated due to aluminium degradation. Finally, data analysis compares aluminum loss and ampacity reduction across models, interpreting how increased fiberglass cracking impacts corrosion rate, ampacity, and conductor longevity.

4.1 Conductor Geometry and Structural Definition

The integrity of the fiberglass layer is crucial to prevent galvanic corrosion. When this layer is intact, it would effectively insulate the aluminum from the carbon fiber core, limiting corrosion development even in humid or salt-rich environments. However, the presence of cracks in the fiberglass layer greatly increases the risk of galvanic corrosion. Even minor crack can allow moisture and ions to penetrate the interface, initiating and accelerating the corrosion process. To further investigate this, artificial mechanical cracks at the fiberglass layer are modelled. The 25% crack model would indicate that 25% of the area at the fiberglass layer has been compromised which allows for the galvanic reaction between the aluminum and core component. The same concept is applied for the conductor model with 50% crack.

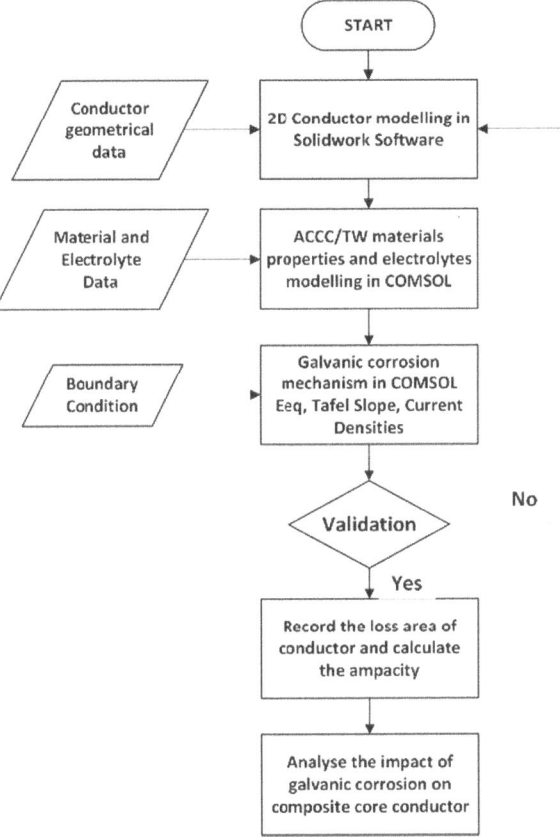

Fig. 3. Flowchart in modelling galvanic corrosion

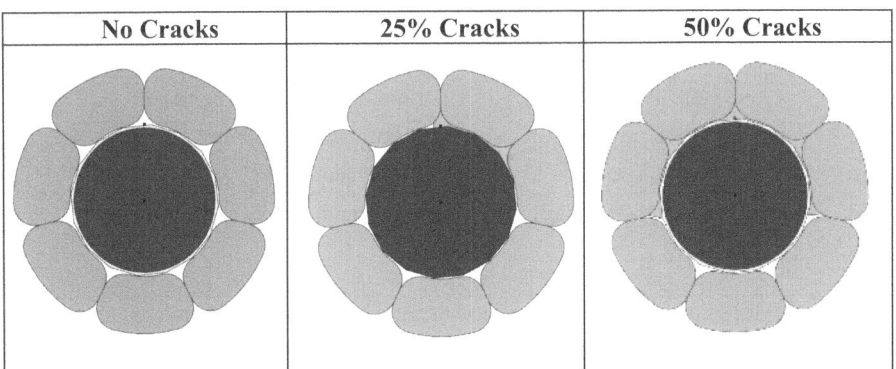

Fig. 4. Two-dimensional model of ACCC/TW conductors

The center of the conductor model represents the carbon fiber composite core, which is encased by a fiberglass insulating layer, and surrounded by aluminum strands shaped into a trapezoidal cross-section. The gaps between the core and the aluminum strands that are highlighted in blue represent the electrolyte. For this investigation, the ACCC/TW conductor with the codename of Hawk (outer diameter of 21.76 mm) is used and Fig. 4 shows the three conductor models used in the analysis. The 25% crack model indicates that the crack allows 25% of the electrolyte that seep through the damaged section simulating the extent of exposure to galvanic corrosion under partial mechanical failure of the fiberglass layer.

4.2 Conductor Geometry and Structural Definition

The galvanic corrosion modelling requires detailed material properties for aluminum, carbon fiber, and fiberglass, each with specific electrochemical and physical characteristics based on literature and experimental data. These parameters are input into the COMSOL model, to calculate the corrosion rate, material loss and changes in electrical and structural properties, using time-dependent simulations to observe corrosion progression in each cracking scenario [16]. The parameters provided in Table 1 are used to model galvanic corrosion processes that involve carbon fiber and aluminum.

Table 1. Input value and function for COMSOL

Parameter	Value	Remarks
Eeq_cat	−0.0605[V]	Equilibrium potential, cathode
i0_cat	0.0547[A/m^2]	Exchange current density, cathode
A_cat	−1.4949[V]	Tafel slope, cathode
Eeq_an	−0.952[V]	Equilibrium potential, anode
i0_an	0.004232[A/m^2]	Exchange current density, anode
A_an	0.1307[V]	Tafel slope, anode
ilim_an	10^2[A/m^2]	Limiting current density, anode
sigma	2.5[S/m]	Electrolyte conductivity
rho_Al	2700[kg/m^3]	Aluminium density
M_Al	0.026982[kg/mol]	Aluminium molecular weight

The Tafel slopes are included to describe the rate of reaction at the anode and cathode surfaces. These slopes show how the corrosion current density increases with overpotential, with distinct values for the anodic (aluminum oxidation) and cathodic (oxygen reduction on carbon fiber) reactions. Tafel slopes value allow for realistic modeling of current density behavior under different environmental and operational conditions.

4.3 Boundary Condition

To accurately model real-world corrosion, the model incorporates boundary conditions that simulate typical environmental factors experienced by overhead conductors. Electrolyte (representing water or moisture with dissolved ions) is applied at the innermost part of the conductor's model to simulate conditions conducive to corrosion. This electrolyte layer allows for ion transfers between anodic and cathodic sites, which is important for galvanic reactions. Boundary conditions are also applied to the aluminum strands (anodic reaction) and the exposed carbon fiber regions (cathodic reaction), allowing COMSOL to calculate the electron exchange and resultant material loss.

In the modelling of ACCC/TW conductors for this study, some assumptions are made to simplify the analysis and focus on the specific parts of galvanic corrosion. External environment parameters such as temperature, electric potential, and ion concentration are not considered directly. These assumptions allow the study to concentrate on the internal corrosion mechanisms. For this modelling, the inner layer carbon fiber that acts as a cathode is assumed to be unaffected by the corrosion process. The model starts with full material integrity in the aluminum strands, with a gradually increasing loss of material as corrosion progresses. The exposure time of the conductors is varied up to 25 months to analyze corrosion over different time frames, simulating both short-term and prolonged exposure.

4.4 Validation of the Model

The developed perfect model is first validated by comparing the results with the experimental and literature data from [16]. The comparison indicates that the model accurately predicts the current density, aluminium cross-sectional area loss and ampacity reduction with deviations within 5–10%. The simulated galvanic corrosion of ACCC/TW obtained through COMSOL agrees with the experiment data validating the accuracy of the computational model. Additionally, the model successfully replicates the electrochemical potential distribution and corrosion progression trends reported in the literature, further supporting its consistency.

5 Results and Analysis

5.1 Geometrical and Structural Changes

Figure 5 and Fig. 6 represent the results from COMSOL numerical computational in terms of the structural changes as the conductors are exposed to galvanic corrosion for up to 25 months. In both figures, the aluminum strands, acting as the anode, show noticeable deformation where the corrosion effects begin to appear on the top-half aluminum strands, indicated by the initial loss of material, proven by a decrease in thickness over 25 months simulation period. Corrosion is distributed across the exposed areas of the aluminum due to increased electrolyte seepage. In contrast, the composite core, serving as the cathode, remains largely unaffected by the corrosion process.

At the beginning of the exposure period of 0 months, the conductor shows minimal signs of corrosion. The cracked area of the fiberglass indicates initial points of exposure

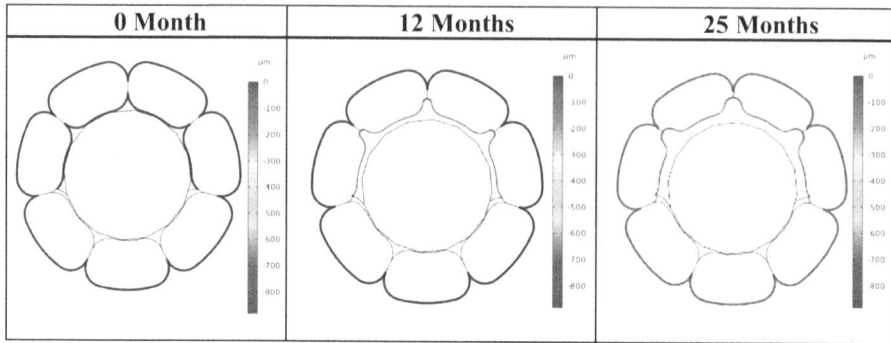

Fig. 5. Structural changes on 50% crack model up to 25 months of exposure of galvanic corrosion

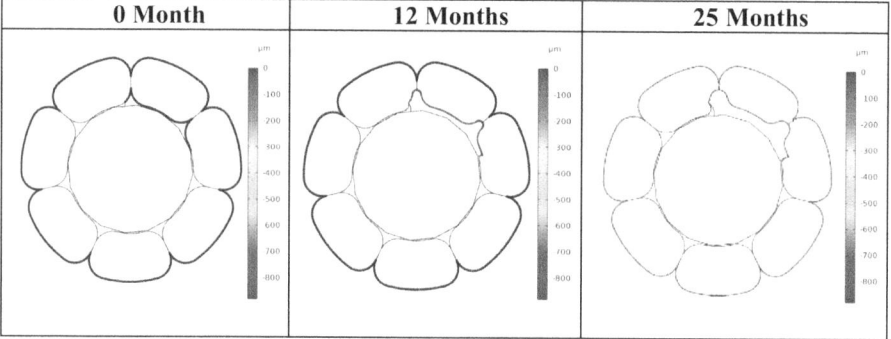

Fig. 6. Structural changes on 25% crack model up to 25 months of exposure of galvanic corrosion

where galvanic coupling could initiate corrosion between aluminum and the composite core due to direct contact at the crack sites. After 12 months to 25 months, corrosion has visibly progressed in the aluminum strands around the cracked region. Even a 25% crack in the fiberglass layer can lead to substantial galvanic corrosion over time. As the exposed aluminum undergoes material loss, the conductor's ability to carry current (ampacity) is likely reduced, alongside the risk of structural weakening.

5.2 Aluminium Area Loss Due to Galvanic Corrosion

The structural changes observed in Fig. 5 and Fig. 6, may be utilized to quantify the amount of aluminium area within the conductor. The results shown in Fig. 7 illustrate the amount of aluminum area loss in the three ACCC/TW conductors models subjected to galvanic corrosion over 25 months of exposure compared with results from [16] that completely removes the fiberglass layer. The measurements were conducted under an electrolyte conductivity of 2.5 S/m, a typical condition that accelerates the corrosion process.

From the figure, it can be concluded that the percentage of crack which also reflect the amount of surface contact between aluminium and the carbon fibre core significantly

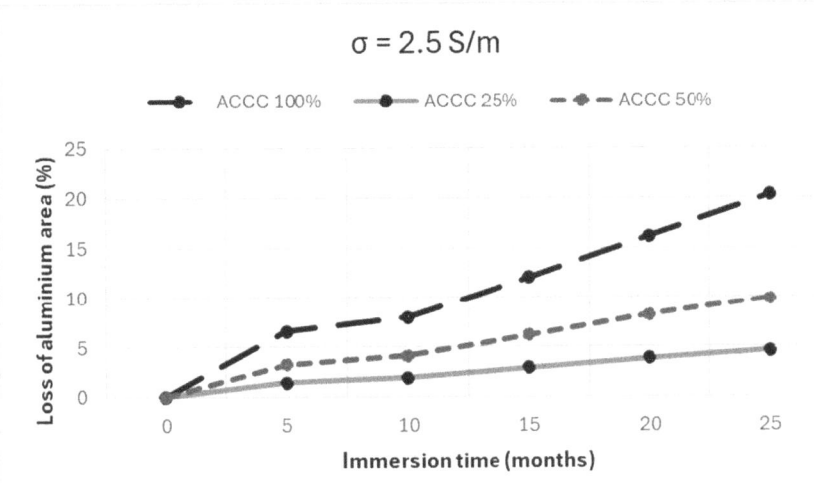

Fig. 7. Percentage of aluminium loss area of conductor models up to 25 months of ezposure

influences the rate of aluminium loss. Conductors with full (100%) contact exhibit the highest percentage of aluminium area loss over time, reaching nearly 25% loss by the 25th month. In contrast, conductors with only 25% surface contact between aluminium and the carbon fiber core show the lowest rate of area loss, achieving 4.57% by the end of the testing period. The reduced contact area limits the electrochemical reactions and slows the corrosion rate. This increase demonstrates that even partial cracking in the fiberglass can expose aluminium to corrosive activity, accelerating degradation in localized areas.

Conductors with 50% contact show intermediate corrosion behaviour, with a gradual increase in aluminium area loss (9.83%) that falls between the 100% and 25% contact cases. The larger crack surface allowed more extensive galvanic coupling between the aluminium and the carbon fibre core, causing accelerated corrosion. This scenario highlighted the vulnerability of conductors with compromised insulation, showing that material loss could lead to structural failure over prolonged exposure.

5.3 Impact of Galvanic Corrosion on ACCC/TW Power Transfer Capability

The power transfer capability, or ampacity, of the conductor is directly influenced by the aluminium cross-sectional area. As corrosion reduces the aluminium area, electrical resistance increases, which affect current flow and reduces the conductor's ampacity. As corrosion advances, the deformation of the cross-sectional area directly impacts the conductor's operational characteristics, particularly its effective resistance. The resistance for each conductor is calculated using the formula:

$$R = \frac{\rho \cdot L}{A} \tag{2}$$

where: R represents the resistance, ρ is the resistivity of the material ($\Omega \cdot m$), L is the length of the conductor (m), and A is the effective cross-sectional area of the conductor (m^2).

In this case, the resistivity of aluminium strands is assumed to be 2.8×10^{-8} $\Omega \cdot m$ with a conductor length (L) 100 m. Based on IEEE Standard 738–2012 [17], the ampacity, I of the conductor is calculated through the heat balance equation:

$$I = \sqrt{\frac{P_c + P_r - P_s}{R}} \qquad (3)$$

where: Where I is the conductor's ampacity, A, P_c is convective heat loss (W/m), P_r is radiated heat loss (W/m), P_s is solar heat gain (W/m), and R is the resistance of the conductor at a given temperature (Ω/m). Assuming the conductor operating temperature at 100 °C, the values used for heat losses and gains are as follows P_r at 39.1 W/m, P_s at 22.44 W/m, and P_c at 81.93 W/m, the resulting values are illustrated in Fig. 8.

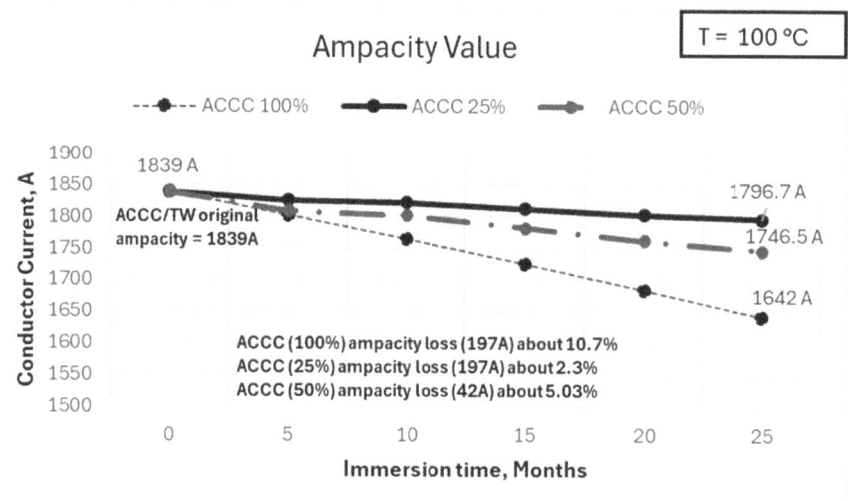

Fig. 8. Comparison of ACCC/TW conductor's ampacity between three scenarios

As the immersion time increases from 0 to 25 months, the ampacity shows a gradual decline across all tested scenarios. This trend indicates that prolonged exposure to galvanic corrosion impacts the conductor's current-carrying capability. For instance, at 100% crack, the ampacity drops to around 1642 A by the 25th month, which represents a loss of approximately 10.7% (or 197 A) relative to the initial ampacity. At 50% crack, the ampacity decreases to around 1746.5 A, reflecting an ampacity loss of about 5.03%. While at 25% aging, the reduction is smaller, with an ampacity of 1796.7 A, corresponding to a 2.3% loss. Heat dissipation is also affected as the conductor's surface area diminishes, leading to localized hotspots that further reduce the current-carrying efficiency. The relationship between aluminium area loss and ampacity reduction highlights the susceptibility of power transfer capability to structural degradation. As cracks in the fiberglass layer grow, the increased exposure accelerates corrosion, causing significant ampacity loss. This reduction in ampacity could result in insufficient power transfer and compromised conductor performance in high-demand situations.

5.4 Future Recommendations

While the developed model provides estimation into galvanic corrosion progression in ACCC/TW conductors, certain environmental factors are not considered in the simulation such as temperature variation, rainfall, pollutants or wind and mechanical stress. The electrolyte properties may change over time due to environmental interactions such as dilution by rainwater or accumulation of contaminants. These changes are not accounted for in the current model but could significantly affect corrosion dynamics whether promoting faster corrosion or slowing down corrosion.

To improve the applicability and accuracy of the model, future studies could combine these environmental variables using multiphysics approaches or extended simulation conditions. For example, introducing temperature-dependent reaction rates and electrolyte conductivity to simulate seasonal variations. As the model assumes static conditions, including humidity and salt deposition effects for coastal or industrial environments can provide a more accurate simulation of corrosion behaviour in ACCC/TW conductors under real-world operating conditions.

6 Conclusion

This study analysed the impact of galvanic corrosion on ACCC/TW conductors, particularly focusing on the aluminium-carbon fiber interfaces, and evaluated the outcomes of varying degrees of crack in the fiberglass layer. Through computational modelling using COMSOL Multiphysics, the analysis demonstrated that the integrity of the fiberglass layer is crucial in preventing galvanic interactions that accelerate aluminium loss and decrease ampacity. As cracks increase (25% and 50%), there is a significant rise in corrosion rate and ampacity reduction due to the greater exposure of aluminium to electrochemical interactions with the carbon fiber core. The findings confirm that even minor damage to the protective fiberglass layer can lead to considerable aluminium degradation, adversely impacting conductor performance over time. The results highlight the necessity for maintaining structural integrity in overhead conductors to ensure reliability and efficiency in power transmission.

Acknowledgements. This research was funded by the Ministry of Higher Education (MOHE) of Malaysia under the Fundamental Research Grant Scheme (FRGS/1/2022/TK07/USIM/02/2).

References

1. Trpezanovski, L.D., Markoski, A.P.: New Bare Conductor for Transmission and Distribution Overhead Lines of the Future, June 2008
2. Calitz, J., Potgieter, J.H.: Corrosion of overhead power transmission conductors in marine environments. Trans.-South African Inst. Electr. Eng. **96**(4), 306 (2005)
3. Håkansson, E., Predecki, P., Kumosa, M.S.: Galvanic corrosion of high temperature low sag aluminum conductor composite core and conventional aluminum conductor steel reinforced overhead high voltage conductors. IEEE Trans. Reliab. **64**(3), 928–934 (2015)
4. Ahmad, Z.: Principles of Corrosion Engineering and Corrosion Control. Elsevier (2006)

5. Parvizi, P., Jalilian, M., Dearn, K.D.: Beyond traditional conductors: aluminium conductor composite core's role in next-generation high temperature-low sag technologies–a review. Electr. Power Syst. Res. **239**, 111251 (2025)
6. Li, Z., Zhao, Q., Yang, B.: Analysis on application of new-type aluminum conductor composite core (ACCC) in power transmission line. 8–6 (2017)
7. Waters, D.H.: Low-Velocity Impact to High-Temperature Low-Sag Overhead Conductors (2016)
8. Alawar, A., Bosze, E.J., Nutt, S.R.: A composite core conductor for low sag at high temperatures. IEEE Trans. Power Delivery **20**(3), 2193–2199 (2005)
9. Burks, B., Armentrout, D.L., Kumosa, M.: Failure prediction analysis of an ACCC conductor subjected to thermal and mechanical stresses. IEEE Trans. Dielectr. Electr. Insul. **17**(2), 588–596 (2010)
10. Gokhale, G., Manish, S.: A study on different conductors for optimisation of power flow/loadability in long and medium transmission line. Int. Res. J. Eng. Technol. (IRJET) 1563–1566 (2017)
11. Lequien, F., et al.: Characterization of an aluminum conductor steel reinforced (ACSR) after 60 years of operation. Eng. Fail. Anal. **120**, 105039 (2021)
12. Riba, J.R., Bogarra, S., Gómez-Pau, Á., Moreno-Eguilaz, M.: Uprating of transmission lines by means of HTLS conductors for a sustainable growth: challenges, opportunities, and research needs. Renew. Sustain. Energy Rev. **134**, 110334 (2020)
13. Zainuddin, N.M., et al.: Review of thermal stress and condition monitoring technologies for overhead transmission lines: Issues and challenges. IEEE Access **8**, 120053–120081 (2020)
14. Håkansson, E.: Galvanic Corrosion of High-Temperature Low-Sag (HTLS) High Voltage Conductors: New Materials—New Challenges (2013)
15. Rico, D.S.R., Heurtault, S., Said, J., Xie, Y., Turmine, M., Vivier, V.: Corrosion of old overhead lines: insights into aluminum-steel galvanic couple using electrochemical techniques. J. Electrochem. Soc. **171**(4), 041504 (2024)
16. Nawawi, M.I., Rahman, S.A., Kopsidas, K.: Assessing the impact of galvanic corrosion on the ampacity of ACSR/TW and ACCC/TW conductors. In: 2024 IEEE 4th International Conference in Power Engineering Applications (ICPEA), pp. 305–310. IEEE, March 2024
17. IEEE Power Engineering Society. IEEE Standard for Calculating the Current-Temperature Relationship of Bare Overhead Conductors IEEE Power and Energy Society. IEEE Std, vol. 738, no. January (2012)

Author Index

A

Ab Rahman, Mohammad Syuhaimi 338
Abd Rahman, Rohizah 182
Abdou Mohamed, Naira 146
Abdul Rahim, Ruzairi 72
Abdul Rahman, Juwairiyyah 338
Abdul Shukor, Syaimak 87
Abdul Wahab, Norhaliza 102
Abdullah, Azizi 87
Abu Hassan, Hasliza 338
Aggarwal, Siddartha 210
Ahmad, Anita 72
Alam, Hilman Syaeful 57
Allak, Anass 146
Aminudin, Ahmad 57
Amran, Mohd Yusuf 171
Arshad, Nurul Wahidah 281
Aw, Suzanna Ridzuan 268, 281

B

Bahafid, Abdessalam 146
Bakar, Mohd Shafie 281
Benelallam, Imade 146
Bhuwana, Jakaisa Riskhalifah 102

C

Chia, Siang Kok 251

D

Daniati, Erna 355
Davidrajuh, Reggie 224, 237
Din, Shahrulnizahani Mohammad 268

E

Elhabib, Mohamed O. 322
Erraji, Zakarya 146

F

Fakhrurroja, Hanif 57
Feng, Yuming 224, 237

G

Gaanoun, Kamel 146
Ghani, Muhamad Fadli 131
Ghazali, Rozaimi 131
Gooi, Wen Pin 268
Gupta, Shrey 210

H

Has, Zulfatman 131
Hor, Xian Feng 268
Huang, Jie 14, 27

I

Idrus Nameh, Sevia Mahdaliza 57
Irianto, Wahyu Sakti Gunawan 355
Ismail, Muhammad Azhar 102

J

Jaafar, Hazriq Izzuan 131
Jamal Mohamad, Farah Aina 72
Jamaludin, Juliza 72

K

Kamal, Iqraq 87
Kamarudin, Nur Diyana 295
Kamisian, Izam 295
Khandelwal, Vansh 210
Kopsidas, Konstantinos 373

L

Le-Khoi, Nguyen 195
Leow, Pei Ling 268
Loh, Shu Ting 295

M

Mai-Cao, Lan 195
Marsono, Muhammad Nadzir 295
Mohamad Ali, Nur Husnina 131
Mohamed, Zaharuddin 322
Mohammad, Mohd Faizzuan 87

Mohd Basri, Mohd Ariffanan 171
Mohd Danuri, Mohd Sharul Nizam 182
Mori, Jun'ichiro 118
Mugunthan, Sivaguru 338
Muslim, Norliana 338

N
Nawawi, Mohamad Izzat 373
Nawawi, Sophan Wahyudi 41
Nguyen-Thanh, Tuan 308
Ntende, Kenneth Edmond 41

O
Ochi, Masanao 118
Othman, Zalinda 87

P
Paraman, Norlina 295
Pham-Tuan, Viet 195
Pusppanathan, Jaysuman 268, 295

R
Rahim, Ruzairi Abdul 268
Rahman, Ab Al-Hadi Ab 295
Rahman, Shahnurriman Abdul 373
Rahmat, Mohd Fua'ad 57
Raisin, Syarfa Najihah 72
Rashidia, Nur Hidayatullah 182
Romdlony, Muhammad Zakiyullah 102
Rusdinar, Angga 102
Rusli, Mohd Shahrizal 295

S
Saari, Mohd Mawardi 281
Sabikan, Sulaiman 41
Sadiah, Shahidatul 295
Sahran, Shahnorbanun 87
Sakata, Ichiro 118
Sakti, Indra 57
Samolej, Slawomir 224, 237
Sani, Nor Samsiah 87

Sarudin, Muhammad Aiqil 281
Shafei, Muhammad Iqbal 373
Shiro, Masanori 118
Shutari, Hussein 322
Singh, Indu 210
Sofia, 57
Soon, Chong Chee 131
Sugiarto, Anto Tri 57

T
Taggu, Amar 159
Tan, Michael Loong Peng 295
Tang, Jia Hui 87
Teyi, Nabam 159
Tran-Hoang-Gia, Lac 195

V
Verma, Abhishek 210
Vo-Tuan, Kiet 308

W
Wahab, Yasmin Abdul 268, 281
Wahid, Herman 322
Wanik, Mohd. Zulkifli Che 251
Wibawa, Aji Prasetya 355
Wong, Li-Pei 251

Y
Yaakub, Mohd Ridzwan 87
Yang, Chenglei 27
Yen, Kin Sam 1
Yeoh, Jocelyn 251
Yu, Sia Yee 281
Yusof, Nurhafizah Abu Talip 281

Z
Zakaria, Shahrudin 41
Zhang, Cheng 1
Zhao, Zixing 14
Zulkifli, Nadiatulhuda 57

The manufacturer's authorised representative in the EU is Springer Nature Customer Service Centre GmbH, Europaplatz 3, 69115 Heidelberg, Germany. If you have any concerns regarding our products, please contact ProductSafety@springernature.com

Printed and bound by CPI Group (UK) Ltd, Croydon, CR0 4YY

26/03/2026

02078935-0013